PETROLEUM CONTAMINATED SOILS Volume I

REMEDIATION TECHNIQUES
ENVIRONMENTAL FATE
RISK ASSESSMENT

PAUL T. KOSTECKI
EDWARD J. CALABRESE

LEWIS PUBLISHERS

Library of Congress Cataloging-in-Publication Data

Petroleum contaminated soils: remediation techniques, environmental fate, and risk assessment/Paul T. Kostecki and Edward J. Calabrese, editors.

p. cm.
Bibliography: p.
Includes index.
1. Soil pollution—Environmental aspects—Congresses. 2. Petroleum products—Environmental aspects—Congresses. 3. Risk assessment—Congresses. I. Kostecki, Paul T. II. Calabrese, Edward J.

TD878.P47   1988   628.5′5—dc19   88-13987
ISBN 0-87371-135-1

LEWIS PUBLISHERS, INC.
121 South Main Street, Chelsea, Michigan 48118

PRINTED IN THE UNITED STATES OF AMERICA

Preface

While petroleum contaminated soils emerged as a major domestic environmental issue in the early 1980s, soil has had the potential for contamination by petroleum products since this country began large-scale production, transportation, storage, and use of petroleum products. In addition, with over one million underground storage tanks in operating and abandoned service stations, farms, industrial sites, or businesses maintaining an automobile or truck fleet, there is no doubt society will face the problems for some time to come. If one also considers the health hazards of some of the constituents of petroleum products, that up to 60% of these tanks may be leaking, and that cleanup costs can amount to well over a million dollars per incident, one can appreciate the magnitude of the social and economic problems posed by petroleum contaminated soils.

Petroleum contamination had been largely perceived as a groundwater contamination problem prior to the early 1980s. However, as public concern and scientific understanding grew, it became clear that the soil component had to be addressed if meaningful solutions were to be found. It also became clear that the problems surrounding petroleum contaminated soils are multidisciplinary in nature and have to be approached with an understanding of the chemical processes involved in petroleum/soil interactions, environmental and public health issues, engineering technologies, and government regulations. Thus, scientists, engineers, and regulators must interact if practical and scientifically defensible solutions are going to be found for the proper handling of petroleum contaminated soils.

This book, the proceedings of the *Second National Conference on the Environmental and Public Health Effects of Soils Contaminated with Petroleum Products,* which took place at the University of Massachusetts, Amherst, September 28, 29 and 30, 1987, provides the common ground for the regulated and regulatory communities to present new information, to exchange ideas, and to focus and refine the nature of the scientific and regulatory problems.

The first step toward any solution is an understanding of the problem. In the case of petroleum contaminated soils, this involves an array of problems. Part I of this book provides the insight and various perspectives on the scientific, economic, and regulatory issues as presented by representatives of the petroleum industry (API) (Chapter 1), state regulatory agencies, California (Chapter 2) and New Jersey (Chapter 4), utilities industry (EEI) (Chapter 3), and the federal government (EPA/Office of Underground Storage Tanks) (Chapter 5).

Understanding why, where, and how quickly petroleum products and their constituents distribute in the environment is essential for informed decisionmaking. Part II provides several chapters on environmental fate (Chapter 6), soil physics

(Chapter 7), state-of-the-art models (Chapter 8), and the application of modeling to regulatory programs (Chapter 9).

Familiarity with the types of remedial technologies available, their environmental effectiveness, technical feasibility, and costs are essential for making sound, cost-effective decisions for the cleanup of contaminated soils. Part III provides information on a wide range of in situ and offsite remedial technologies that have been used, as well as some that show promise. More specifically, this section presents an overview of existing technologies (Chapters 10 and 18), as well as methodologies such as stabilization (Chapter 11), vitrification (Chapter 12), soil washing (Chapter 13), land treatment (Chapter 14), asphalt batching (Chapter 15), incorporation into bituminous concrete (Chapter 16), and low temperature stripping (Chapter 17).

Risk assessment provides the basis for rational decisionmaking and there are several risk assessment models which can be used for petroleum contaminated soils. In Part IV the theory behind the rationale for using risk assessment (Chapters 19 and 20) is provided, as well as information for use in the health hazard and exposure components (Chapters 22 and 23) of the risk assessment. Several models are critiqued and alternatives are offered (Chapter 21). The utility of the models is demonstrated by proposing an implementation plan for state agencies (Chapter 24).

Contamination of soils by petroleum products has become a major concern in all sectors of our society. Billions of dollars will be spent in the next few years to clean up and minimize future occurrences. This book brings together researchers and others charged with correcting the problem in a statement and discussion of the multifaceted nature of this problem and a presentation of the solutions.

Paul T. Kostecki, a Senior Research Associate/Adjunct Assistant Professor in the Environmental Health/Science Program at the University of Massachusetts, Amherst, received his PhD from the School of Natural Resources at the University of Michigan in 1980. He has been involved with risk assessment and risk management research for contaminated soils for the last five years, and is coauthor of *Remedial Technologies for Leaking Underground Storage Tanks* and coeditor of *Soils Contaminated by Petroleum Products.* Dr. Kostecki's yearly conference on petroleum contaminated soils draws hundreds of researchers and regulatory scientists to present and discuss state-of-the-art solutions to the multidisciplinary problems surrounding this issue.

Edward J. Calabrese is a board certified toxicologist who is professor of toxicology at the University of Massachusetts School of Health Sciences, Amherst. Calabrese has researched extensively in the area of host factors affecting susceptibility to pollutants, and is the author of more than 240 papers in scholarly journals, as well as 10 books, including *Principles of Animal Extrapolation, Nutrition and Environmental Health,* Vol. I and II, and *Ecogenetics.* He has been a member of the U.S. National Academy of Sciences and NATO Countries Safe Drinking Water committees, and most recently has been appointed to the Board of Scientific Counselors for the Agency for Toxic Substances and Disease Registry (ATSDR).

Contents

PART I
DEFINING THE PROBLEM

PART II
ENVIRONMENTAL FATE AND MODELING

PART III
REMEDIAL OPTIONS

PART IV
RISK ASSESSMENT/RISK MANAGEMENT

PART I

Defining the Problem

CHAPTER 1

Soils Contaminated by Motor Fuels: Research Activities and Perspectives of the American Petroleum Institute

Bruce J. Bauman

The American Petroleum Institute (API) is a trade association representing the domestic petroleum industry, serving over 200 corporate and nearly 5000 individual members. In addition to engaging in traditional trade association activities, API conducts an active health and environmental research program keyed toward addressing the pressing health and environmental issues of the day. API and its members have a keen interest in contaminated soils, and are aggressively pursuing topical research on this important issue.

The conference in Amherst focused on the environmental and public health effects resulting from soil contamination by petroleum hydrocarbons. This chapter will consider the impacts of petroleum products, with particular emphasis on motor fuels (such as gasoline and diesel) and heating fuels. Enormous volumes of these fuels are required by our society on a daily basis (e.g., daily average consumption of gasoline in the United States approaches 500 million gal., as shown in Table 1). Despite the best efforts of both the petroleum industry and the regulatory community, it is inevitable that there will be inadvertent releases to the environment. A number of these releases will require site restoration in order to mitigate the impacts of soil and groundwater contamination.

Because these fuels are commonly stored in underground storage tanks (UST), regulatory activities in that arena have contributed to the increased interest in problems of soil contamination that may be encountered during the removal or

Table 1. Daily Average Supply of Selected Petroleum Fuels, 1986.

| GASOLINE | | FUEL OIL | | AVIATION |
Unleaded	Leaded	Distillate	Residual	Jet Fuels
4,656,000	2,096,000	2,796,000	889,000	1,186,000

(Units are in barrels; 1 barrel = 42 gallons)

DAILY TOTAL = 11,623,000 BARRELS (488,166,000 GALLONS)

Source: Petroleum Supply Annual, Energy Information Administration, Office of Oil and Gas, U.S. Dept. of Energy, Washington, D.C.

replacement of USTs.[1] As tank replacement and soil remediation programs continue over the next decade in response to these new regulations, there will be a need for responsible and effective public policy toward this type of soil contamination. Open conferences such as the one held in Amherst will play an important role in the development of that policy.

Interestingly, concern with soil contamination and its potential impacts has lagged behind that of groundwater contamination. However, over the last several years, especially as appreciation of the importance of contaminant behavior in the vadose zone has grown, there has been a parallel growth in efforts to evaluate and mitigate the environmental risks that may be posed by the presence of residual soil contaminants. Where there is a migration route by which these contaminants may result in an exposure of a human population, it is important that the potential pathway for exposure be evaluated carefully. For soils contaminated by motor fuels, there are three potential pathways for such exposure: transport of soluble components to groundwater as water percolates through the soil; migration of vapors through the soil and into structures or the atmosphere; and direct contact with particles of contaminated soil, e.g., by inhalation or ingestion. Decisions regarding the need for corrective action, and the type and extent of cleanup required, should be based on a site-specific evaluation of the potential risks posed by any of these potential pathways.

However, many efforts toward accurate and rational assessments of these risks are frustrated by lack of an adequate understanding of the dynamics of the interaction of motor fuels and soils, and of the toxicological mechanisms of potential short-term, low-level exposures. This is primarily a result of the complex nature of motor fuel chemistry and environmental behavior, coupled with the equally complex nature of the chemistry and physical properties of soils. Unfortunately, there are few individuals who possess the education and experience that would provide them with a thorough knowledge of the basic properties of both soils and fuels. As a consequence, in some cases, cleanup goals may be established that are very costly, yet provide only a minimal incremental reduction in the net risk to the environment and to public health.

These are complex, technical issues. Because they have developed so recently and rapidly, there has been little time to analyze this environmental problem in the detailed manner necessary to construct adequate solutions. The net result of the current situation is an environmental decisionmaking system that is often inefficient and sometimes ineffective in addressing the potential problems caused by soil contamination. Currently all involved parties (i.e., industry, regulators, and the public) are struggling to develop effective solutions and policies for the issues discussed above. It is clear that this issue will continue to grow as UST regulatory requirements impact a larger proportion of the UST population. For these reasons it is important to acquire a solid scientific understanding of the risks posed by petroleum contamination of soil, which can then be employed to develop a rational regulatory policy.

The following paragraphs will address in limited detail the nature of the inter-action of soil and motor fuels, the variable nature of the risks associated with motor fuel contaminated soil, and a suggested approach for managing sites with contaminated soils. At the end of the chapter, a brief review of recent and current research being conducted by API will be presented.

MOTOR FUEL BEHAVIOR IN SOIL

This chapter will briefly examine the transport and behavior of motor fuel contaminants in a general, conceptual way. Given its pervasive use in our society, gasoline provides the most fitting example of a contaminating motor fuel. Special emphasis will be given to the processes of fuel retention and transport, volatilization, and biodegradation. Current issues concerning sampling and analysis of soils contaminated by petroleum products will also be discussed.

Retention and Transport

When motor fuel leaks from its storage vessel and enters the soil, gravitational forces act to draw the fluid in a downward direction. Other forces act to retain the fuel, which is either adsorbed to soil particles or trapped in soil pores. The amount of fuel retained in the soil is of primary importance, for it will determine both the degree of contamination and the likelihood of subsequent contaminant transport to groundwater due to leaching effects. Little formal data on this phenomenon have been documented in the literature, but soil particle size, moisture content, and fluid viscosity have been identified as the regulating parameters.[2]

Similarly, there have been few published studies regarding motor fuel solubilization and transport (i.e., leaching) from soils.[3] The general solubility and speciation of common fuels such as gasoline and fuel oil has rarely been documented.[4-8] The solubility of gasoline in water is typically in the range of 50–150 mg/L, and reported values for #2 fuel oil or diesel, between 0.4 and 8.0 mg/L. Few studies have examined the potential for solubilization and transport of soil-adsorbed fuels through the vadose zone to the water table.[9,10] Additional information on these processes would be beneficial for providing guidance for assessing the potential for contaminant migration from motor fuel-contaminated soils.

Volatilization

Due to the volatile nature of gasoline (its vapor pressure is typically between 60 and 105 kPa, depending on the season and the geographic area of use), it will tend to enter the vapor phase.[11,12] Since gasoline is a complex mixture of hundreds of compounds, the composition of these vapors will reflect the composition of the compounds within the gasoline that have relatively high vapor pressures. Those compounds with higher vapor pressures (e.g., butane, propane, benzene, toluene, xylene) will be preferentially volatilized, and consequently, their concentration in the fuel remaining in the soil will be diminished.

It is interesting to note that as soon as the source of the release has been stopped, these natural volatile forces are acting to reduce the concentration of the gasoline in the soil (i.e., hydrocarbon molecules experience a spontaneous phase change from liquid to vapor, and thus can escape to the atmosphere). This volatility can be both an asset and a liability. It can be utilized to effectively accelerate removal of the gasoline from the soil through the use of vacuum extraction (i.e., soil venting) either in situ, or in excavated piles of contaminated soil.[2,13-15] Simple mechanical aeration (e.g., turning a pile of soil regularly) will also stimulate the volatilization of gasoline from the soil.[16] While physical and chemical principles dictate that this volatilization will occur, it is important to recognize that soil textural properties (e.g., sandy, silty, clayey) and other natural factors that influence vapor flow in soils (e.g., temperature, soil moisture content, surface wind movement, relative humidity) will affect the overall rate at which volatilization will occur.

The liability side to the volatile nature of gasoline is the potential for the buildup of explosive vapors in subsurface structures, or the potential for direct human exposure to the vapors. At some sites these hazards may be nonexistent. At others it may be necessary to actively control vapors, and/or to remove residual gasoline from the soil to eliminate vapor generation. As with decisions concerning the need for corrective action to protect groundwater quality, a decision as to what level of risk is presented by the vapors in the soil should include a careful and knowledgeable assessment of the nature and extent of the contaminant, the properties of the particular site, and potential routes of exposure.

Biodegradation

Another process, somewhat slower than volatilization, will occur simultaneously with the volatile loss of the gasoline. Soil microbial organisms will begin to selectively consume the hydrocarbon compounds that comprise the fuel, transforming them into carbon dioxide, water, microbial biomass, and humus. Biodegradation of petroleum hydrocarbon has been well documented in the literature.[17-20] The several dozen species of microbes that effect the degradation are naturally present in all soils.[17] While their natural populations may be relatively small in some (especially subsurface) soils, the introduction of petroleum compounds that act as a hydrocarbon source for microbial metabolic processes can stimulate the growth of that diverse population of soil bacteria and fungi.

Certain types of waste treatment processes exploit this beneficial microbial soil property by carefully managing the degradation of organic waste material through soil application. Land treatment of petroleum refining wastes has been employed successfully at many locations for over 20 years.[21] In comparison to those types of high molecular weight wastes which typically contain a high percentage of long chain and branched hydrocarbons, gasoline (along with other motor fuels) contains predominately short chain compounds that are more readily biodegradable.[17,20] Again, the rate of these reactions will be site-specific, and, as with volatilization, may be accelerated artificially. A number of separate studies of various petroleum oils have shown mineralization rates in surficial soils of 0.02 to more than 0.4 g of hydrocarbon per kg of soil per day, with 0.09 to 0.14 g/kg soil/day representing a reasonable average.[17] Considering that 100 mg/kg petroleum hydrocarbon in soil equals 0.1 g hydrocarbon per kg of soil, it is clear that soils have the capability to mitigate the impacts of motor fuel spills or leaks through microbial degradation. It is known that the rate of subsurface degradation will be slower, usually due to lower concentrations of oxygen, or smaller populations of microbes. Several studies have examined the potential for acceleration biodegradation rates where the availability of oxygen is limited.[22,23] However, over time, degradation of the petroleum hydrocarbon will inevitably occur, with or without artificial inputs.[24]

A further observation regarding volatilization and biodegradation is that both processes tend to selectively remove the lighter end hydrocarbon compounds. This is important, since this lighter fraction includes the monoaromatic hydrocarbons benzene, toluene, and xylene (BTX). These compounds have relatively high vapor pressure (benzene—12.7 kPa, toluene—3.8 kPa, and xylenes—0.9 to 1.2 kPa), and so they are easily volatilized from the soil compared to other gasoline components. They also are considered to be relatively easy for microbes to degrade, although it must be recognized that at high concentrations they are toxic to those same microorganisms.[17] Because these are the same compounds most commonly targeted as the soluble and toxic compounds of concern, it is fortuitous that natural processes also act on these compounds to bring about their early depletion in the soil.

The point to be made about the potential for natural processes in soil to remove petroleum hydrocarbon is that while the rates at which these removal processes occur may vary, it is clear that motor fuels can be considered nonpersistent contaminants in the soil (i.e., their concentrations in soils can be reduced through a variety of methods, both natural and engineered).[20] These phenomena distinguish motor fuels from other types of contaminants that are not volatile and are much more resistant to degradation (e.g., classes of chlorinated hydrocarbons). Therefore it is very important that these concentration-reducing processes be incorporated into the evaluation of the nature of the risk present at a site.

Sampling and Analysis

A critical component in characterizing the nature of soil contamination is the determination of the concentration of the contaminant(s) present. Procedures for analysis of gasoline range hydrocarbons in soil, together with sampling and handling techniques prior to analysis, are currently a source of much debate.[25] Whether measuring parameters such as total petroleum hydrocarbons (TPH), specific compounds such as benzene, toluene, and xylene (BTX), or soil vapor, arguments arise as to which methods are most accurate, which are most appropriate and meaningful in relation to risk characterization, and what are the real detection limits of the methods.

Analysis of soil liquid extracts is complicated by the variability of the matrix solution that is extracted from the soil. This is because soils typically contain a variety of naturally occuring organic compounds, as well as inorganic substances, that interfere with the detection capabilities for the petroleum hydrocarbons of interest.[25] It is not always possible to eliminate these interferences, or even recognize when they may be present. In addition, it is difficult to relate concentrations measured in soil extract solutions to the actual concentrations that would occur in soil under the normal ranges of moisture content found in soils.

A related and perhaps more serious problem exists for those methods currently used to determine soil contaminant concentrations in the field. Since gasoline and other motor fuels are volatile liquids, many of these field methods measure the concentration of contaminant vapors in or above the soil rather than the mass of the contaminant adsorbed to the soil. Such vapor measurements can be made by a variety of instruments (e.g., combustible gas indicator, organic vapor analyzer, portable gas chromatograph) with varying levels of specificity and accuracy. Measurements are typically obtained by sampling the vapor present in the headspace of a bottle or jar containing a volume of contaminated soil; creating a hole in the contaminated soil, and lowering a vapor sampling probe into the hole; or simply holding a vapor sampling probe close to the surface of the contaminated soil. The accuracy and precision of these methods have not been determined, and there is no nationally accepted standard methodology. The existence of naturally occuring gaseous hydrocarbons in soil can also contribute to measurement errors.[26] Because of these limitations, such methods have proved to be most useful for rapid onsite qualitative assessment of the presence and nature of soil contaminants.

However, these methods are also sometimes used as semiquantitative techniques for estimating levels of soil contamination, or even for establishing standards for corrective action despite the lack of accepted standardized methods or any demonstrated correlation between vapor concentration and soil contaminant concentration.[27,28] Further, there is an absence of significant literature that provides guidance as to the relationship between motor fuel contaminant levels in soil and their potential migration to the water table (i.e., what can be considered acceptable levels of residual contamination).

Given all the sources of variability in trying to make field measurements, the concentrations reported by any method must be cautiously evaluated as to their validity, meaning, utility, and possible limitations. The instruments used need to be carefully calibrated, maintained, and operated. Again, careful and reasoned interpretation of the results is necessary if meaningful decisions are to be made. There is a need for development of more standardized methods that will provide accurate, reliable, and significant analytical results. It is hoped that this topic will become the focus of future research projects. API has begun to address this problem area in its research efforts, both in the field and in the laboratory.[29] These projects will be briefly summarized at the end of this chapter.

CORRECTIVE ACTION STANDARDS FOR CONTAMINATED SOILS

The technical processes discussed above provide important input for decisions regarding the need for corrective action at sites where motor fuel contamination exists. A fundamental concern of many readers will be how to define an appropriate strategy for such sites. Of primary interest is the determination of what can be considered ''safe'' levels of these classes of hydrocarbons in soil, i.e., the determination of cleanup standards. During the last three years, movement has occurred at the state level toward development of ''guidelines'' or ''standards'' that establish general cleanup target concentrations for fuels or fuel constituents.[30] There has been no consensus as to what levels of what contaminants represent unacceptable health or safety risks. This is due partially to the newness of this environmental issue and the resultant lack of relevant technical assessments, and partially to the difficulties in understanding the nature and consequences of soils contaminated by motor fuel.

The numerical standards or guidelines that have been established generally suffer from being the product of ''educated guesses'' as to what these levels should be. Such numerical standards are typically based on limited field data generated by analytical techniques of questionable accuracy, or based on modifications of existing water quality standards that may have no relevance to soil contamination. Examples of some of these numerical standards that have been used (formally or informally) or are under consideration in various regulatory jurisdictions, include either 10, 100, or 1000 ppm total petroleum hydrocarbon; 100 and 500 ppm cumulative benzene, toluene, xylene and 60 ppb benzene (a very conservative standard which is below analytical detection limits).[30,31] In its proposed regulations for USTs, the EPA has indicated a concentration of 100 ppm total petroleum hydrocarbon in the soil as evidence that a release has occurred, although it offers no substantive documentation for selection of that value.[1]

API questions the propriety of any such standards when they are rigorously applied, either for groundwater (where they have received wide application), or especially for soil. During an assessment of the need for corrective action at sites with contaminated soil, there is no generic single number of relevance that can be universally applied to all sites as a target concentration for cleanup efforts. Rather, there are many important numbers (e.g., the concentration of contaminants, their extent, the depth to groundwater, the distance to populations or subsurface structures) that collectively must be evaluated to estimate the level of risk present at any given site. Based on this cumulative risk, an appropriate cleanup target can be established.

RISK ASSESSMENT AND MANAGEMENT

Environmental risk characterization involves the assessment of three key components: (1) the nature and extent of the contamination; (2) the pathways by which the contaminant(s) may migrate to expose any given population of interest; and (3) a recipient population(s) that potentially could be exposed to the contaminant(s). It can be argued that if any one of these three components is not present at a site, then there is no risk. Indeed, many current techniques of site corrective action focus on risk reduction by either reducing/eliminating the source of the hazard, or containing the hazard so that migration will not occur.

At those sites where there is no potential for exposure of a population to the hazard, then remedial action may not be necessary. This is expecially true if migration (and thus exposure) is unlikely and if the contaminant is considered to be nonpersistent (i.e., will diminish with time). As discussed earlier, because of their degradability and volatility, motor fuels can be considered to be nonpersistent soil contaminants. This lack of persistence is a key factor to be evaluated when characterizing risk at a site where a release has occurred.

While many site assessments tend to focus on the characterization of the type of contaminant and its concentration, it is most important to consider the potential for *exposure* to the contaminant.[32] The concept of risk (e.g., to a human population) implies that there is some potential for exposure to a defined hazard. Estimating potential exposure requires both a knowledge of possible receptors (i.e., human beings) in the area around the contaminated site, and an understanding of the mobility of the contaminant and potential routes of migration that could carry the contaminant to the receptors.

Most retail motor fuel facilities and their USTs are located in urban or suburban areas, and thus there usually are potential receptors in the area that could be exposed to the contaminant should a release occur. Under these circumstances it is most important to examine the likelihood of migration of the contaminant toward potential receptors, where exposure might then occur. It is this estimate of the probability of actual exposure to the contaminant that should factor heavily in the ultimate decision for the level of remedial activity that will be required for the contaminated soil.[32]

When examining the potential for exposure to contaminated soil, it is found that when compared to the other major environmental media subject to contamination (i.e., water and air), soil can be considered relatively immobile, and so it is far less likely for the soil itself to be transported to a point of exposure. Further, unlike either water or air, soil is not commonly inhaled or ingested by humans. Direct exposure to soil can usually be avoided through limiting access to the site where the contamination has occurred, and by covering the soil to prevent atmospheric transport. For these reasons, it is often comparatively easy to control direct exposure to petroleum contaminated soils.

For the same reasons, it is generally acknowledged that contaminated soils may be considered intrinsically less hazardous than similarly contaminated air or water. However, it is also known that such soils can present serious environmental and public health hazards in those situations where contaminant levels are high and exposures (either direct or indirect) are likely. In most cases it is the potential for the contaminated soil to serve as a source of contaminant release to another media (e.g., air or groundwater) that creates the need to consider remedial action. Leaching of soluble contaminants to groundwater, where they may travel to public or private water supply wells, or release of volatile contaminants to the atmosphere, where they may be inhaled, are the two most common sources of environmental degradation caused by gasoline contaminated soil. Another hazard is posed by the potential for the accumulation of explosive hydrocarbon vapors in subsurface structures. If these contaminant transport processes can be controlled, the level of risk presented by the contaminated soil can be reduced or eliminated.

When subsurface movement of vapors offsite is determined to present a hazard, such vapors can be effectively controlled through use of soil venting.[33] In addition to its benefits for vapor control, soil venting has been shown to be an efficient means for removal of volatile hydrocarbons and reducing their concentrations in soil. [2,13-15] Recently, concern has been expressed regarding potential atmospheric adverse effects to local air quality from venting of volatile motor fuel hydrocarbons. While this may be a valid concern at some sites, it is also necessary to consider that in all but very extreme cases, the process of removing residual gasoline from the soil will take a relatively short time, and a relatively small mass of liquid will be vented to the atmosphere (small in terms of net mass loading to the atmosphere from all sources in the local or regional air basin). These factors mean that there will be only a short-lived environmental impact during the site restoration process. The risks from this short-term impact must be compared to the risks and costs posed by other practical remediation alternatives.

The other major pathway of contaminant migration from contaminated soil is through dissolution of residual hydrocarbon in the soil by water that has infiltrated into the soil and is moving vertically to the water table.[10] Once they have reached groundwater, contaminants can subsequently travel offsite in the direction of groundwater flow. At many service station sites this may be an improbable occurrence, because the soil is commonly covered by a concrete or asphalt pad of very low permeability which prevents rainfall infiltration and percolation through the soil. As a result it is unlikely that consequential amounts of water might move through the soil and leach contaminants to groundwater. If the zone of contaminated soil does not extend to the water table, it is unlikely that groundwater will come in direct contact with contaminants. In these situations, contamination has been effectively isolated from the hydrogeologic system, and migration is improbable. Thus potential exposure to the contaminants is eliminated, and it can be argued that there is no need for formal remediation measures. This isolation of the contamination, and the likelihood of gradual chemical and microbial degradation, combine to provide an effective remedial solution.

This is not to suggest that all motor fuel contaminated sites can be addressed by the simple containment process described above. At some sites contamination levels may be too high, or groundwater too shallow, for this approach to be effective. Therefore, it is important that before deciding on a course of corrective action, the potential risks present at the site be clearly defined and evaluated. As emphasized in the preceding paragraphs, much of the potential risk can be attributed to the likelihood of exposure to the site contaminants. If potential exposure (direct or indirect) to the soil is negligible, either because there is no population in the vicinity, or because it is possible to prevent migration of the contaminants and the associated potential for exposure, then potential risk has also been, in essence, eliminated.

When the nature of the risk (both existing and potential risk) has been determined, an appropriate corrective action plan can be developed. At some sites, the potential risks may be low enough that there is no need to actively engineer the removal of the contaminants from the soil. Other sites may require that remedial actions of varying levels be taken. For example, one site may have a high concentration of TPH, but because it is relatively isolated from groundwater (e.g., covered or capped by concrete or some other impervious barrier) and is remote from human populations, it may not require any direct treatment. Another site may have lower TPH concentrations, but is located directly above a drinking water aquifer. In this case there would likely be a need to actively reduce the concentration of contaminants in the soil to a level that minimizes the potential risk for groundwater quality degradation. For each site, exposure scenarios can be developed to evaluate potential risks which are unique to site characteristics, and to determine the most appropriate type of remedial response.[32]

This discussion has emphasized the importance of assessing potential exposure to contaminated soil. Regarding direct exposure to contaminated soil, it is obvious that one of the primary objectives in minimizing potential exposure is to restrict access to the soil. It is interesting to note that at many sites this can be best accomplished by leaving the soil in place. This avoids the potential exposure that would occur during and after the excavation of the soil.

In situ remedial methods such as soil venting or natural biodegradation and attenuation are attractive, proven alternatives to soil excavation that deserve consideration when deciding on the most appropriate cleanup strategy for a site. Implementation of in situ methods is especially logical when there will also be a need for groundwater remediation at the site. Finally, there is a growing trend in federal environmental programs [e.g., Comprehensive Environmental Response, Compensation, and Liability Act (CERCLA)] toward promotion of treatment approaches that deal with the contamination onsite.[34]

The selection of an appropriate corrective action strategy for motor fuel contaminated soils needs, by the very nature of the problem, to be a site-specific process.[33] There are a variety of technologies that can be used to reduce potential risks present at contaminated sites. At sites where corrective action is needed, preferred techniques are those that employ onsite contaminant removal (either in situ or from excavated soil), and allow for the subsequent reuse (e.g., backfilling) of the treated soils wherever practical.

However, at some locations site conditions may dictate soil removal and disposal of contaminated soil, a remedial technique that is currently the source of much discussion. In some state and local jurisdictions it is currently a requirement, or may be the primary remedial technique.[30] In others, it is to be used only when other methods are impractical. In the draft underground storage tank regulations, the EPA has proposed that all "visibly" contaminated soil be excavated and properly disposed of.[1] While clearly there are situations where landfilling of the contaminated soil may represent an appropriate cleanup solution, it should not be considered as an exclusive or mandatory solution.

Its principal limitation, and an important one, is that it can represent an expensive form of transferring a potential environmental hazard (and its resultant long-term liability) from one location to another, with little reduction in net risk. It may have the added undesirable result of filling up limited waste disposal volume at permitted land disposal facilities. However, if adequate landfill capacity can be found close to the contaminated site, this soil disposal option may represent an effective and economical solution. In some areas, landfill operators may encourage such soil disposal, because of the scarcity of the cover soil needed for landfill operations. Therefore, it is desirable that excavation and landfill disposal of motor fuel contaminated soil be preserved as a cleanup alternative.

SUMMARY

In this chapter we have discussed the general nature of the processes that affect motor fuel in soils, and have emphasized the natural propensity for contaminant depletion in soils which in turn mitigates potential environmental risks. The importance of careful definition and evaluation of the potential risks that are uniquely associated with each site has also been emphasized. It has been demonstrated that at any given site, any potential risks are the result of the:

- nature and level of the contamination
- site-specific conditions which influence possible routes of contaminant exposure
- proximity of potential receptors (e.g., human populations) to the contaminated soil

The unique potential risk at a given site will in turn require that careful consideration be given to the variety of remedial strategies that can effectively reduce or control that risk.

It is important to appreciate that the inherent diversity among sites dictates that a regulatory program possess sufficient flexibility to allow for the design of site-specific cleanup goals and technical strategies to achieve them. This can best be achieved through an evaluation of the potential environmental, public health, and safety risks present at a site, as opposed to strict assignment of numerical standards as cleanup objectives.

Working together within this framework, industry and the regulatory community can provide practical solutions to the complex problem of soil contamination by petroleum hydrocarbons. API will continue its research efforts to provide an improved understanding of soil contamination and its mitigation. The regulatory community can facilitate the development of innovative cleanup techniques by encouraging careful experimentation with new methods, and by streamlining the permitting process required for such projects. Perhaps more importantly, by continuing efforts in the area of prevention of petroleum hydrocarbon releases to the environment, it will be possible to maximize the protection of soil and groundwater quality.

API RESEARCH

As noted earlier, API has an active environmental research program that includes several studies on contaminated soil. A summary of some these projects is provided in the following paragraphs.

Sampling and Analysis of Soils for Gasoline-Range Organics (Rocky Mountain Analytical Laboratory)

Those who have been active in the area of site assessment for soils are aware that there are inherent limitations in the current sampling and analytical procedures for gasoline contaminated soils. This serves as a continual source of frustration to efforts to obtain accurate characterization of the extent and the level of contaminants present. Recently, API initiated a study to examine ways to improve existing sampling and analytical techniques. This work has included a literature review and a workshop involving technical experts on this issue. Two procedures (involving both field sampling and laboratory analyisis) have been identified as deserving careful evaluation. Two rounds of field sampling will take place in the coming months and a series of comprehensive laboratory analytical studies will be conducted. An eventual goal of this effort is to establish a nationally recognized, standardized procedure for these important field and laboratory procedures.

Short-Term Fate and Persistence of
Motor Fuel in Soil (Radian Corporation)

One of the most commonly used methods for reducing levels of gasoline in contaminated soils is to facilitate the natural volatilization of the fuel from the soil. While the effectiveness of this method is widely recognized, there is little published data from scientific studies. API is conducting a pilot scale study involving small piles of soil (four cubic yards) that will be mixed with gasoline to simulate a typical contaminated soil. Four different aeration treatments will be monitored for six weeks. The treatments are a control, a soil pile that is turned over twice a week, a soil pile that is constantly heated to 100°F, and a pile that is constantly under a slight vacuum. Soils will be analyzed for total hydrocarbons and for common monoaromatics. All soil piles will also be monitored for air emissions. The results of this study will help to quantify the rate and mass of hydrocarbon losses from excavated soils.

Natural and Nonpoint Source Benzene, Toluene, and
Xylene in Soils (Environmental Science and Engineering, Inc.)

This study includes a review of literature to accumulate data on natural and nonpoint source concentrations of BTX in soils that have not been directly impacted by motor fuel spills. A second phase of this work will entail sampling soils at nine field sites where these nonspill BTX compounds might be expected to occur (e.g., along highways, in high-organic content soils, etc.). The objective of the project is to document the natural, common ranges of concentrations of these compounds in soils.

Assessment of the Impact of Residual Hydrocarbons in
Soils on Groundwater (University of Waterloo)

An important question to be answered during the establishment of cleanup goals at contaminated sites is, "At what level of residual petroleum hydrocarbon concentration is there a potential for continued leaching of contaminants that may significantly degrade groundwater quality?" During this study, cores of soil of varying levels of residual petroleum hydrocarbon contamination will be obtained from field sites. These core samples will then be used in laboratory column leaching studies to characterize the nature of the leachates from soils with different levels of residual hydrocarbon contamination.

Field Evaluation of Vacuum Extraction of
Gasoline from Excavated Soils (Roy F. Weston, Inc.)

This project, to be conducted early in 1988, is a field-scale evaluation of the use of vacuum extraction (i.e., soil venting) to accelerate the volatilization of gasoline from excavated piles of soil. Previous research by API and others has shown venting to be an effective remedial technique for soils in situ. There is an increased interest in application of this technology for onsite treatment of soils that have been excavated. This study will provide a more complete characterization of the necessary design parameters, and the general effectiveness of this method.

The Effects of Multiple Vapor Recovery Wells During Subsurface Venting of Hydrocarbon Vapors
(Reidel Environmental Services, Inc., and Radian Corporation)

API has been very active in exploring the utilization of vacuum extraction of volatile motor fuel hydrocarbons from soil, initiating research on this topic back in the late 1970s. This particular project is an extension of the research reported in an API publication which examined the effects of venting with a single extraction well, as a means for removal of gasoline from subsurface soils.[14] This phase of the study evaluates the effectiveness of multiple vapor wells, their area of influence, and the effects of various rates of evacuation. A final report should be available in early 1988.

Another related project is the development of a field manual for soil venting that draws on the knowledge gained from these and other studies. The manual will provide guidance for the design, construction, and evaluation of soil venting systems. This manual should be completed in early 1988. Further research on in situ soil venting will also be initiated in 1988.

In addition to the above research directly related to soil, API has an active groundwater research program that often involves studies which indirectly encompass soil topics. For example, a recently completed study conducted at the University of Waterloo examined the ability of soils to attenuate low levels of dissolved, gasoline-derived contaminants in groundwater when it was applied to surface soil and percolated through the vadose zone. This would be analogous to a field situation where air stripper effluent would be discharged to an infiltration gallery in order to further remove the last traces of organic contamination by means of natural soil attenuation. In this study, the applied water was sampled at various depths as it percolated through an instrumented column of soil in the field. Natural removal of these low-level organics was observed to occur in the vadose zone, demonstrating the natural attenuation capacity of the soil. Additional study of vadose zone attenuation of motor fuel contaminants is planned in 1988.

This comprehensive and diverse environmental research program has been and will continue to be a valued source of technical information regarding motor fuel contamination of soil and groundwater. All API research reports are available to the general public.

ACKNOWLEDGMENT

The author would like to gratefully acknowledge the helpful comments and review of Dorothy Keech, Chevron Oilfield Research, during the development of this chapter.

REFERENCES

1. *Federal Register,* Vol. 52, No. 74. (April 17, 1987), pp. 12662–12786.
2. Hoag, G. E., and M. C. Marley. "Gasoline Residual Saturation in Unsaturated Uniform Aquifer Materials." *J. Environ. Eng.* 112:586-604 (1986).
3. American Petroleum Institute. "Soils Impacted by Motor Fuels: A Literature Review." Report of the Marketing Department (1987).
4. American Petroleum Institute. "Literature Survey: Hydrocarbon Solubilities and Attenuation Mechanisms." Publ. No. 4414 (1985).
5. American Petroleum Institute. "Laboratory Study on Solubilities of Petroleum Hydrocarbons in Groundwater." Publ. No. 4395 (1985).
6. Environment Canada. "A Study of the Solubility of Oil in Water." Technology Development Report EPS-4-EC-76-1 (1976).
7. Guard, H. E., J. Ng, and R. B. Laughlin, Jr. "Characterization of Gasolines, Diesel Fuels, and Their Water Soluble Fractions." Report for the Naval Biosciences Laboratory, Naval Supply Center, Oakland, CA 94625 (1983).
8. Smith, J. H., J. C. Harper, and J. Jaber. "Analysis and Environmental Fate of Air Force Distillate and High Density Fuels." Report for the Engineering and Services Laboratory, Tyndall Air Force Base, FL 32403, ESL-TR-81-54.
9. Dietz, I. D. N. "Large Scale Experiment on Groundwater Pollution by Oil Spills—Interim Results," *H₂O,* 4:77-80 (1978).
10. Schwille, F. "Migration of Organic Fluids Immiscible with Water in the Unsaturated Zone, in *Pollutants in Porous Media,* D. Yaron et al., Eds. (New York: Springer-Verlag, 1985).
11. "Interim Report: Fate and Transport of Substances Leaking from Underground Storage Tanks." Vol. I–Technical Report. Office of Underground Storage Tanks, U.S. EPA (1986).
12. "Interim Report: Fate and Transport of Substances Leaking from Underground Storage Tanks." Vol. II–Appendices. Office of Underground Storage Tanks, U.S. EPA (1986).
13. American Petroleum Institute. "Forced Venting to Remove Gasoline Vapor from a Large-Scale Model Aquifer." Publ. No. 4431 (1984).
14. American Petroleum Institute. "Subsurface Venting of Hydrocarbon Vapors From an Underground Aquifer." Publ. No. 4410 (1985).
15. Marley, M. C., and G. E. Hoag. "Induced Soil Venting for Recovery/Restoration of Gasoline Hydrocarbons in the Vadose Zone," in *Petroleum Hydrocarbons and Organic Chemicals in Ground Water—Prevention, Detection and Restoration, 1984 Conference and Exhibition* (Worthington, OH: Water Well Journal Publishing Co., 1984).

16. American Petroleum Institute. "Short Term Fate and Persistence of Motor Fuels in Soil." Draft report for the Marketing Department (1987).
17. Bossert, I., and R. Bartha. "The Fate of Petroleum in Soil Ecosystems," in *Petroleum Microbiology,* R. Atlas, Ed. (New York: MacMillan Publishing Co., 1984).
18. de Kreuk, J. F. "Microbiological Decontamination of Excavated Soil," in *Contaminated Soil: First International TNO Conference on Contaminated Soil,* J. W. Assink and W. J. van den Brink, Eds. (Boston, MA: Martinus Nijhoff Publishers, 1986).
19. Dragun, J. "Microbial Degradation of Petroleum Products in Soil," in *Petroleum Contaminated Soil: Public and Environmental Health Effects,* E.J. Calabrese and P. T. Kostecki, Eds. (New York: John Wiley & Sons, in press).
20. Soczo, E. R., and K. Visscher. "Biological Treatment Techniques for Contaminated Soil." *Resources and Conservation* 15:125–136 (1987).
21. American Petroleum Institute. "The Land Treatibility of Appendix VIII Constituents Present in Petroleum Industry Wastes." Publ. No. 4379 (1984).
22. American Petroleum Institute. "Feasibility Studies on the Use of Hydrogen Peroxide to Enhance Microbial Degradation of Gasoline," Publ. No. 4389 (1985).
23. American Petroleum Institute. "Field Study of Enhanced Subsurface Biodegradation of Hydrocarbons Using Hydrogen Peroxide as an Oxygen Source." Publ. No. 4448 (1987).
24. American Petroleum Institute. "Literature Survey: Unassisted Natural Mechanisms to Reduce Concentrations of Soluble Gasoline Components," Publ. No. 4415 (1985).
25. American Petroleum Institute. "Proceedings: Sampling and Analytical Methods for Determining Petroleum Hydrocarbons in Groundwater and Soil." Publication of the API Health and Environmental Sciences Department (1987).
26. Van Cleemput, O., and A. S. El-Sebaay. "Gaseous Hydrocarbons in Soil," in *Advances in Agronomy* Vol. 38. (San Diego, CA: Academic Press, 1985).
27. Holbrook, T. "Hydrocarbon Vapor Plume Definition Using Ambient Temperature Headspace Analysis," in *Petroleum Hydrocarbons and Organic Chemicals in Ground Water—Prevention, Detection and Restoration, 1987 Conference and Exhibition* (Worthington, OH: Water Well Journal Publishing Co., 1987).
28. Robbins, G. A., J. T. Griffith, J. D. Stuart, and V. D. Tillinghast. "Use of Headspace Sampling Techniques in the Field to Quantify Levels of Gasoline Contamination in Soil and Groundwater," in *Petroleum Hydrocarbons and Organic Chemicals in Ground Water—Prevention, Detection, and Restoration, 1987 Conference and Exhibition* (Worthington, OH: Water Well Journal Publishing Co., 1987).
29. American Petroleum Institute. "Manual of Sampling and Analytical Methods for Petroleum in Groundwater and Soil," Publ. No. 4449 (1987).
30. Kostecki, P. T., and E. J. Calabrese. "A National Survey of State Regulatory Approaches to Dealing with Soil Contaminated with Petroleum Products," in *Petroleum Contaminated Soils: Public and Environmental Health Effects,* E. J. Calabrese and P. T. Kostecki, Eds. (New York: John Wiley & Sons, in press).
31. Stokman, S., and R. Dime. "Soil Cleanup Criteria for Selected Petroleum Products," in *Proceedings of the National Conference on Hazardous Wastes and Hazardous Materials,* (Silver Spring, MD: Hazardous Materials Control Research Institute, 1986).
32. Tucker, W. A., and C. Poppell. "Method for Determining Acceptable Levels of Residual Soil Contamination," in *Proceedings of the National Conference on Hazardous Wastes and Hazardous Materials* (Silver Spring, MD: Hazardous Materials Control Research Institute, 1986).

33. "Underground Storage Tank Corrective Action Technologies," Office of Solid Waste and Emergency Response, EPA/625/6-87-015 (1987).
34. Skinner, J. H. "Innovative Technology for Superfund," in *The Fifth Annual Hazardous Materials Conference* (Wheaton, IL: Tower Conference Management Co., 1987).

CHAPTER 2

A State's Perspective of the Problems Associated with Petroleum Contaminated Soils

John J. Hills

INTRODUCTION

With the substantial increase in production of automobiles in the United States during the 1940s and 1950s came an unprecedented demand for petroleum by-products to fuel the high number of new automobiles.

To meet this demand, fuels were transported to retail and wholesale distribution points by rail, highway, and underground transmission lines, and stored in aboveground as well as underground tanks. These practices, in turn, created a significant potential for accidental discharges onto land of the petroleum by-products.

The discharges that were a potential problem in the 1950s have become a reality in the 1980s. The purpose of this chapter is to identify, from a regulatory perspective, the various environmental and public health problems that have resulted from soil being contaminated with petroleum by-products.

As you may be aware, the Environmental Protection Agency (EPA) is currently developing regulations requiring construction and monitoring requirements for underground tanks storing hazardous substances. In 1984 similar laws were enacted in California, and by the latter part of that year California was knee-deep in the problems associated with the enforcement of a statewide underground tank regulatory program.

Identifying the number and location of the tanks was one of the prime objectives of the program, but it was not difficult to accomplish. Enforcing the new construction requirements was another of the initial objectives, and that also was not difficult to implement. However, we eventually did encounter a major problem in implementing the program in California. That problem was addressing the issue of underground tanks that had leaked petroleum by-products. Although I was aware of the fact that Southern California is a heavily urbanized area, I was surprised to learn that there were over 10,000 underground tanks within our jurisdiction, with approximately 80% of these tanks storing petroleum-related products. From our inventory it was calculated that we had a potential underground storage capacity of 80 million gallons of petroleum products.

It is not my intention to report, "Here are the problems we encountered in California, and this is how we solved them." The reason why I cannot make this statement is because it is not entirely true. We did, in fact, encounter many problems, but not all of the problems we encountered that were associated with petroleum contaminated soil have been resolved. The purpose of this chapter is to identify the various environmental and public health problems we encountered as a result of petroleum contaminated soils, and the difficulties we faced in assessing the contamination, establishing cleanup levels, and evaluating mitigation technologies.

Many states will be implementing underground storage tank regulatory programs as the federal regulations are adopted, and will probably encounter problems and develop concerns similar to those we experienced. In an effort to help provide consultants, regulators, and companies with petroleum contaminated soil problems with the tools that they need to effectively assess, evaluate, and mitigate these problems, I will discuss our experience.

THE PROBLEM

To begin, I would like to identify three major issues that Orange County initially faced in dealing with the problem of the effectiveness of the methodologies being utilized to assess the nature and extent of soil contamination. Subsurface investigation for petroleum contamination was not a new field, but suddenly, almost overnight, many companies had developed a "fail-safe" approach to accurately assessing the extent of contamination. With limited data and information, individuals were attempting to delineate the exact amount and location of petroleum contamination present in the soil at various sites. The problem with this approach was that these individuals attempted to generalize and oversimplify what often is a very complex problem. Additionally, the methodologies being utilized for obtaining soil samples and the laboratory methods and techniques did not, in many cases, provide an accurate picture of the site contamination.

Another issue that we faced was our need to develop cleanup standards for the soils contaminated with petroleum by-products. At that time there was limited information and toxicity data available addressing environmental and public health issues associated with petroleum contaminated soils. Due to the lack of data, we reacted to the problem by requiring that the contamination be removed to zero levels. This approach, which I will discuss later, further complicated our cleanup problems.

The third issue that we faced was evaluating the mitigation methodologies that were available and viable, apart from soil removal and disposal. During this time a number of companies began marketing treatment systems for petroleum contaminated soils. Because of limited financial and personnel resources, it was extremely difficult to evaluate these methods for effectiveness. As a result of the limited technologies available at the time, and our policy of requiring petroleum contaminated soils to be cleaned to zero, more and more responsible parties were removing the contaminated soil and disposing of it at hazardous waste landfills. However, three additional problems associated with this approach were identified early in the program—high costs, limited availability of hazardous waste landfills, and long-term liability.

The high costs associated with the transportation of hazardous waste, and fees charged by hazardous waste landfills, became a major financial burden to companies with large volumes of petroleum contaminated soil. Additionally, it should be noted that because of the limited number of existing and future hazardous waste disposal facilities, and the high overhead associated with this type of operation, costs can be expected to rise significantly in the future.

There are no hazardous waste landfills in many areas of the country, and the limited number of hazardous waste landfills that are available should be utilized for the disposal of those types of hazardous waste for which treatment technologies are not available. Federal and state waste reduction programs are placing more and more responsibility on the generators of hazardous waste to recycle or treat contaminated soils that are amenable to recycling or treatment. As we saw hundreds of tons of petroleum contaminated soil being excavated and hauled from various sites, it soon became apparent that alternative treatment technologies were necessary.

The liability for the disposal of petroleum contaminated soil at hazardous waste landfills remains with the generator of the waste indefinitely. Should adverse environmental or public health impacts associated with the landfill develop in the future, the generator of the waste could be required to bear a portion of the financial burden of the costs associated with mitigating the landfill.

THE SOLUTION?

As a regulator, these issues plagued our agency for the first six months of the program as we began encountering more and more petroleum contaminated sites. In response to these issues, I developed what is know as a "blinder" mentality or "tunnel vision." This approach allows just one remediation method and one acceptable cleanup level, regardless of the site conditions. The cleanup level was zero and the remediation method was to remove the petroleum contaminated soil and haul the soil to a hazardous waste landfill. This approach was developed in part due to the fact that there were limited references to guide me in addressing the issues, and as a regulator, I felt that I could not rely on the information or perspectives provided by consultants, engineers, and contractors familiar with assessing and mitigating petroleum contaminated sites. The approach lasted only a short while until we realized that we were creating more problems than we were solving with this policy. Petroleum contamination had migrated beneath multistory commercial buildings, under roads, highways, and major intersections, and hazardous waste landfills were receiving record amounts of petroleum contaminated soil.

About this time in 1985 I attended a conference on problems related to petroleum contaminated soil, sponsored by the University of Massachusetts. I attended the conference because the program topics addressed the very issues that our agency was facing. My attendance at the conference greatly increased my understanding of the problems, and provided a number of potential solutions to the issues associated with petroleum contaminated soils.

From the engineers and geologists who presented papers and attended the conference, I obtained a better understanding of the applications and limitations of the various assessment and mitigation technologies available.

From the regulators who were present at the conference, I learned not only of their frustrations in dealing with the issues, but also of their perspectives of the actual and potential public and environmental health considerations associated with petroleum contaminated soil.

From the representatives of industries responsible for the cleanup of petroleum contaminated soil, I gained a better understanding of the difficulties associated with dealing with a problem that is waiting for technologies to catch up with it.

And finally, from research scientists actually involved in dealing with this problem, I gained what turned out to be an invaluable perspective, which is that only through a practical approach and a cooperative effort with the various disciplines involved would we be able to effectively address and resolve the issues.

After returning to California from the conference, the problem that I thought was difficult soon began to grow out of control. It was during this period that our agency developed what we considered to be innovative and progressive policies relating to the problem, many of which were influenced by the information and data that I obtained at the University at Massachusetts conference. Looking back to 1985, the changes that we initiated in our cleanup policies were critical to the development of our program for managing the cleanup of sites contaminated by petroleum by-products. For, over the next two years, we encountered over 600 sites that were contaminated with petroleum by-products.

CURRENT PERSPECTIVE

As a result of the 600-plus cleanups of petroleum contaminated soil we have overseen to date, we have concluded that a remedial action plan for petroleum contaminated sites is not complete unless the cleanup of soil contamination is addressed. There are those who believe that contaminated soil need not be addressed as part of a site cleanup, and that only groundwater cleanup should be required. We disagree with this approach because of the following five problems that we feel are associated with petroleum contaminated soils:

1. *Soil with significant petroleum contamination can act as a source of contamination for groundwater supplies.* Early in the program, we encountered a number of sites where groundwater remediation had been completed, only to discover at a later date that the groundwater had been recontaminated. It was later discovered that the source of contamination was petroleum hydrocarbons that were contained in the unsaturated zone.

2. *Significant concentrations of petroleum by-products in soil have been found to permeate PVC water lines and contaminate drinking water supplies.* Approximately six months ago our agency responded to a water quality complaint at a convenience store in Santa Ana, California where a number of customers had complained that the water had a "gasoline" taste. The water was tested and found to have high concentrations of benzene, toluene, and xylene. Through a subsequent investigation it was determined that underground tanks at a gasoline station adjacent to the store had leaked, and the excessive concentrations of gasoline in the soil had permeated the store's water supply line.

3. *Vapors from petroleum contaminated soils may migrate laterally and/or vertically and collect in underground utility vaults, sewer lines, and basements, creating serious fire or explosion hazards.* Although there are relatively few homes with basements in southern California, there are many underground utility vaults. Because of numerous instances of petroleum products and vapors accumulating in these vaults, a number of utility companies require employees to monitor the vaults with instruments prior to entry in order to determine that the vault is safe to enter both from fire safety and health exposure concerns.

4. *Surface soils contaminated by petroleum by-products with nonbiodegradable additives such as lead are a potential source of ingestion by small children.* Our agency became involved with a state Superfund site in Westminster, where we were asked to investigate an excessively high lead value in a soil sample obtained from a preschool playground area. The analysis indicated a lead level of 3800 ppm. For those of you who are familiar

with the playing habits of small children, you are probably aware of the fact that they are constantly placing dirty objects as well as dirty fingers into their mouths. In doing so, they unknowingly are ingesting small amounts of soil, a phenomenon known as pica. Upon questioning the operator of the preschool, it was learned that leaded gasoline was routinely distributed in the playground area to control weed growth in the area, and that in doing so, a potential mechanism for contaminant ingestion had been created.

5. *Those constituents of gasoline with high vapor pressures may migrate upward through the unsaturated zone presenting a potential source of contamination for surface waters and certain food crops being grown in the immediate proximity of the contaminated soil.* Although our experience with this problem has been limited to contamination of surface waters, we perceive this to be a significant potential problem.

As stated earlier, it is for these reasons that we require parties responsible for subsurface contaminations to address soil cleanup as part of the overall remediation of petroleum contaminated sites. However, we feel that we must act in a practical and realistic manner by evaluating each site on a case-by-case basis in relation to these concerns when determining site cleanup levels.

THE CHALLENGE

This completes my summary of California's perspective on the problems associated with petroleum contaminated soil. However, I want to direct a specific challenge to each of several groups, in order for them to more effectively address the problems I have identified.

I encourage the consultants to develop their technical expertise on the subject in order to increase their knowledge and be able to provide their clients with the most effective and responsible approach to deal with the problem.

I challenge the industry representatives to familiarize themselves with the technologies and options available for assessing and remediating petroleum contaminated sites that they may be responsible for, in order that they may, from a financial and liability standpoint, be able to effectively participate in resolving the problem.

And finally I encourage the regulators to evaluate whether they have fallen, as I did, into the blinder or tunnel vision syndrome. If so, they should evaluate their position on the regulatory issues I have identified, realizing they are difficult, but not insurmountable, and should be addressed from a realistic and practical standpoint and approached in a cooperative effort with all of the parties involved.

CHAPTER 3

Electric Utility Perspective of the Cleanup of Petroleum Contaminated Soils

William A. Kucharski

Electric utilities were affected by the passage of Subtitle I just like every other industry in the country. While we had expected that underground storage tanks (USTs) would become regulated, we were surprised to learn the breadth of issues that would come to affect our operations. We acknowledged early that the petroleum industry had superior knowledge, greater numbers of tanks, and more experience associated with USTs than we as an industry ever wanted to have. It did not take us long to discover that electric utilities had similar but limited areas of coincident concern with the petroleum industry.

Underground storage tanks are a small part of our operations, and as an industry we have approximately 12,000 tanks. More than 50% are older than 10 years and 85% have between 500- and 20,000-gallon capacities. Eighty percent are steel, fewer than 50% have internal protection, and fewer than 15% are cathodically protected. Eighty percent of our tanks contain gasoline or diesel. The probability that the electric utility industry will encounter contaminated soil cleanup in the next few years is greater than one. For this reason, I would like to discuss a process for negotiating an acceptable cleanup. I will first, however, describe how the electric utility industry structured itself to respond to the issue of UST regulations.

HISTORY

The electric utility industry is recognizing that the economic well-being of its service areas, as well as the cost of electricity, is of critical concern to its own health. For this reason, the industry has opted to become more involved in federal and state regulatory issues. When Subtitle I was enacted in November of 1984, the utility group that deals with Resource Conservation and Recovery Act (RCRA) issues, the Utility Solid Waste Activity Group (USWAG*), prepared to respond to this legislation by creating a new Leaking Underground Storage Tank (LUST) committee, which I have chaired since its inception. The committee organized two subgroups, a performance standards subcommittee and a remedial action task force. The LUST Committee was created in March of 1985, and one of the first things to accomplish was defining a goal. We settled on the aim of achieving flexible and reasonable UST regulations. In addition to this primary goal, the committee had to anticipate which regulatory and operational issues were going to be raised, identify information that would be helpful in mitigating impacts on our industry, and finally, obtain and disseminate technical information to our industry on UST management options.

To prepare our industry for the new UST requirements, we have had eight studies completed over the past two and one-half years. They are:

- a generic response plan
- a new tank installation and specification guide
- a review of leak detection devices
- a review of cleanup methodologies
- a revised survey of underground tanks in the industry
- a review of secondary containment methodologies
- a decision tree model for use in tank management
- a review of hydrogeologic susceptibility to tank leakage.

These projects are all complete and are being used by electric utilities throughout the nation. In addition, more than 140 pages of comments were submitted to the Environmental Protection Agency (EPA) on the proposed UST regulations published in April of 1987.

The final UST regulations are scheduled to be published in the spring of 1988, and if the cleanup rules are finalized as proposed, cleanup will be site by site. This "flexibility" is a mixed blessing, in our opinion. While we have not had overly stringent and unreasonable cleanup levels imposed, we have not had

---

*USWAG is an informal consortium of Edison Electric Institute (EEI), American Public Power Association (APPA), National Rural Electric Cooperative Association (NRECA), and approximately 75 electric utility operating companies. EEI is the principal national association of investor-owned electric power and light companies. APPA is the national association of rural electric cooperatives. Together, USWAG members represent more than 85% of the total electric generating capacity of the United States and service more than 95% of the nation's consumers of electricity.

reasonable levels provided either. This lack of firm guidance, therefore, grants all of us the privilege of fighting over what is reasonable, time and time again. Implementation of this privilege leads me to my major concern—how do we "negotiate" a cleanup?

First, I would like to share a thought process that an official from the 3M Corporation came up with in the mid-1970s, which I plagiarized from the *Environment Reporter* and have used ever since. I call it Environmental Reality.

1. Environmental *issues* are emotional.
2. Environmental *decisions* are political.
3. Environmental *solutions* are technical.

I have added the following empirical observation to this axiom: whoever controls any two of these points will prevail in an environmental negotiation.

In my opinion, 95% of all industry regulatory failures have resulted from our training and belief that only the technical solution is necessary in a negotiation. As science/engineering professionals, we are trained and educated to find analytical answers to all problems, and we easily fall into the trap of believing that everyone thinks "our" way, and that all parties have similar constraints and objectives.

Turning a multidimensional problem into a single-dimensional solution affects regulatory agencies as well. Let us examine the basic plight of the regulatory agencies: they have a thankless job. If the industry being regulated does not sue, the professional environmental interest groups will. No decision an agency can make will ever please everyone. I therefore submit that the axiom of Environmental Reality affects a regulatory agency as much, if not more, than it does any industry. Recognizing that even regulatory agencies have multiple goals and constraints is conceptually very important.

Can we negotiate the cleanup of an individual petroleum contaminated soil site, meet the goals of protecting human health and the environment, and also execute this in a cost-effective manner? The answer is yes, and I will attempt to describe a process by which these goals can be achieved.

To negotiate satisfactorily, we must be cognizant of three major requirements:

1. We must be prepared.
2. We must be credible.
3. We must maintain the initiative.

PREPAREDNESS

In order to deal successfully with any environmental issue, one should be an "expert," someone who knows more than his/her audience. When a leak or a potential leak has been discovered, it is crucial that we *understand* not only what the problem is, but also what the applicable law and regulation requires. Note that I mentioned both the law and the regulations. As an aside, let me say that I review many proposed regulations. When I find a rigid, impossible, or plainly ridiculous phrase in a proposed regulation, I invariably find that these words have been taken verbatim from the law. Regulating by legislation leaves the presumed authors of the regulations, by definition, with no flexibility or options, but the agency must obey the law.

We must know what an agency can do and what it must do before we can know what areas of negotiation are open to us. We need to know where the agency has no options. Only then can we place the leak into the perspective of these regulatory obligations. If an agency is constrained by state laws to the cleanup of all constituents down to 1/100th of a drinking water standard or to zero, we must find another channel of negotiation other than arguing that "the standard is stupid." In short, we should not break our pick on a clear loser.

We must be prepared to define what cleanup goal we expect to achieve and which cleanup methodology we expect to utilize, as well as why each was chosen. The EPRI/USWAG/UMASS/WESTON review of cleanup alternatives describes many of the possible onsite and offsite cleanup methods available, and as we learned at the Amherst conference, more and better methodologies are now being developed. It is crucial, however, to have done our homework to ensure that site-specific political and emotional considerations are factored into all cleanup proposals. For example, if the local newspaper has just published a series of articles on the hazard of benzene, be prepared to address this issue—whether relevant to your situation or not.

Whatever else we do, we must provide an agency with data that it can use to support a final cleanup decision. We must be prepared to discuss the alternatives we have examined and the rationale for our decisions. While liability considerations make a nontechnical cleanup decision foolish for a viable company, we should, in addition to providing technical data, be prepared to discuss the economic considerations as well as demographics, populations at risk, and potential exposure pathways related to the cleanup proposal. The agency will ultimately have to justify all final cleanups, and the agency will do so—with our input or without it. Certainly, with an input is better.

Another aspect of being prepared is that we must start collecting data and reviewing alternatives before a spill or leak or tank removal occurs. This was the motivation behind the Generic Response Plan that the UST committee of USWAG commissioned as one of its first projects. We tried to define the variables and identify those pieces of information that each tank location should have collected prior to an incident. In summary, if we are well prepared we will have a better chance of achieving approval for a cost-effective cleanup. This preparation, however, must be joined with the other two ingredients—credibility and initiative.

CREDIBILITY

We must be credible. Credible means that we must be honest and straightforward. This does not mean confessing all past sins; neither does it mean forgetting to mention problems. Acquiring a reputation for being forthright and honest does not happen overnight. It takes time and a lot of effort to create and nurture, and we have to work continually to keep it. If there is one truism related to credibility, it is that it cannot be bought after we have discovered a problem.

We have to have been willing to cooperate, and to exchange or provide data and test results, etc., when we had no problem, in order for the credibility factor to work for us when we do have a problem. The result of lack of credibility is invariably a legal confrontation and the goal is to avoid, whenever possible, a legal confrontation. The short-term solution of tricking an agency can possibly work—once. I have found no long-term advantage to misleading an agency, however. I should point out that the agency, whichever it is, has the same obligation to be honest and credible. If cooperation and mutual solutions to real problems are to occur, each side of the operation must be willing to trust the other.

This means that an agency ought not to ask for tons of data when a decision has already been made, and it should not promise what it cannot deliver.

Above all, we should recognize that the regulatory agency is simply trying to do the best job it can. In general, industry has more experienced and better technical resources available than do most regulatory agencies. As I pointed out earlier, this may cause more problems than it solves, especially if one-dimensional technical considerations are allowed to be the only basis of a cleanup position. Obviously, the technical expertise available must be used intelligently, and all data and cleanup alternatives should be presented in a nonpatronizing but technically correct way. We should view our job to be providing enough data so that the agency will be technically secure in supporting our proposal. Do not forget the political/emotional constraints, however. Keep in mind that in political terms, the concerned parent in the back of the room may have a more powerful voice in influencing a cleanup decision than our PhD in Soil Chemistry. The more straightforward we are, the more answers or data we provide before a question is asked, the better position we create—the more credible we become. We must have contacted, talked to, and developed rapport with agencies, citizen groups, and other potential critics long before we developed a problem. Trying to talk to such groups only when we have a problem sends a very negative message.

Being credible means being trusted. I can think of no cleanup scenario that is perfect, no cleanup that has no risk or exposure or downside. If a proposal to an agency is only peaches and cream, it will be difficult to have created an image of a credible report. By identifying the negative as well as the positive aspects of a cleanup in the proposal, I believe we encourage more support than criticism from an agency. After all, we deal with relative risks and hazards. While we know that there are no absolutes, political requirements dictate that we communicate this reality effectively. One way is the discussion of the problems and risks associated with a chosen cleanup, relative to the other potential technologies we did not choose. The use of nontechnical communication is a basic part of creating an atmosphere of trust. This examination of risks, expenses, and alternatives can be placed in the category of emotional preparation or political necessity. In essence, we are saying "Yes, we *cared* enough to examine all aspects of our proposal." Providing both the risks and benefits of a cleanup is a necessary and critical part of developing credibility. If we are not trusted as a company, the best technical work becomes useless; it is imperative that we create and maintain a reputation for good and credible work.

INITIATIVE

In order to have any chance of influencing a regulatory decision, such as a satisfactory cleanup, we must have control of the process. This means simply that we need to act first, to get out in front of the situation, and not to get into a position of *reacting* to the problem. It has been my experience that if we are prepared, do credible work, and anticipate agency requirements, everyone wins. The agency has enough problems with companies and situations where there is no chance of cooperation. If we are reasonable, competent, and showing progress, the probability of achieving a reasonable cleanup is enhanced. While a truculent agency or inspector can make attempts to cooperate very difficult, I believe that most regulatory agencies allow good companies to do good work.

How does one maintain initiative? Well, we must know where we want to go, and we must take a risk by doing something before we have approval. If we know how an agency must determine an appropriate cleanup, we can minimize the risk of doing unnecessary work. For example, if a leak is suspected, mobilize a consultant or company crew at once. Do not wait until the agency requires this action. Have the field people develop a preliminary work plan as soon as possible. When the agency is contacted, attempt to set up a meeting and then tell the agency what is being done. It is surprising how much cooperation can be achieved if you are doing what has to be done before an agency requires it of you. Additionally, this approach will require the agency to review your plan. Unless an agency has more experienced and better trained personnel than you do, you do not want to be in the position of asking the agency what to do next. If you are prepared, you will be in the position of having anticipated what the agency will need in order to make a decision and you will be ready before the agency is. When the time for discussion arrives, it is highly desirable that the basis for the discussion be *your plan*.

Remember that an agency probably has a lot more open cases/spills and problems than we do. Ours is just one case out of many, and the more initiative we take, the less initial work that agency has to do. Does this mean we get to run over the agency? Of course not. When we couple the preparation, which defines our work goals and procedures, with credible work, we find a cooperative, professional, and potentially efficient program. To achieve such a result is a primary goal of the regulatory agency. Again, both sides should expect the *best,* rather than the *worst,* from each other.

Believe it or not, the approach I have just described does work. It may not work all of the time, such as when an emotional outcry triggers a political decision that results in a nightmare. In that case, all the data, science, and facts in the world can be useless. The environmental arena is unique; it is unique because it has no certainty associated with it—none. All we can deal with is probability. So when you set out to negotiate a cleanup, recognize that you can do everything right, and lose. Conversely, you can ignore the odds and maybe never get caught. (Or you could win the lottery.) The negotiating formula I have described is one that will maximize, in my opinion, the probability of achieving a successful, cost-effective cleanup.

I have presented a formula for negotiating a cleanup. One could reasonably ask a simple question: Why bother? What does it matter how we clean up a petroleum spill or leak? The answer is dollars—BIG dollars. I described earlier the research projects that the UST Committee of USWAG commissioned.

Figure 1 refers to the site risk cost curves produced by Dames & Moore. This study was very difficult to produce, mainly because we were unable to define what level of cleanup would be acceptable under the new UST rules. The goal was to have a set of curves produced that would compare cleanup costs for a generic leak into several generic hydrogeologic regimes.

Whether the cost variations between the hydrogeologic scenarios are precise, or even accurate, is not so much the issue as indicative of what unreasonable cleanup goals could cost.

The study assumed that a tank leaked at a constant rate of 0.1 gal/hr. Varying depths to groundwater were assumed, based on the hydrogeologic structure. High, medium, and low cleanup costs were developed for each type of site. Air stripping with air pollution controls, soil venting, interceptor trenches, scavenger/wells, and soil excavation were some of the technologies proposed to accomplish the cleanup. I would like to quote the final assumption and caveat for this report:

"Cleanup levels and costs can often be negotiated. Negotiated settlements are not considered in this report."

The results for medium level cleanups under various hydrogeologic conditions are shown in Figure 1. The X axis reflects the length of time a leak has been occurring at 0.1 gal/hr; the Y axis, the resultant cleanup cost. For the cleanup of a five-year leak, the costs range from $320,000 to $2.7 million, depending upon the site.

This is why we negotiate.

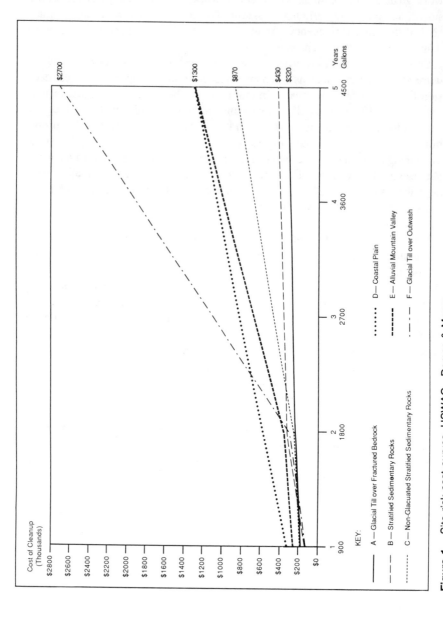

Figure 1. Site risk cost curves, USWAG—Dames & Moore.

If there is a final message, it is that environmental negotiations are multifaceted and more than technically based. No amount of data, even when it is correct and peer reviewed, will guarantee a satisfactory negotiation with an agency.

CHAPTER 4

Problems Dealing with Petroleum Contaminated Soils: A New Jersey Perspective

Robert K. Tucker

The New Jersey experience reinforces the perspective of the complex nature of soil contamination. Petroleum products significantly contribute to such contamination; New Jersey's giant petrochemical industry, second largest in the nation, dominates the economy of this state. Tourism is New Jersey's second leading industry; in the relationships between these two activities, ironies abound. Travel is fueled by the products of petroleum; the industry provides the raw materials for plastics and synthetics, and the energy for refining metals for autos, trains, and planes that transport people to New Jersey's vacation spots, and for many of the useful products people use everyday, even on vacation. Yet tar balls, oil slicks, and plastics washing up along our shores, as well as contamination of drinking water, landfills, and dirty air, are cited as factors that discourage visitors to the Garden State.

What to do with waste products of our highly technological society and how to deal with pollution of soil, air, and water have become the most important environmental questions of the 1980s.

The magnitude of chemical use, and consequently the amount of waste products involved, is staggering. Petroleum use in 1985 in the United States was on the order of 30,000,000 barrels per day. Consider the sheer volume of plastic production—more than 20 million tons a year in the U.S. By 1976, the cubic volume of plastic manufactured in the U.S. had surpassed that of steel, copper, and aluminum combined.

Because of soil contamination and subsequent groundwater problems, land disposal is no longer a viable option in New Jersey. Our state is in desperate straits regarding domestic landfill capacity; in some parts of New Jersey, tipping fees have gone up an order of magnitude over the last couple of years. Finding sites for hazardous waste disposal or treatment is substantially more complicated and difficult.

New Jersey's status as the most densely populated of U.S. states and also one of the most highly industrialized has intensified contamination problems. We've become one of the nation's premier laboratories for investigating the problems and studying remedial solutions for toxic chemicals.

Work carried out in New Jersey makes it more and more apparent that we must view contamination problems from an overall perspective which accounts for their complexity and effect on all media. While it is appropriate to focus on a particular aspect, such as soils contamination, or to deal with particular components, we must not lose sight of the interactions and multimedia implications. Although a chemist might look at soil contamination from a compound-specific point of view—simply studying behavior of benzene, for example—alternative ways of approaching the problem include ecological perspectives or an overall systems approach.

When we speak of petroleum products, not only are we dealing with the components of petroleum, but also the many additional compounds for which petroleum serves as the starting material. Thus, logically, we should consider halogenated solvents, pesticides, plastics, and other synthetics when we discuss soil contamination by petroleum products.

Our regulatory strategies depend on how we define the sets of problems and specify boundaries.

Consider some additional complications:

- Petroleum products become contaminated in use. Crankcase oils, for example, pick up metals and also have been shown to become mutagenic, exponentially with time of use, when run in vehicles.
- Used oils have become media for disposal of other toxic chemicals. A notorious example is the addition of PCBs to fuel oils sold on the residential heating markets in New York and New Jersey.
- Additives to gasoline for scavenging lead, DBE and DCE, produce many additional halogenated volatile organics and have the potential for dioxin formation.
- Lead, in addition to its use as an additive to gasoline, has been a substantially increased environmental contaminant through auto battery use.
- Petroleum products have been used as dispersal media for pesticides, or as substantial components of pesticides, as illustrated in a California study of pesticide use:

Chevron oil, a nonselective weed oil, is a broad-spectrum herbicide commonly applied to alfalfa as a seed harvest aid; it contains predominantly alkyl-substantial benzene and naphthalene derivatives, along with some aliphatics (C9–C19). Beacon oil, a typical selective weed oil, commonly applied to fields containing young carrots (three frond stage) for weed control, consists primarily of aliphatics (C9–C19). For Beacon oil, the field data convincingly show that over 90% of the residue evaporated within three hours after application. For Chevron oil, about 45% apparently became airborne as drift during application, and 90% of the foliage surface residue evaporated in 0.44 to 0.75 hours after application. Evaporation behavior of mixture changes with time. More volatile components evaporate early, resulting in a decreased rate of the mixture as its vapor pressure decreases.

Direct soil contamination from leaking underground tanks provides substantial amounts of petroleum pollutants. However, direct disposal of used oils on the ground by individuals or industries; seepage from landfills, illegal dumps, unlined pits, ponds, or lagoons; and spills from transport accidents or even auto accidents all make contributions to the soil burden.

Indirect contamination through aerial transport plays a role in widely spreading pollutants, also contributing to soil residues.

Complications of dealing with soil contamination include problems with sampling, detection, and analysis; determining fate and transport of individual components or mixtures; and evaluating actual exposure in order to be able to undertake valid risk assessments.

New Jersey's experience with strategies for dealing with soil contamination has led to the realization that we must emphasize prevention. Without major efforts at source reduction, the problems will overwhelm us.

New Jersey has a substantial data base with which to begin understanding soil contamination and its interactions with other environmental media. The New Jersey Department of Environmental Protection (DEP), and particularly its Division of Science and Research (DSR), has pioneered statewide studies of toxics in air, surface water, groundwater, drinking water, and at waste sites around the state.

WATER

In the mid-1970s DSR began statewide water surveys. Our original groundwater study pointed out the problems with halogenated solvents, a finding subsequently confirmed to be almost ubiquitous nationwide. It was determined that 16.6% of New Jersey wells had volatile organics (VOs) greater than 10 parts per billion (ppb) and 3.1% greater than 100 ppb. Lead was also found to be a problem, detected in 48.8% of wells and above drinking water standards in 13 of 670 wells tested.

The United States Geological Survey conducted a more focused study on a single coastal plain aquifer, the Potomac-Raritan-Magothy, in southern New Jersey. They found benzene in 8.6% and toluene in 6.6% of wells in the outcrop area of the aquifer. Halogenated solvents TCE and PCE, were frequently detected.

DSR followed up the groundwater study with tests of all types of drinking water supplies, analyzing for an expanded list of chemicals. Aromatics such as benzene, xylenes, toluene, and halogenated organics were detected most often.

The DSR research and survey findings galvanized the state legislature into passage of amendments to the N.J. Safe Drinking Water Act. This legislation, P.L. 1983, c.443 (A-280), called for the establishment of Maximum Contaminant Levels (MCLs) based on one in a million excess risk for carcinogens, and mandated testing on a regular basis by every supplier in the state.

The first round testing, reported in 1985, showed the halogenated VOs TCE (5.5%), PCE (4.9%), and 1,1,1-TCA (4.9%), most frequently detected. Benzene was the aromatic most frequently occurring, reported in 1.1% of samples at a range of 1.4 to 5.0 ppb. The second round is little different overall, although the range for benzene, 0.24 to 33.0 ppb (found in 1% of samples), is a little greater.

New Jersey now has the most comprehensive monitoring program in the nation for volatile organic chemicals and other organics in drinking water. This monitoring has been performed prior to any federal monitoring requirements, and consequently more is known about drinking water quality in New Jersey than in most other states.

AIR

DSR's Airborne Toxic Elements and Organic Substances project was designed to simultaneously measure atmospheric levels of more than 50 toxic and carcinogenic chemicals within three urban population centers and one rural area. Conducted in the summers of 1981 and 1982, and winters of 1982 and 1983, the extensive analysis included examination of seasonal variations, pollution episodes, interurban differences, and source composition. Parameters measured included: extractable organic matter, PAHs, VOCs, mutagenic activity (Ames), trace elements, inhalable particulate matter, and individual compounds. The aromatic compounds—benzene; toluene; o-, m-, p-xylenes; ethylbenzene; and styrene—generally had the highest concentration and greatest frequency of the 26 compounds measured. The results of this extensive project are described in a book edited by Lioy and Daisey.[1]

SOILS

The Division of Science and Research has been supervising a statewide study of soils for background levels of contamination. In addition to chemical analysis, more than 80 locations have been studied for land use, soil type, and surrounding activities which might contribute to any anomalies. Soil samples have been characterized for general soil parameters, including texture, percent organic matter, pH, and cation exchange capacity. Thirty-four of the soils sampled represent the major soil types found in New Jersey, while others were collected from a variety of land use categories, including golf courses, state and local parks, and farmland. These analyses will provide comparisons with the extensive data the state now has on the many severely contaminated locations that have been located and analyzed during the last several years under Superfund, the Environmental Cleanup Responsibility Act (ECRA), the Resource Conservation and Recovery Act (RCRA), etc. Assisting DSR in this study have been the U.S. Soil Conservation Service and Rutgers University.

POLYNUCLEAR AROMATIC HYDROCARBONS (PAH)

The Division of Science and Research has also conducted extensive multimedia analyses on particular contaminants. Of special interest in the area of soil contamination is our work on PAHs. In the early 1980s we carried out a study of PAHs in soil, air, and water around a foundry in Mahwah. More recently, we have coordinated an extensive Total Human Environmental Exposure Study for benzo(a)pyrene, including measurements in outdoor air, indoor air, water, food, and soils, as well as in blood and urine of subjects in our study area.

We have a better data base for concentrations of PAH in water, air, and sediments than in soils. In part, this may be due to analytical difficulties and difficulties in interpretation of results. We have also spent considerable research effort in environmental fate and transport studies, especially hydrocarbon transport into aquatic ecosystems. Hydrocarbon compounds attach to particles and settle out of the water column quickly, but may be resuspended and contribute to lateral transport in the bottom nepheloid layer. We still need to know more about transport mechanisms, and also about the range of point and nonpoint sources.

Vogt et al.[2] compared PAHs in soil and air around industrial sources and found patterns different than in relatively unpolluted soil. Pattern recognition methods and correlation analysis may allow interpretation of mechanisms that govern the distribution of PAH between soil and air.

New Jersey DEP's Division of Science and Research will continue its research program to measure contaminants in soils and other media, and to understand transport and other interactions. Research in New Jersey has led to legislation, regulation, and vigorous enforcement of environmental laws. A recent example is our push to prevent leaks from underground tanks.

UNDERGROUND STORAGE TANKS

There are approximately 150,000 underground storage tanks in New Jersey. It is estimated that approximately one-third of these are discharging hazardous substances into the environment. The DEP conducted a study of the contamination caused by underground storage tanks subject to New Jersey's ECRA, N.J.S.A. 13:1K–6 *et seq*. The study indicated that 36% of the tank systems (at 46% of all sites under ECRA's jurisdiction) exhibited evidence of contamination from overfills, spillage, discharges from leaking tanks or piping, or generally poor management practices. Improperly installed, maintained, removed, or abandoned tanks are generally the cause of these releases. Such releases have the potential to cause severe harm to human health and the environment.

NEW JERSEY LEGISLATION FOR UNDERGROUND STORAGE TANKS

The New Jersey Underground Storage of Hazardous Substances Act (N.J.S.A. 58:10A–21 *et seq*.) is patterned after the federal "Hazardous and Solid Waste Amendments of 1984 to the Resource Conservation and Recovery Act" (42 U.S.C. 6901 *et seq*.). The federal act provides for the delegation of the federal program to the individual states. Both the federal and the state statutes require the registration of facilities, technical standards for new tanks, leak detection programs, remedial action for sites found to be contaminated, and financial responsibility. In addition, the state act, unlike the federal act, provides for a permit program for new tank installations, for DEP approval of tank closures, for the annual reregistration of all facilities, and for a loan program to ease the economic burden for owners who replace their tanks.

The New Jersey Underground Storage of Hazardous Substances Act (N.J.S.A. 58:10A–21 *et seq*.), passed in 1986, requires the DEP to adopt comprehensive regulations to carry out the following purposes, as set forth in the opening paragraph of the Act:

> The Legislature finds and declares that millions of gallons of gasoline and other hazardous substances are stored prior to use or disposal, in underground storage tanks; that a significant percentage of these underground storage tanks are leaking due to corrosion or structural defect; that this leakage of hazardous substances from underground storage tanks is among the most common causes of groundwater

pollution in the State; and that it is thus necessary to provide for the registration and the systematic testing and monitoring of underground storage tanks to detect leaks and discharges as early as possible and thus minimize further degradation of potable water supplies. The Legislaturefurther finds and declares that with the enactment by the United States Congress of the "Hazardous and Solid Waste Amendments of 1984" Pub. L. 98-616 (42 U.S.C. 6991) it is necessary to authorize the Department of Environmental Protection to adopt a regulatory program that permits the delegation of authority to carry out the federal act, but also recognizes the need of this State to protect its natural resources in the manner consistent with well-established principles.

The DEP has divided the legislative requirements into three sets of regulations. The first set, Administrative Regulations, spells out general information, registration requirements and procedures, fees, and penalties. The second set, Technical Regulations, includes new tank performance standards, monitoring and testing requirements, installation and closure requirements, other criteria which the Department considers necessary for the prevention of releases from underground storage tanks, and corrective action criteria governing appropriate action in the event a release occurs. The third set of regulations provides for a loan program authorized in the legislation, and includes criteria for issuance and terms and conditions of loan approvals. The loan program is designed to ease the economic burden for owners who replace their tanks.

New Jersey state regulations incorporate the federal exemptions for flow-through process tanks, wastewater treatment tanks, and electrical equipment. The U.S. Environmental Protection Agency (EPA) has deferred imposing a notification requirement on five other types of tanks: sumps, used oil tanks, bulk storage tanks, radioactive waste tanks, and hydraulic lift tanks. The USEPA suggests that there is insufficient information at this time to determine whether regulation of these types of tanks is necessary. New Jersey intends to impose some technical requirements on these five types of tanks. The extent of the requirements is not defined at this time; however, the registration of these tanks will allow the NJDEP to identify the nature and extent of these types of tanks and their potential impact on groundwater. Registration requirements of the New Jersey state act require that the owner or operator of each underground storage tank in the ground on or after September 3, 1986 register the facility with the NJDEP. The purpose of the tank registration is to accumulate a data base of all the regulated underground storage tanks in the state that are subject to the state act. The owner or operator shall annually certify, on the New Jersey Underground Storage Tank Annual Certification Form, the operational status of the facility. In addition, each owner or operator is required to maintain reconciled inventory control records. If the inventory control records indicate that a release has occurred, or if any person discovers that a release has occurred, the owner or operator must, within 24 hours, notify the NJDEP.

The owner or operator of any tank which is abandoned but left in the ground must register the tank with the NJDEP. Unless the tank is totally removed from the ground and the site fully examined for possible contamination, or the tank is properly abandoned under procedures to be provided in the technical regulations, the tank will be considered to be in use and the owner or operator will be required to annually certify the tank with the NJDEP.

Failure by an owner or operator of an underground storage tank to comply with any requirement of the state act may result in the imposition of penalties as set forth in N.J.S.A. 58:10A–1O, which authorizes fines up to $50,000 per day per violation.

NEW JERSEY EXAMPLES OF SOIL CONTAMINATION

Specific examples of soil or groundwater problems show why we in New Jersey are so concerned about soil contamination. These examples have been taken from a report by DEP's Division of Hazardous Site Management.[3]

ROCKAWAY TOWNSHIP WELLS

Rockaway Township has a population of approximately 20,000. The township water supply system serves 12,000 of these residents and consists of three pumping wells (numbers 4, 6, and 7) located in a wetlands region on the northeast side of Green Pond Road. In November 1979, trichloroethylene (TCE) was detected in all three wells at levels causing concern for public health. Despite efforts to eliminate it, contamination worsened. In October 1980, di-isopropyl ether and methyl tertiary butyl ether were detected in well #7 at levels of 100 ppb and 40 ppb, respectively. These results forced the township to declare a water emergency, and discontinue use of all three wells. Investigation determined that several sources were contributing to the contamination of these wells. A Shell Oil Company gasoline station approximately 1,100 ft from the well field was confirmed as one source. In 1980 the owners, in conjunction with the township, provided funds for the installation of a treatment system consisting of an air stripping unit and an activated carbon filtration unit to treat the contaminated water, and thus allow the township to resume use of this supply.

The significant levels of toxic organics found in the Rockaway Township wells raised concern about its safety as a supply of potable water. Moreover, the township wellfield draws from the Valley Fill aquifer, a sole source aquifer which represents the only supply within the township capable of meeting the community's water needs. Water-bearing zones in the surrounding fractured bedrock formations are very low-yielding.

LONE PINE LANDFILL

Lone Pine, a 45-acre landfill, operated for about 20 years until the late 1970s.

During that time, in addition to municipal garbage and septage, more than 17,000 drums of chemical waste and several million gallons of bulk liquid chemical waste were disposed of in the landfill. Several contaminated groundwater plumes in both the shallow Vincentown and the deeper Red Bank aquifers appear to migrate from the landfill in a northerly direction toward and into the Manasquan River.

There is considerable leachate seepage, especially after rainstorms. Contaminated surface water runoff flows into adjoining wetlands to the north and then into the Manasquan River. Volatile organics and other contaminates have been detected in the river water column and sediments downstream from Lone Pine. Cleanup is imperative, since a potable water reservoir is slated for construction with an intake downriver on the Manasquan from this landfill. Benzene has been measured in surface water of the Manasquan at 25 ppb. Substantially higher levels, 1940 ppb in groundwater and 2900 ppb in soil, occur at the landfill. Levels of toluene up to 4700 ppb in groundwater and 80,000 ppb in soils have been found. Benzene and toluene have been detected in air samples, but at low levels.

LIPARI LANDFILL

Of 97 New Jersey sites on the National Priorities List, Lipari Landfill is ranked first. Lipari Landfill is a private inactive landfill located in Gloucester County. This 16-acre site is an old sand and gravel pit that was converted into a solid waste disposal facility. The site is surrounded by fruit orchards. Between 1958 and 1981, domestic and industrial wastes including methanol, benzene, toluene, xylene, isopropanol, butanol, bis (2-chloroethyl) ether, beryllium, and mercury, were disposed of at the landfill. It is reported that several thousand drums and several hundred thousand gallons of hazardous chemical wastes are presently on site.

A significant amount of leachate is entering an adjacent creek which empties into Alcyon Lake. The Kirkwood aquifer which underlies the site may become contaminated. Potable groundwater wells serve a population of approximately 20,000 area residents. There is a severe air pollution problem in the vicinity of the site due to the volatilization of chemicals.

BRIDGEPORT

Ranking among the top 10 New Jersey Superfund sites, Bridgeport Rental and Oil Storage Services, Inc., operated a facility in Gloucester County for waste oil storage and recovery, storage tank leasing operations, and illegal dumping operations. The site contains approximately 18 large tanks ranging in size from 75,000 to 750,000 gal, in addition to other smaller vats and stills comprising a total of 90 tanks overall, and a 12.7 acre waste oil and wastewater lagoon. The lagoon was reportedly formed by previous sand and gravel mining operations. The average depth is approximately 12–18 ft, with reports of two holes that may be as deep as 60 ft. Preliminary estimates indicate the lagoon contains 50,000,000 gal of contaminated liquids, six to eight in. of free-floating waste oil, and three to four ft of an oily sediment. Sampling of the lagoon indicated high levels of polychlorinated biphenyls (PCBs), oil and grease, heavy metals, phenols, and benzenes. Commercial waste-handling activities are presently prohibited at the site by court order.

The characteristics of the lagoon are such that it has contaminated local groundwater, surface water, and soil. Of particular concern is the oil layer floating on the surface of the lagoon sediment, which contains PCBs at an average concentration exceeding 500 ppm.

Because the lagoon level rises with each rainfall, overflow and dike breaches have caused some lagoon oil and water to contaminate areas east and northeast of the lagoon. Groundwater pollutants emanating from the site have contaminated several domestic wells west and northwest of the site and several other residential wells in this area are threatened.

BURNT FLY BOG

Also among the top 10 sites on New Jersey's National Priorities List, Burnt Fly Bog is an area of approximately 1700 acres in Monmouth County. During the 1950s and early 1960s a portion of the site was used for the storage of waste oil in approximately five unlined lagoons. Discharges from these lagoons have contaminated an area of at least 17 acres. In addition, an area of approximately 10 acres known as the Westerly Wetlands is also suspected to be contaminated. There are currently four lagoons onsite (one containing liquid oil and contaminated water, underlain by oil sludges, and three that have been backfilled with soil), as well as a mound of contaminated material known as the ''Asphalt Pile,'' and approximately 350 exposed and partially buried drums.

The site is currently a groundwater discharge area for the Englishtown aquifer. The ground water flows to the surface and drains into Deep Run. Analyses of the groundwater and surface water sediments indicate contamination with oil and various organic chemicals. The Englishtown aquifer may be affected, due to lack of an impervious clay layer beneath the lagoon area.

FORT DIX LANDFILL

The Fort Dix Landfill encompasses approximately 126 acres at the southwestern boundary of McGuire Air Force Base (MAFB). The natural geologic setting surrounding the landfill is a sandy coastal plain with a mixture of pine, oak, and scrub vegetation. The area is drained by numerous creeks. In the past, the landfill was used for the disposal of municipal wastes that were generated at Fort Dix. From the 1970s to the 1980s, MAFB disposed of chemical wastes in the Fort Dix Landfill. The hazardous substances disposed of at this site include petroleum products, thinners (methyl ethyl ketone), strippers (methylene chloride), and paints. The landfill was closed in July 1984. The recently filled landfill area is covered with a sand cap and sprayed with an erosion-retarding agent.

The U.S. Army conducted groundwater sampling around the landfill which indicated the presence of chloroform; 1,1,1–trichloroethane; methylene chloride; toluene; and 1,1–dichloroethene. Approximately 7,300 people within a three-mile radius depend upon groundwater for their potable water supply. The closest potable well is on a domestic farm, situated 3,500 ft south of the Fort Dix Landfill.

Methylene chloride and trichloroethane were also detected in surface water samples taken from Cannon Run, downstream from the landfill, indicating the possibility of downstream contaminant migration from the landfill. In addition, Cannon Run originates adjacent to the landfill and flows into Rancocas Creek, which is about one mile south of the site and is used for recreation.

FAA, ATLANTIC CITY

The Federal Aviation Administration (FAA) owns a site located on a 5,059-acre parcel of land. The FAA property is located to the west of the Garden State Parkway with Galloway Township to the north and Egg Harbor to the south. This area is widely used for aviation activities, including a base for the Air National Guard, the Atlantic City Airport, and the National Aviation Facilities Experimental Center (NAFEC). Jet fuel is stored onsite and used extensively in varying capacities, including training and experimental and operational tasks.

During the period of 1943 to 1958 the Navy operated a landfill at the site, which is now covered by a hanger, a parking lot, and a ball field. Other areas and operations of concern include a salvage yard, an abandoned fuel farm, and a photo lab.

As a result of a study commissioned by the Atlantic City Municipal Utilities Authority (ACMUA) to assess potential pollution threats to current and proposed ACMUA wellfields, five potentially long-term problem spots were discovered on the FAA grounds. Aquifer characteristics were defined and it was determined that groundwater, which ranges from 3 to 23 ft below the surface at test areas, is part of the Upper Cohansey aquifer, and that only shallow localized contamination exists below the five problem sites. Surface waters exist onsite with the largest being the 250 million gal Kuehnle reservoir.

Conclusions from the ACMUA study indicate that two of the five sites pose a threat, and that there is a greater probability of horizontal contaminant migration to surface water than vertical contaminant migration to the depths of the

Cohansey aquifer, where potable groundwater is obtained. A system of eight shallow (Upper Cohansey) observation wells and eight deep (Lower Cohansey) observation wells ring the new ACMUA wellfield to monitor groundwater quality and pumping effects.

UNIVERSAL OIL PRODUCTS

The Universal Oil Products (UOP) Chemical Division site in Bergen County is located along State Route 17 on a relatively flat 85-acre tract of land within the coastal wetlands management area of the Hackensack River Basin. The site is bordered on the southwest by Berry's Creek, which joins the Hackensack River about 3.5 miles downstream. One of its tributaries, Ackerman's Creek, flows through the property. In 1960 UOP purchased the property from Truebeck Laboratories, who had used the site since 1955 as a recovery facility for solvents and waste chemicals. A waste treatment plant was constructed in 1956 and further expanded wastewater holding lagoons in 1959. UOP adopted the facility to manufacture specialty organic chemicals (primarily benzyl alcohol and amyl salicylate). Based upon manifest records, it is estimated that approximately 4.5 million gal of waste solvents and solid chemical wastes were dumped into the unlined lagoons located on the eastern part of the property. By 1971, neither the onsite treatment system nor the two lagoons were in use. UOP terminated its operations at the site in 1979 and razed the plant in 1980. The contaminants found on site include chromium, arsenic, lead, benzene, chlorobenzene, 1,1,2,2–tetrachloroethane, trichloroethylene, vinyl chloride, toluene, and other organic and inorganic chemicals.

BORNE CHEMICAL COMPANY

Borne Chemical Company in Union County operated a now-inactive lubricating oil manufacturing plant located on a five-acre tract of land bordering the Arthur Kill Waterway in a mixed residential and industrial section of the city of Elizabeth. In addition to petroleum processing and blending, the plant also manufactured products used in the leather tanning industry, tints for the textile industry, and oil additives. The facility includes several bulk storage tanks, warehouses, and a railroad car loading facility. The company also operated a leasing business for bulk storage tanks. Sampling at the site has shown polychlorinated biphenyls (PCBs) present in the waste oil contained in the bulk storage tanks.

BEL-RAY COMPANY

The Bel-Ray Company site in Monmouth County encompasses approximately 29 acres in a semiresidential area. The company, which has been in operation for about 25 years, manufactures petroleum and synthetic-based specialized lubricants. In the past, Bel-Ray manufactured oils and lubricants which contained polychlorinated biphenyls (PCBs). During the company's operating history, the facility accumulated an undetermined number of drums that contain waste solvents, waste oils, and off-spec products. These drums are stacked on their sides for storage onsite. Some of the drums have developed leaks, causing the contents to spill directly onto the ground. The site also has approximately 38 aboveground oil and solvent storage tanks with capacities ranging from 1,000 to 20,000 gal. Only some of these tanks are diked. In addition, the facility has an unlined storage lagoon with a 125,000-gal capacity which collects rainwater used for cooling. This lagoon also collects oil-contaminated runoff, as well as process spillages which occur both inside and outside the building. Currently the only treatment the wastewater receives is from a small belt skimmer which is not satisfactory for a lagoon of this size.

There is contamination of groundwater as well as surface water from the unlined lagoon. Analysis of the wastewater in the lagoon revealed benzene, toluene, PCBs, and chloroform. The unlined lagoon discharges via a pipe through a marshy area into Jumping Brook, a tributary of Shark River. Remnants of oil have been observed in the woods and in the marshy area past the discharge pipe.

SWOPE OIL AND CHEMICAL COMPANY

The Swope Oil and Chemical Company north of Pennsauken on a one-acre site in an industrial park, surrounded by warehouses and railroad right-of-ways, is 1.2 miles from the Delaware River and 0.8 miles from Pennsauken Creek. Swope Oil and Chemical Company operated a chemical reclamation operation from 1963 until December 1979. Some of the products processed at the site include hydraulic fluids, paints and varnishes, solvents, oils, plasticizers, and printing inks. Waste sludges were discharged to an unlined lagoon. Contaminated material was ponded within a diked tank farm and in an exposed drum storage area. Reports indicate that past discharges from the lagoon and onsite spills entered drainage ditches and storm drains which discharged into Pennsauken Creek. There were 30 storage tanks ranging in size from 3,000 to 20,000 gal, and a 4,000- ft^2 unlined lagoon onsite.

There is a considerable amount of soil contamination, including high concentrations of toluene. The unlined wastewater lagoon contains various hydrocarbons, phenols, and metals which have contributed to the groundwater contamination. The groundwater is at a depth of 80 to 100 ft. Several wells, including a municipal water supply well, are within one mile of the site. The site lies within a recharge area for the Potomac-Raritan-Magothy aquifer system, which is one of the most important sources of potable water in Camden County. Air pollutants may be generated by vapors which are emitted into the atmosphere by chemicals in the lagoon.

SCIENTIFIC CHEMICAL PROCESSING, INC.

Scientific Chemical Processing, Inc. in Bergen County is an inactive waste processing facility situated on a six-acre tract of land directly across from the Meadowland Sports Complex. The company was formerly involved in the recovery and recycling of various by-product and waste solvents and other chemicals. The site is bordered on the northeast by Peach Island Creek, which joins Berry's Creek, a tributary of the Hackensack River, and is located within a coastal wetlands management area. Groundwater is found near the surface. Currently about 375,000 gal of hazardous substances are stored onsite in 44 tanks ranging in capacity from 3,000 to 20,000 gal; 15 2,000- to 7,000-gal tank trailers; and numerous 55-gal drums and other containers. These materials are broadly classified as crude oils, fuel oils, paint sludges, solvent residues, water-latex mixtures, phenolic resins, and other unidentified chemicals.

An ongoing investigation of reported hazardous waste spills and suspected violations of governmental regulations revealed extensive soil contamination and suspected groundwater contamination. The investigation indicated frequent spillage due to poor housekeeping procedures and an inadequate maintenance program. Some of the drums are rusted and leaking, and a product-sheen is visible in rainwater runoff. A petroleum-like discharge was found leaching from the banks of Peach Island Creek, while contaminated surface water flows to the creek as well. Direct discharges of hazardous substances to Peach Island Creek and the municipal sewer system were also known to occur. Local surface water bodies are used for recreation and industrial water supplies and could be affected by surface water contamination. Air quality is threatened by volatile organic solvents, and there is a danger of fire and explosion. Extensive soil contamination at the site could be aggravated as the integrity of the vessels continues to degrade. The aquifer in the area is shallow, and groundwater contamination is strongly suspected.

RENORA, INC.

Renora, Inc. in Middlesex County was a hazardous waste collector/hauler operating from 1977 to 1980. The site of its operations is bordered on one side by
the Mill Creek, and the other sides by the New Jersey Turnpike South and the
Conrail railroad tracks. Presently onsite are as many as 1,100 containers (55-gal
and 5-gal) and 18 tankers and tanks containing solvents, resins, oils, greases,
and other hazardous substances. Some of the tanks, tankers, and drums are deteriorating and leaking their contents onto the soil. Several ponds of hazardous
materials have accumulated on site.

Soil at the site has been contaminated by leakage. It is suspected that area groundwater may be contaminated. Groundwater below this site probably discharges
to Mill Brook.

QUANTA RESOURCES CORPORATION

The Quanta Resources Corporation site in Bergen County is bordered on the
east by the Hudson River, between the George Washington Bridge and Lincoln
Tunnels, and consists of an inactive facility formerly used for storage, reprocess-
ing, and reclamation of waste oil. The site is also bordered to the north by a fresh
produce distribution warehouse, and a fertilizer distribution facility to the west.
Condominiums atop the Palisades overlook the site.

Within the facility there are four tank farms, an oil/water separator, a process
building, and an office trailer. The 61 aboveground tanks range from 20,000 to
600,000 gal and have a combined capacity of 9 million gal. Ten underground
tanks have an estimated capacity of 40,000 gal. Many of the steel plate tanks
have wooden roofs which, in some cases, are partially or totally collapsed, and
allow rainwater to enter. The contents are stratified and consist of a layer of heavy
sludge at the bottom, an oil layer, and water. Tank farms B and C hold polychlorinated biphenyl (PCB) contaminated oil, with concentrations as high as 260+
ppm.

The Hudson River in this area is a very important shipping lane, contains many
marinas, and has been cited as an important striped bass habitat. Pathways from
the site, amounts and concentrations, and proximity to the public create the potential for substantial problems. In addition to PCBs, benzene, toluene, phenol, ethyl
benzene, trichloroethane, lead, cyanide anthracene, and chloroform have been
found at the site. Some storage tanks have been found to have low flash points,

increasing the possibility of fire and explosion, which could create toxic fumes. Dikes have been placed around the storage areas, but are inadequate. Soils in the tank farm areas are grossly contaminated, and drainage from the site has resulted in releases of oil to the Hudson River. Seven such spills were recorded by the USEPA alone, between February 1982 and September 1983. Contaminate booms/sweeps in place are not effective due to the lack of maintenance. The deteriorating tanks onsite could produce a major spill.

IMPERIAL OIL COMPANY

Imperial Oil Company, Inc. in Monmouth County is an active oil blending facility. Several different companies have operated out of the facility in the past. It is believed that this oil reprocessor discharged waste products to a nearby stream. There are two additional suspected contaminated areas adjacent to the Imperial Oil Company site. A waste site at the facility is contaminated with PCBs, arsenic, lead, petroleum hydrocarbon, and other synthetic chemicals. The soil surrounding both the waste pile and the oil/water separator is contaminated. Sediments in a nearby stream and groundwater beneath the site are contaminated, and have migrated offsite, causing further pollution of the surrounding area. Floating PCB-contaminated oil has been detected in the groundwater.

BARRIER CHEMICAL INDUSTRIES

Soil sampling by the New Jersey Department of Environmental Protection in 1978 at the former site of Barrier Chemical Industries, Vernon Township in Sussex County, revealed high concentrations of several volatile organic compounds, including 800 ppm of benzene. Groundwater in the area is contaminated. This site was formerly used as a chemical formulation plant to produce degreasers, sewer cleaners, soaps, disinfectants, insecticides, and floor waxes. This two-acre tract is adjacent to a wetlands area, less than a quarter of a mile from Wawayanda Creek. According to residents in the area, the company dumped chemicals directly onto the ground for several years before manufacturing operations were terminated at this site.

AMERICAN CYANAMID COMPANY

The American Cyanamid Company site is a 575-acre active industrial facility adjacent to the Raritan River and the Elizabethtown Water Company. The facility encompasses a number of individual disposal sites that include 27 lagoons, both active and inactive, and inactive landfills. Approximately 800 chemicals, including dyes and textile chemicals, organic pigments, rubber chemicals, pharmaceuticals, and intermediate chemicals are produced. During the company's more than 50 years of operation here, an unknown quantity of chemical wastes were buried at the site. The company uses unlined lagoons for treatment and storage of wastewater and sludges. Sludge lagoons were allowed to reach their capacity and were then covered. An incinerator was put into operation in 1979 for the disposal of newly produced sludge.

The lagoons are a potential source of ground- and surface water contamination due to percolation and mixing with storm water. The groundwater beneath the site is severely contaminated with organic chemicals. The potential spread of contamination into nearby wells and surface water is of concern. The Elizabethtown water supply intake is within 2,000 ft of American Cyanamid's settling lagoon. There are also at least 20 private wells in the immediate area. Groundwater here is part of the Brunswick aquifer and is the state's second largest source of drinking water.

REFERENCES

1. Lioy, P. J., and J. M. Daisey. *Toxic Air Pollution: A Comprehensive Study of Non-Criteria Air Pollutants* (Chelsea, MI: Lewis Publishers, Inc., 1987).
2. Vogt, N. B., F. Brakstad, K. Thrane, S. Nordenson, J. Krane, E. Aamot, K. Kolset, K. Esbensen, and E. Steinnes, "Polycyclic Aromatic Hydrocarbons in Soil and Air: Statistical Analyses and Classification IV: SIMCA Method," *Environ. Sci. Technol.* 21:35–44 (1987).
3. New Jersey Department of Environmental Protection. 1987 Site Status Report on Hazardous Waste Remediation in New Jersey; Site Specific Information.

CHAPTER 5

Federal Underground Storage Tank Regulations and Contaminated Soils

Richard A. Valentinetti

The proposed federal regulations for underground storage tanks (UST) reflect the scientific uncertainty of many aspects of the contaminated soil problem, and the ways of solving this problem.[1] In the past, government, private sector cleanup contractors, and the academic community have directed too little attention to the issue of scientific uncertainty surrounding the contaminated soil problem. The number and quality of attendees at the soil contamination conference held at Amherst in September 1987 indicated that this is no longer the case.

I will address the philosophy of the federal UST program, summarize the proposed requirements for contaminated soil, discuss central issues relating to the scientific uncertainty associated with remediation options, and present an overview of our continuing research objectives. Before discussing the specific provisions of the Environmental Protection Agency (EPA)-proposed regulations that relate to soil contamination, an appreciation of the general philosophy behind the program is necessary.

To begin with, there are approximately two million tanks of various sizes, ages, and construction existing in the federal universe, excluding onsite heating oil tanks, which are not subject to the regulations. The sheer number and diversity of these tanks makes the program difficult for a centralized federal organization like EPA to implement effectively. As a result, this program must be decentralized as much as possible and placed at the lowest governmental unit for operations, so that reports of gasoline detected in the drinking water of private homes can be handled by a state, county, or municipal government organization. In order to allow local government organizations to implement the program, the Office of Underground Storage Tanks (OUST) must provide flexibility in the regulations, and recognize that many different solutions exist for the same problem.

The OUST office director, Ron Brand, believes that a "franchise approach" to the UST program will provide local government organizations with the flexibility and support they need to implement the UST program. As he views it, the UST program will become the McDonald's of the EPA environmental programs. As the owner of a franchise, one needs to commit basic resources and comply with certain requirements of the franchiser. In return, the franchiser provides to the franchisee the basic tools and assistance to ensure his operation achieves a higher level of performance. The franchiser also should highlight the success of other franchisees, in order to increase performance in a specific area. As it applies to government, the "franchise approach" means that local government acts as "franchisee," committing resources and complying with requirements, while the OUST office acts as "franchiser" providing support, but allowing autonomy to make decisions based on local conditions and knowledge. The key element of the success of this approach is providing flexibility to local government organizations. This flexibility is reflected in the approach that the proposed regulation takes to contaminated soils.

In the proposed EPA regulations, the issue of contaminated soils is considered as part of the corrective action process for petroleum spills. The corrective action process is broken down into four major areas, including:

- initial response action
- data gathering
- developing and receiving approval for cleanup plan
- implementing plan and monitoring results

The initial response action is a critical element in the process. Unlike many of the other remedial action programs, the owner or operator is required to perform immediate work at the site to mitigate fire and safety hazards, and to control the migration of product through the environment. In addition, the owner or operator is required to remove, to the maximum extent possible, free product from the groundwater.

In the initial response action phase of the proposed regulations, the cleanup of contaminated soil is addressed in the requirement that the owner or operator "remove and properly dispose of visibly contaminated soil from the excavation zone." In testimony presented at the public hearings and written comment submitted during the comment period, commenters expressed concern that this requirement revealed a preference for removal and offsite disposal of large amounts of soil. As I indicated at the public hearings, this preference was an attempt to codify a standard operating practice of removing highly contaminated soil adjacent to the leaking tank, but was not intended to require that the entire zone of contaminated soil be excavated, removed, and properly disposed of. I anticipate that the final regulation will clarify the agency's intention.

The other area of the regulation that may require management of contaminated soil is in developing the corrective action plan. The regulation proposes that contaminated soil be addressed as part of a corrective action plan.

In neither area of the regulation does the agency set a specific cleanup level for contaminated soil. The agency's position is that corrective action decisions should be made on a site-by-site basis in order to provide flexibility to the implementing agency.

More important, however, the agency's ability to establish a standard cleanup level is hampered by the lack of sufficient data supporting one consistent level. State and locally established cleanup levels for contaminated soil are inconsistent among one another because they are often established by officials of different branches of government, such as the public safety department, health department, or environmental agency officials, who may have different objectives for developing regulation. To confuse matters even more, standard methods do not exist for collecting soil samples prior to a field analysis for a certain constituent. Clearly, the agency does not have all the answers.

I feel that it is important to divide the problems associated with soil contaminated by petroleum into the following categories:

- removal and treatment of soils
- in situ treatment of soils

In my opinion, the issues, problems, and solutions surrounding these two methods of treating contaminated soil are so vastly different that we should not continue to discuss treatment of soil contamination as a single issue. It is in fact two distinct problems. I am concerned that public citizens and government officials in decisionmaking positions recognize that there are two distinct problems.

Much of our past experience has been with the removal of contaminated soil—the "muck and truck" approach—with little attention devoted to treatment. We must recognize that in most cases, some removals, hopefully low mass and confined to the immediate area of the tank, will occur. But the first question we should ask is "What do we do with the dirt?" (attributed to Henry Lord, Liberty Engineering Services). This portrays a serious problem. Once you dig the dirt up, what are your options?

The first question for narrowing the options of handling contaminated soil should be, "Do we dispose of it on- or offsite?" If one can find an offsite facility, such as a landfill, that does not charge hazardous waste disposal rates, then offsite disposal is the preferred option. However, many sanitary landfills want to avoid potential liabilities of disposing of contaminated soil; consequently, this option is disappearing.

The next question concerns enhanced volatilization. "Do we place soil in the parking lot or the back forty to allow hydrocarbons to volatilize, and how do we enhance this volatilization?'' I am not aware of much data that indicates how to optimize methods for enhanced volatilization, such as rototilling, soil turning, or air injection, or what impact this process will have on air quality.

A third question that will affect the option to land dispose soils contaminated with petroleum is ''Are these soils considered hazardous?'' The answer is ''It all depends.'' It depends on where you are located. The OUST office is aware of some jurisdictions that have set a low level of contamination for determining if soil is hazardous, in order to justify removing the soil and shipping it to another jurisdiction. Other jurisdictions have threshold levels for determining if a soil is hazardous, but have adopted provisions for handling soils contaminated with petroleum as special wastes that have different disposal alternatives than ''real'' hazardous wastes.

At the federal level, the answer to the question of whether soil contaminated with petroleum is considered hazardous is also ''It all depends.'' It depends on what one measures for, and what the future EPA policy will be on the no-migration issue for land disposal, as well as the use of an expanded Extraction Procedure (EP) toxicity test. At the present time, soil contaminated with petroleum is considered hazardous if it is ignitable, or if it contains leaded fuel. If the contaminated soil fails an EP toxicity text for lead, it would be considered hazardous. EPA has proposed to expand the EP toxicity test so that it will consider the ''capture of organic constituents.''[2] The OUST office has initiated a project to determine at what level soils contaminated with petroleum would fail the modified test. Because other EPA offices share our concern with the applicability of the proposed test methods and regulatory levels for organics in contaminated soils, an internal EPA workgroup has formed to resolve this issue.

Another issue that complicates the handling of contaminated soil is potential recontamination resulting from fluctuating groundwater levels.

In addition to land disposal of contaminated soils, other treatment processes, such as an asphalt batch plant, have been developed to fulfill the requirement for treating hazardous wastes. We still need to determine their applicability to treating soils contaminated with petroleum by pilot testing, bench scale analysis, and free testing.

During the short time I have been involved with the UST program, I have grown to understand that the area for which we have the least amount of information and in which we can make the greatest impact is in situ treatment of soils. To date, most remediation efforts have focused on eliminating the primary source of contamination, the tank, and the highly contaminated soil adjacent to the tanks; and pumping and treating the groundwater to an acceptable level. If groundwater were used as a drinking water source, the level would range from 1 to 10 ppb of benzene. It was soon evident that this approach was unworkable, not because the air stripper or carbon filter could not reach these levels, but because, due to the continual source of contamination from the remaining petroleum product absorbed into the unsaturated zone, the pump and treatment system would have to continue operating for 15–20 years to reach the required cleanup levels.

The American Petroleum Institute soil column study that revealed 850 soil washings were necessary to extract most of the petroleum product from the soil provided further evidence that a secondary source of contamination, such as absorbed petroleum, would continue to release petroleum for a long time. If the UST program is to succeed, we must investigate the various methods for releasing absorbed product from the unsaturated zone, so that cleanup can be accomplished within a reasonable time period.

Although the proposed federal regulations are necessary to initiate cleanup of soils contaminated with petroleum, EPA/OUST is enthusiastic about pursuing any activities that will enhance the cleanup of contaminated soil "outside the regulations." Earlier in this chapter, I introduced one function of the federal government as providing basic assistance and tools to its franchisees. The greatest need for this approach is in the area of contaminated soils, because so little information currently is available. With the assistance of our Office of Research and Development (ORD), we have initiated the following two-pronged attack on this problem:

- Focus on better understanding the physical and chemical mechanisms that retain and release petroleum that has been absorbed into the soil.
- Evaluate the capability for innovative technologies to achieve clean up of transforming existing treatment levels to achieve cleanups to the same low level.

The objectives of this first project will be (a) to characterize the mechanism of hydrocarbon migration, retention, and transformation in the saturated, as well as unsaturated, zones; and (b) to use this information to improve removal technologies, especially in insitu settings. These objectives could also be described as fate and transport as it applies to corrective action.

The results of the research involved in this first project will be used to accomplish the following:

(1) Identify the requirements for remobilizing constituents of petroleum motor fuels from distinct physiochemical phase settings.
(2) Examine existing corrective action technologies to determine their relative effectiveness in removing petroleum products from the various physiochemical phase settings.
(3) Prepare a research report on the findings and recommendations for further work, highlighting technologies with a potential for rapid and significant improvement, while identifying physiochemical settings with minimal potential for significant improvement.
(4) Prepare a handbook for federal, state, and local screening and selecting appropriate response technologies.

The second ORD effort will be to evaluate some innovative technologies in the field, somewhat similar to the Superfund Innovative Technology Evaluation Program (SITE program). Both EPA and technology developers share a strong, common interest in the new cleanup requirements for Superfund sites. EPA is actively seeking developers of innovative technology for participation in this unique new program. Through the SITE program, EPA will assist in commercially developing and demonstrating at hazardous waste sites the use of promising new technologies. The UST program will try to build on the technology development and experience of SITE's program.

ORD will evaluate one technology, a vacuum extraction process, during early 1988. Evidence indicates that this process is capable of collecting hydrocarbons from the unsaturated zone, but little evaluation of its level of performance has been done. The goal of this work assignment is to determine transport mechanisms and the controlling factors for the extraction of gasoline components from the subsurface zones through the use of the vacuum extraction technology. The application of long-term corrective action technologies requires further refinement of the mechanisms for the removal of leaked products from underground storage tanks. The information developed from this assignment will provide an understanding of the characteristics of subsurface zones and vacuum extraction technology design.

In addition, two sites will be investigated in cooperation with Terra Vac and the state of Florida. Our site (Bellview), which as an ongoing project has had a substantial reduction in hydrocarbons, will continue to be used to get additional information on measurements, to practice debugging the field equipment, and to receive initial information on the vacuum extraction technology. A full evaluation will be conducted at the old 441 site in Oscala. At this site, measurements will be taken before, during, and after the process has been in operation.

REFERENCES

1. 52 *Federal Register* 12652, April 17, 1987.
2. 51 *Federal Register* 21648, June 13, 1986.

PART II

Environmental Fate and Modeling

CHAPTER 6

Modeling Petroleum Products in Soils

Linda Eastcott, Wan Ying Shiu, and Donald Mackay

INTRODUCTION

For reasons more eloquently expressed in other chapters, there is a compelling incentive to understand and describe, in quantitative terms, the behavior and effects of petroleum in soils. It is relatively easy to describe the fate of petroleum in soils in merely qualitative terms. For example, it is clear that volatile petroleum products such as gasoline experience considerable loss by evaporation; normal alkanes are subject to fairly rapid biodegradation; aromatic hydrocarbons, particularly those of lower molecular weight, are very susceptible to dissolution into water and may thus cause contamination of water supplies in the locality. It is, however, much more difficult to make rigorous, scientifically justifiable statements in which these processes are described in quantitative terms. In very few cases has it been possible to state, for example, that in a period of one year 20% of a particular mass of oil spill was lost by evaporation, 2% by dissolution, and 30% by biodegradation, and that 3% was altered by photolysis.

There have been numerous studies and reviews of this issue, including the proceedings of the previous conference in this series,[1] work sponsored by the American Petroleum Institute,[2,3] and reports by our group.[4-8] There are also numerous reports on oil biodegradation in soil; for example, Lee and Ward,[9] Jobson et al.,[10] Walker and Colwell,[11] Westlake and Cook,[12] and Westlake et al.[13] More exhaustive references can be found in other papers in this series.

The ultimate goal of our work is to enable quantitative statements of past and future behavior to be made for field situations in which actual spills have occurred. This should ensure that appropriate remedial actions can be taken. In some cases it may be justifiable to take no remedial action, but this should be done for good reason, i.e., it can be demonstrated that this is the best course of action. Unfortunately, in field situations it is very difficult to gather quantitative data of this type. There is usually an inadequate data base on the amount of oil spilled; its location, horizontally and vertically; and the prevailing conditions of exposure to evaporation, dissolution, and biodegradation. The spill is probably highly heterogeneous in nature, and thus it is difficult to compile a mass balance. These difficulties can be eliminated by performing laboratory experiments in which controlled amounts of oil are spilled under closely defined and monitored conditions. Of course the disadvantage of this approach is that there is no assurance that these conditions are representative of those which will occur in the real environment.

It is believed, however, that by a combination of complementary laboratory studies and field studies, it should be possible to develop an improved capability of predicting the fate of petroleum in soils.

In this chapter we describe and discuss one approach toward developing this capability, namely, the study of the fate of oil in laboratory soil microcosms or "soil boxes." This approach has the advantage that more information can be obtained with greater accuracy and precision, and modeling is easier.

There are two distinct intellectual challenges in this work. The first is to design and conduct experiments, and interpret the resulting data to obtain an understanding of the dominant processes and their approximate rates. We describe and discuss here some experimental work, the objective of which was to provide quantitative data for oil spill fate in soils. The two oils studied were a crude oil and a diesel fuel, along with a synthetic hydrocarbon mixture. The data obtained from these studies are being used to develop a general model of petroleum fate in soil.

The second task is to formulate a mathematical model which will consist of equations founded in physical and chemical reality, and which describe only the more significant processes. The various parameters in this model must then be determined, as well as their dependence on variables such as temperature, soil type, oxygen status, and availability of nutrients.

In summary, our purpose is to discuss experimental and modeling approaches and give some illustrative results in the hope that others may follow this approach and thus contribute to an increased understanding of oil behavior in soil.

EXPERIMENTAL

We describe here only the general approach and procedures in our work. More complete details and results can be found in the project reports.

The experimental work was undertaken in soil boxes of dimensions 50 cm × 50 cm × 50 cm. The soil was a homogeneous mixture of garden topsoil and sand, to give an organic carbon content of approximately 2%. Each box typically contained 136 kg or 100 L of soil and was maintained at constant temperature conditions. The top 10 cm of the soil, representing about 25 L or 34 kg of the mixture, was thoroughly mixed with 400 mL or 340 kg of oil to obtain a 1% by mass mixture. A diagram of the setup is shown in Figure 1. In the crude oil experiments, four boxes with varied treatments were used. The base case contained 1% by mass of Norman Wells crude oil in the soil and was watered at a rate of 2 L/week. A second box contained a smaller amount of oil, i.e., 0.5% by

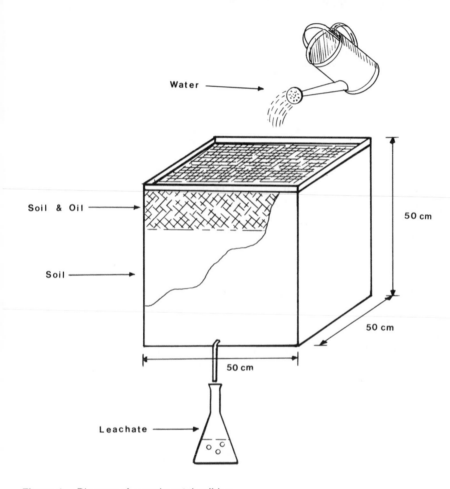

Figure 1. Diagram of experimental soil box.

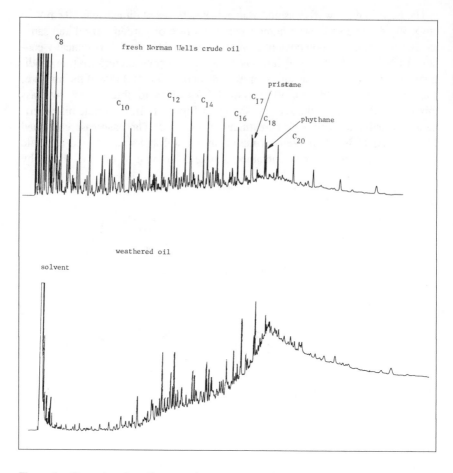

Figure 2.   Examples of capillary gas chromatograms of fresh (above) and heavily degraded (below) crude oil showing loss of volatile and degradable hydrocarbons.

mass, and was also watered 2 L/week. In a third box, an attempt was made to sterilize the soil and inhibit or eliminate biodegradation by adding a 1% by mass solution of $HgCl_2$ in 2 L of water per week. Sterilization has proved to be an impossible goal at this low biocide concentration. It is believed that the best that can be achieved, at this scale of experiment with the facilities available to us, is to hinder biodegradation by adding mercuric chloride directly to the soil at 2% by mass of soil. We refer to this box as one with "hindered biodegradation." In the final box, a quantity of nitrogen-phosphorus fertilizer was added frequently to the oil, in addition to regular watering, in an attempt to stimulate biodegradation. The boxes were run for four months, and soil and leachate sampled at regular intervals.

Analysis involved direct capillary column gas chromatography of the samples and, in addition, the separation of the samples into saturates, aromatics, polars, and asphaltenes, and separate GC identification of the first three fractions. Examples of the gas chromatograms of the fresh and weathered Norman Wells crude oil are given in Figure 2. The normal alkane peaks decrease considerably over the four months. The prominent peaks remaining after four months include the isoprenoids and other microbiologically recalcitrant hydrocarbons, which are relatively resistant to degradation. The unresolved mass or "hump" under the peaks gradually changes shape over the four months. The change in hump shape with biodegradation is clearly seen at the bottom of Figure 2, showing a chromatogram of the parent oil after three months. The fertilized box, showing the greatest biodegradation, had a different hump shape than the other boxes. The relative size of the hump can probably be used as an indicator of the extent of biodegradation.

The water soluble fraction of the fresh Norman Wells crude oil is shown in Figure 3 with the major components indicated. It is clearly very different in composition.

In a study of diesel fuel, a similar approach has been taken. However, 10 boxes have been used, the conditions being listed in Table 1. Some specimen results are shown in Figure 4, which are gas chromatograms of the fresh diesel fuel and weathered samples at 1% by mass in soil for three and six months, showing change in n-alkane to isoprenoid ratio and emergence of the "hump."

Table 1. Experimental Conditions of Soil Boxes Used in Fuel Oil Study.

	Boxes
1. Diesel fuel:[a]	(1% by wt. oil in soil as are all the boxes)
2. Diesel fuel:	fertilized
3. Diesel fuel:	2% $HgCl_2$ in soil—"hindered"
4. Diesel fuel:	tilled once a week
5. Diesel fuel:	flooded[b]—anaerobic + H_2O_2 at 1% of oil mass/week
6. Diesel fuel:	flooded—anaerobic
7. Synthetic hydrocarbon mixture:[c]	1% by wt. oil in soil
8. Synthetic hydrocarbon mixture:	fertilized
9. Synthetic hydrocarbon mixture:	+2% $HgCl_2$ in soil—"hindered"
10. Synthetic hydrocarbon mixture:	tilled once a week

[a]Each box watered 2 L/wk.
[b]Flooded boxes are watered 6 L/wk.
[c]Consists of 12 normal alkanes and 12 aromatics + hexachlorobenzene and pristane as an internal standard.

The results from these and other studies show clearly that oils, such as crude oil and diesel fuel, are subject to evaporation, dissolution, and biodegradation. However, it is difficult to be quantitative about the relative roles for each process.

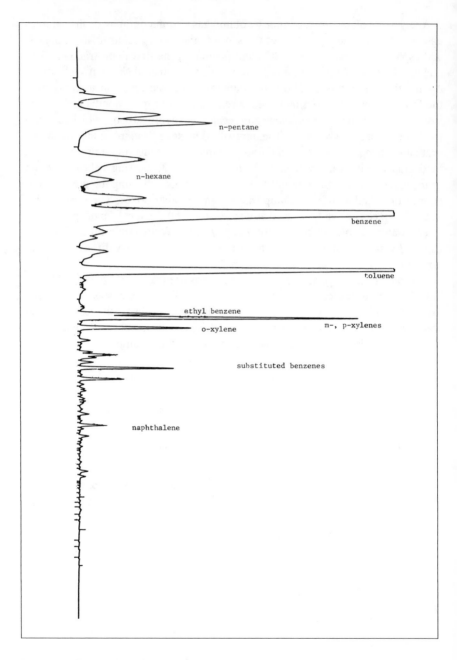

Figure 3.   Gas chromatogram of the water soluble fraction of a crude oil showing dominance of the more soluble monoaromatics.

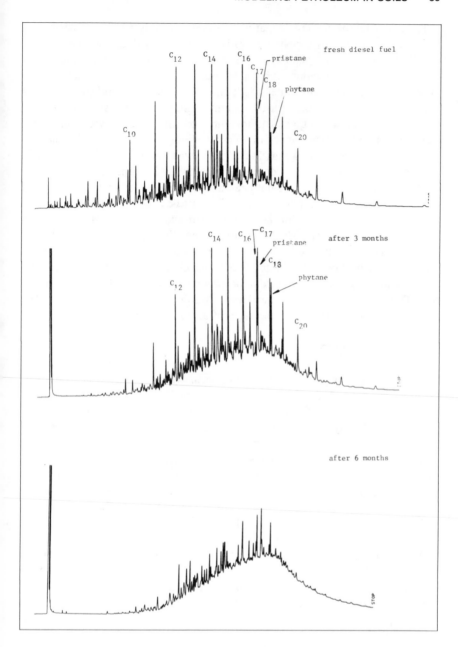

Figure 4. Specimen gas chromatograms of fresh diesel fuel, and weathered samples showing change in n-alkane to isoprenoid ratio and emergence of the "hump."

The major reason for this is that it is difficult to study all three processes in isolation. We can study evaporation in isolation by tray evaporation or "stripping" of the fresh oil. Such experiments have been done in our laboratories to test the effect of evaporation on composition. It is also possible to expose oil to dissolution or contact with water by "plating" the oil on a solid support (as is done with the liquid phase on a GC column) and then flow water over the oil for a prolonged period of time. The amount of oil which is dissolved in the water can be measured, and the change in oil composition can be assessed by analysis of the oil after exposure to water. In this case, evaporation is eliminated. Details of these techniques have been discussed recently by Maijanen et al.[6] and Shiu et al.[8] Unfortunately, we have not been able to devise and conduct an isolated biodegradation experiment because the oils must be exposed to some water and to air. Any attempt to control the excess of air and water is likely to result in biodegradation conditions which are unrepresentative of those which prevail in natural soils. The only feasible approach appears to be to try to calculate the extent of biodegradation by "subtracting out" the deduced amounts of evaporation and dissolution which are believed to occur simultaneously with the biodegradation. This is a somewhat difficult and unsatisfactory procedure, but it can be done if one has a model describing the rates of all three processes.

Model Development

If the spill consisted of a single component with well-established physical-chemical and biodegradability properties, it would be relatively easy to assemble a fate model on lines similar to those which have been used for agricultural chemicals. The difficulty with oils is that they consist of a large number of components, only a few of which have been well characterized, and their individual and collective properties are in considerable doubt. The logical approach, when faced with such a situation, is to break the oil down into a manageable number of pseudocomponents. One can then assign properties to these pseudocomponents and calculate the fate of each, making some assumption as to how components influence the fate of each other, i.e., devising "mixing rules," such as Raoult's law.

A fundamental modeling problem arises at this point. Oil or oil components can migrate through the soil matrix by two general pathways. First is bulk flow of an oil phase which infiltrates the soil under the influence of gravity and capillary forces. Second, the oil components can dissolve in air or water and migrate in these phases by diffusion or bulk flow. We are of the opinion that successful mathematical models can be devised to treat either case in isolation, but the combination is likely to be excessively complex. A justification for separating these pathways is that when bulk oil flow does occur, it results in little or no component separation (i.e., benzene travels as fast as hexane) and the transport rate is usually fast relative to that of the dissolution rate. We thus suggest a two-stage model, first a bulk flow model which applies until the oil is rendered immobile, then second, a diffusion/dissolution/transformation model which applies over the longer term to the immobilized oil. We do not treat the first type here, but focus on the second, which is quite compound-specific and results in substantial changes in oil composition with time.

Our ability to model the fate of oil under these conditions is largely controlled by the availability of experimental data, which in turn is controlled by analytical capability. It is thus useful at this point to digress to consider the contribution and constraints of analytical methodologies in controlling the selection of pseudocomponents and limiting the use of models.

ANALYTICAL METHODS

Current analytical methods, such as GC, HPLC, and spectroscopic analyses, do not give a total description of the composition of the oil.[14-16] They provide only a partial picture of the amounts of some of the components of the oil. We are thus inherently limited in our ability to probe the oil's composition. We can refer to the available data as "analytically accessible" information. If an oil is spilled it is likely that we will have "analytically accessible" information for the original or parent oil. We can expose the oil to migration, evaporation, dissolution, and biodegradation in a soil environment, sample it, and then obtain new sets of "analytically accessible" information about the oil at various times.

If a model is developed to describe this process, then the best that can be accomplished is that the model will accept the "analytically accessible" information about the parent oil, process this information by some numerical scheme, and then generate by computation a set of data which is consistent with the "analytically accessible" information for the exposed oil. Because it is much easier to model oil fate than it is to obtain analytical data or conduct experiments, models have a tendency to produce vast amounts of data which cannot be confirmed experimentally.

We have concluded that the best analytical approach is to subject the oil to preparative-scale high pressure liquid chromatographic (HPLC) separation into the groups of saturates, aromatics, and polar compounds[17] (after removing the asphaltenes gravimetrically), and separate the latter three fractions into various groups based on volatility by capillary gas chromatography. This enables us to break down the saturates into normal, branched, and isoprenoid hydrocarbons, and assign approximate relative proportions. In theory it should be possible to digitize the gas chromatographic traces of the original oil, calculate its changing composition, and regenerate the gas chromatograph of the weathered oil.

Comparison could then be made of the observed and calculated group GC traces and the model adjusted to give the observed behavior. This is presently not possible. But we have taken a step in this direction.

DATA PRESENTATION

The GC traces can be converted into histograms in which the ordinate is the amount of material, corresponding to the peak height or area, and the abscissa is retention time on the GC column, grouped into a number of categories which we term "elution groups." These correspond to peaks lying between designated normal alkanes. The resulting histograms reconstruct or approximate the GC trace. Such histograms can be prepared for the parent oil or for the saturate or aromatic fraction of the oil, as are shown in Figure 5A and 5B for crude oil after one month. The distribution of components in the oil is then characterized by the relative amounts in the various elements of a matrix consisting of columns of various chemical class (n-alkanes, branched alkanes, isoprenoids, aromatics, and polars) and elution groups defining the rows (conveniently divided by the normal alkanes), as shown in Table 2. By identifying some of the hydrocarbons which are present in each element, we can assign to each element physical chemical properties of volatility, water solubility, molecular weight, molar volume, and ultimately, it is hoped, biodegradability. It is then relatively easy to assemble a computer program which will calculate the change in amount of each element as a function of time as oil is exposed to defined conditions of evaporation, dissolution, and biodegradation. The model produces a new composition matrix which, using spreadsheet software, is converted into a histogram. The histogram can be compared with the observed GC traces.

There are some difficulties in this process. It is necessary to treat the hydrocarbons which are present in the "hump" of the GC trace. The hump is made up of unresolved peaks and therefore must be included in the mass balance. We have developed a method of assigning these unresolved hydrocarbons to the various groups, but this introduces a considerable uncertainty.

Further, it is probable that some of the chemicals contributing to the GC trace and especially to the "hump" are polar compounds formed as a result of biodegradation of the oil. The present system does not treat this effect adequately, because we are unable to calculate the extent of this phenomenon.

An advantage of this system is that the model can be used to test compositional changes of the oil when exposed to isolated evaporation and dissolution conditions, as discussed earlier. Figures 6 and 7 show the results of such experiments, and they are generally in satisfactory agreement. It is apparent that the model describes evaporation and dissolution fairly well. The principal difficulty and primary task is to define a biodegradability for each of the elements in the matrix.

Table. 2 Matrix of Crude Oil After One Month in Soil.

Final Mass Percent

Class elution group	n-Alkane		Br-Alkane		Isoprenoid		Aromatic		Polar		Totals
	hump	peaks	hump	peaks	hump	peaks	hump	peaks	hump	peaks	
6	0.0000	0.0000	0.0000	0.0000	0.0000	0.0000	0.0000	0.0000	0.0000	0.0000	0.0000
7	0.0000	0.0000	0.0000	0.0000	0.0000	0.0000	0.0000	0.0000	0.0000	0.0000	0.0000
8	0.0000	0.0000	0.0000	0.0000	0.0000	0.0000	0.0000	0.0000	0.0000	0.0000	0.0000
9	0.0000	0.0000	0.0000	0.0000	0.0000	0.0000	0.0000	0.0000	0.0000	0.0000	0.0000
10	0.0000	0.0742	0.0000	0.0236	0.0000	0.0000	0.0000	0.0151	0.0188	0.0094	0.1411
11	0.0000	0.3620	0.0000	0.6639	0.0000	0.0000	0.0000	0.3617	0.2484	0.1242	1.7602
12	0.0210	0.6250	0.0948	1.5210	0.0000	0.0000	0.2422	1.2974	0.6412	0.2699	4.7125
13	0.0377	0.7195	0.6099	1.4450	0.0970	1.1092	0.7954	1.2728	0.9072	0.4536	7.4473
14	0.0788	0.6885	1.0704	1.0150	0.0676	0.5137	1.1260	1.8134	1.0295	0.5148	7.9177
15	0.1067	0.7810	1.2367	1.0111	0.1180	0.9940	1.6046	1.7658	1.0780	0.5390	9.2349
16	0.1339	0.4412	1.7136	0.9593	0.0000	0.0000	2.7493	2.4546	1.0961	0.5480	10.0960
17	0.0673	0.4134	1.7451	0.3052	0.1206	0.6603	2.0479	1.0856	1.1027	0.5513	8.0994
18	0.0805	0.3725	1.6561	0.5069	0.1444	0.7199	1.9092	1.8267	1.1051	0.5525	8.8738
19	0.0942	0.3241	1.9106	0.5902	0.3370	0.6259	2.5697	1.6933	1.1059	0.5529	9.8038
20	0.0674	0.5061	2.2097	0.4883	0.0000	0.0000	2.2486	1.8037	1.1062	0.5531	8.9831
21	0.1610	1.2050	8.0568	0.3891	0.0000	0.0000	10.6786	0.7296	1.1064	0.5532	22.8797
Totals	0.8485	6.5125	20.3037	8.9186	0.8846	4.6230	25.9715	16.1197	10.5455	5.2219	99.9495

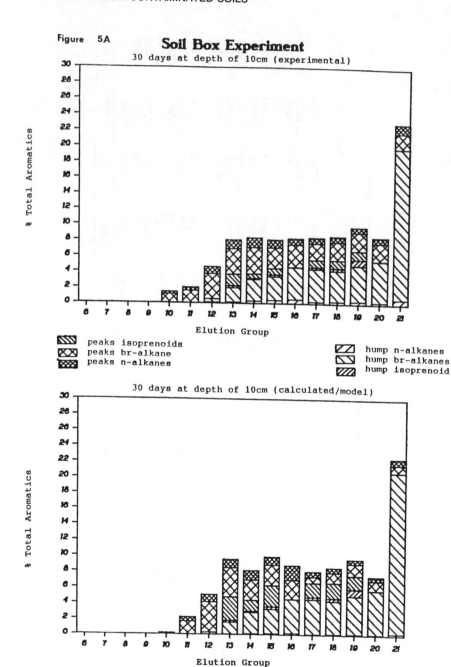

Figure 5A — Soil Box Experiment

30 days at depth of 10cm (experimental)

30 days at depth of 10cm (calculated/model)

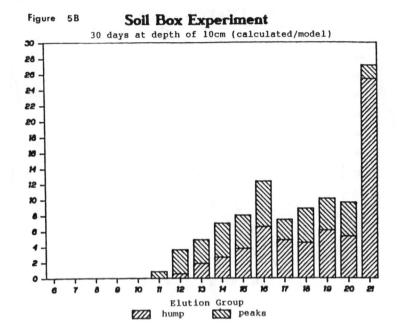

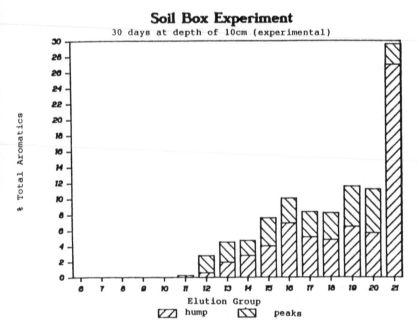

Figure 5A and 5B.   Histograms comparing the experimental and calculated saturated and aromatic fractions of a crude oil after one month.

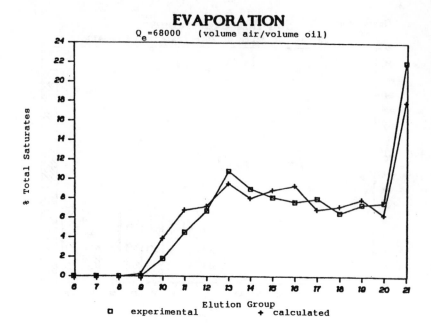

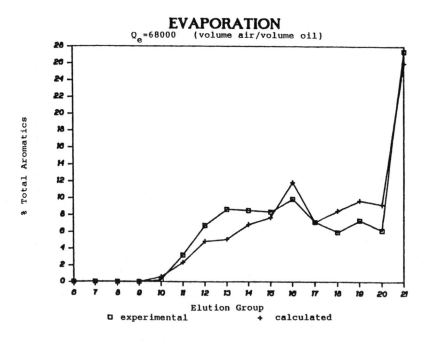

Figure 6. Line graph of evaporation by gas stripping results of the saturated and aromatic fractions of a crude oil at air to oil volume ratio of 68000 (Reference 4).

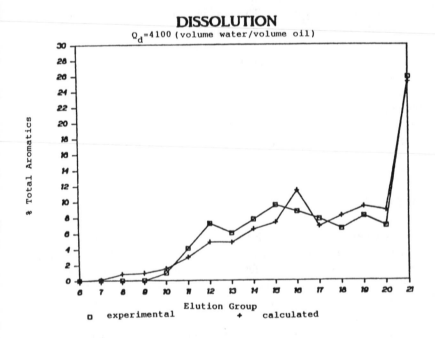

Figure 7. Line graph of dissolution by the generator column results for the saturated and aromatic fractions of a crude oil at water to oil volume ratio of 4100 (References 7 and 8).

This can be done by "backing out" a suitable biodegradation rate constant by subtracting from the total observed compositional change the expected evaporation and dissolution changes. Fortunately, dissolution is usually responsible for a negligible change in oil composition, but of course it can have profound effects on water quality.

A major difficulty is ascertaining, in the soil volume, the extent of heterogeneity in process rate. We suspect that there is a considerable variation in the rate of evaporation and biodegradation from place to place within the soil. Occasional sampling can thus give a misleading impression of the overall rate of conversion. We refer to this colloquially as the "Danish Blue Cheese" problem. If samples of that cheese were taken of perhaps 1 mm³ size, it could be deduced that there was either very little, or very intense microbial activity, because the activity is very heterogeneous. Fortunately, that heterogeneity is visible. On the contrary, we have no visible or mental impression of the heterogeneity of oil biodegradation. Regretfully, oil degrading microorganisms are not a visible blue. We believe that it is essential to characterize this effect, both in the laboratory and in the field.

A further problem is the effect of oil concentration. We find that at concentrations of 1 or 0.5% of oil by volume, the degradation rate is fairly independent of oil concentration. However, as oil concentration rises, the first order degradation rate must fall and the oil biodegradation half-life must rise. Ultimately, of course, when the oil reaches saturation conditions in the soil, i.e., 30–50% oil, biodegradation virtually ceases. This is an important consideration in real spills. The location of the point at which biodegradation starts to be adversely affected by the amount of oil present is not well established. The effect arises because of the toxicity of the large amount of oil to the microbial population and possibly other factors, such as alteration of the water environment, the amount of air, and the amount of nutrients relative to the amount of hydrocarbon. It should be possible, we believe, to develop equations expressing biodegradation rate as a function of variables such as hydrocarbon type (i.e., element in the matrix), total concentration of oil, amounts of oxygen and nutrients, and exposure or acclimation time.

CONCLUSIONS

We have described in qualitative terms the nature and results of ongoing experiments and our present thoughts on the feasibility of modeling petroleum in soils. We believe that in the next few years it should be possible to put into place models which will describe the behavior of hydrocarbons in soils under relatively dilute conditions, i.e., when hydrocarbon is present at concentrations of a few percent and is relatively immobile. A more difficult problem will be that of coupling this to models containing information on the flow of hydrocarbons and on the factors affecting biodegradation rate.

In summary, we believe that a suite of models is emerging describing the fate of oil in soils. Some will treat the migration or flow of the bulk phase under the action of gravity and hydraulic and capillary forces. Others, such as that described here, will treat the gradual decay of hydrocarbons over a period of months as they are subject to evaporation, dissolution, and biodegradation. The purpose of these models is not just to describe the science of chemical fate in soil (although this is justification enough), but to contribute to an improved ability to describe, respond to, and mitigate the fate and effects of hydrocarbons in real spill situations.

ACKNOWLEDGMENTS

The authors are indebted to the Association of American Railroads and the Petroleum Association for the Conservation of the Canadian Environment for support of work described in this chapter.

REFERENCES

1. Calabrese, E. J., and P. T. Kostecki, Eds. *Soils Contaminated by Petroleum: Environmental & Public Health Effects* (New York: John Wiley & Sons Inc., 1988).
2. Brookman, G. T., M. Flanagan, and J. O. Kebe, "Laboratory Study on Solubilities of Petroleum Hydrocarbons in Groundwater," API Publ. No. 4395, American Petroleum Institute, Washington DC, August, 1985.
3. Brookman, G. T., M. Flanagan, and J. O. Kebe, "Literature Survey: Hydrocarbon Solubilities and Attenuation Mechanisms," Report of Health and Environmental Sciences Department, American Petroleum Institute, Washington, DC, August, 1985.
4. Stiver, W., and D. Mackay. "Evaporation Rate of Spills of Hydrocarbons and Petroleum Mixtures," *Environ. Sci. Technol.*, 18, 834–840 (1984).
5. Mackay, D., and G. E. Hoag. "A Perspective on the Behaviour of Chemicals Spilled in Soil," in *Proceedings of the Technical Seminar on Chemical Spills*, Montreal, 1986.
6. Maijanen, A., A. Ng, W. Y. Shiu, and D. Mackay. "The Preparation and Composition of Aqueous Solutions of Crude Oils and Petroleum Products," report from the University of Toronto for PACE (Petroleum Association for the Conservation of the Canadian Environment), Ottawa, 1984.
7. Billington, J. W., G. L. Haung, F. Szeto, W. Y. Shiu, and D. Mackay. "Preparation of Aqueous Solutions of Sparingly Soluble Organic Substances: I. Single Component Systems." *Environ. Toxicol. Chem.* 7, 117–124 (1988).
8. Shiu, W. Y., A. Maijanen, A. L. Y. Ng, and D. Mackay. "Preparation of Aqueous Solutions of Sparingly Soluble Organic Substances: II. Multicomponent Systems—Hydrocarbon Mixtures and Petroleum Products," *Environ. Toxicol. Chem.* 7, 125–137 (1988).
9. Lee, M. D., and C. H. Ward. "Biological Methods for the Restoration of Contaminated Aquifers," *Environ. Toxicol. Chem.*, 4, 743–750 (1985).
10. Jobson, A., F. D. Cook, and W. S. Westlake. "Microbial Utilization of Crude Oil," *Appl. Microbiol.*, 23, 1082–1089, (1972).

11. Walker, J. D., and R. R. Colwell. "Biodegradation Rates of Components of Petroleum," *Can. J. Microbiol.*, 22, 1209–1213 (1976).
12. Westlake, O. W. S., and F. D. Cook. "Biodegradability of Northern Crude Oils," Report ALUR 74-75-81, Department of Indian and Northern Affairs, Ottawa, 1984.
13. Westlake, D. W. S., A. M. Jobson, and F. D. Cook. "*In situ* Degradation of Oil in a Soil of the Boreal Region of the Northwest Territories," *Can. J. Microbiol.*, 24, 254–260 (1978).
14. Mackay, D. and W. Y. Shiu. "Aqueous Solubilities of Weathered Northern Crude Oils," *Bull. Environ. Contamin. Toxicol.*, 15, 101–109 (1976).
15. Mackay, D., W. Y. Shiu, A. Chau, J. Southwood, and C. I. Johnson. "Environmental fate of Diesel Fuel Spills on Land," Report for the Association of American Railroads, Washington, D.C., 1985.
16. May, W.E., S. P. Wasik, and D. H. Freeman. "The Determination of the Aqueous Solubility of Polynuclear Aromatic Hydrocarbons by a Coupled-Column Liquid Chromatographic Technique," *Anal. Chem.* 50, 175–179 (1978).
17. Energy Resources Separations Manual, Waters Associates Inc., Milford, Massachusetts, 1978.

CHAPTER 7

Movement and Retention of Organics in Soil: A Review and a Critique of Modeling

Daniel Hillel

In this chapter, I shall attempt to sketch out the movement and retention of organic contaminants in soil, to describe the phenomenology in general terms, to say something about the mechanisms, and then to sound a critique of modeling.

As a starting point, the fate of a mixed organic material applied to the soil surface, or introduced into the soil profile, obviously depends on an extremely interesting and complex combination of interactive processes relating to the nature of the material, the manner of its application, the fundamental nature of the soil, and its transient state at the time and place of interest. The chain of sequential simultaneous processes may include:

1. volatilization of the lighter components at the soil surface with consequent, possibly temporary, pollution of the atmosphere, or transfer via the atmosphere
2. runoff over the soil surface, driven by gravity and affected by the surface configuration and possibly by rainfall, with consequent contamination of nearby surface waters or concentration in surface depressions
3. adherence of the heavier components to the soil surface, possibly resulting in clogging and hydrophobization of the soil
4. infiltration into the soil proper
5. downward and lateral flow within the unsaturated zone of the soil profile
6. detention above layer interfaces within the profile

7. retention in soil pores and attachment to soil grains, as well as to soil organic matter
8. volatilization and vapor diffusion within the soil and out of the surface, or possibly movement of that vapor (if its vapor density is greater than that of air) downward within the unsaturated zone, and possibly redissolution in the groundwater at the water table
9. chromatographic separation of components within the profile, resulting in a selective migration of lighter and less viscous components
10. partial dissolution of soluble or emulsifiable components within the water phase of the soil
11. degradation resulting from both nonbiological (i.e., chemical) and biological processes
12. internal drainage or leaching from the soil, either within or alongside the water phase toward the water table
13. mounding over the water table and creep over it, possibly including convergence in cones of depression or drawdown regions at wells or streams
14. penetration into the aquifer of the soluble and denser components
15. dispersion and further migration within the aquifer and eventual appearance in the water supplies.

From the complexity of these catenary processes, we begin to have an inkling of the difficulty of the problem that we face and that we are trying to handle and control. The significance of the unsaturated vadose zone arises from its possible function as a protective buffer over the groundwater aquifer.

The migration, retention, and transformation of mobile materials in the soil is affected by the nature of the soil, of course, and among the properties and conditions of the soil which affect the flow regime are the following:

1. soil type, or, to simplify, *soil texture*. It is obvious that migration in coarse texture (i.e., in sandy and gravelly soils) is generally faster than in fine texture (silty or clay) soils, which are more likely to attenuate and retain contaminants and prevent them from reaching groundwater. Furthermore, such dynamic parameters as permeability, diffusivity (both to the vapor phase and to the liquid phase), and hydrodynamic dispersivity are all texture-dependent.
2. the vertical uniformity or nonuniformity of the soil. *Layered soils* are more likely to retard migration in the profile than are uniform profiles. Since the latter are rare, we almost always must deal with layered conditions in the profile.
3. the *configuration of the soil layers*. Horizontal layers within concave troughs or depressions are more likely to detain perched bodies of the contaminating fluid, whereas slanted or sloping layers may direct the contaminant toward wells or springs serving water supplies. Standing at the surface of the soil, we very seldom know the configuration of the subterranean layers.

4. the *depth to the water table.* This determines the opportunity time and space for the retention and degradation of the contaminant prior to its penetration into the aquifer. Because the retentivity of soil profiles is limited, any particular quantity applied or spilled on the soil will migrate only so far, and if we are lucky, then that distance of travel prior to total retention will be less than the distance to the water table. At other times, that distance is not less than the water table, so we can either get or avoid getting penetration of the groundwater, depending on the depth of the water table and the retentive properties of the soil.

5. the *structure of the soil* is another important consideration. The presence of macropores, for instance, cannot be duplicated in the laboratory. It is a common feature of field soils that some have fissures, cracks, or channels, possibly due to roots or burrowing animals. These can serve as preferred pathways, allowing transient streams of pollutants to spurt rapidly toward the groundwater, thus bypassing the greater volume of the vandose zone and evading its filtration and degradation mechanisms. Even in the absence of such channels, the structural aggregation of soils induces faster flow through the inter-aggregate macropores than within the intra-aggregate micropores where potential pollutants might otherwise be retained.

6. in addition to the above, there is the possibility of *unstable flow,* a seemingly anomalous phenomenon only recently recognized. Apart from flow in macropores or fissures which are detectable preexisting features in the soil profile, there is an occasional tendency for flow to concentrate in tongue-like streams or convergent currents (often called "fingers") which generally begin at the transition from fine-textured to coarse-textured layers, again bypassing or short-circuiting the greater volume of the vadose zone and allowing direct transmission of contaminants to the water table. Such streams do not follow any particular discernible feature in the profile, but are spontaneous constrictions of the flow field that seem to occur whenever the flow velocity accelerates with distance.

7. *soil moisture.* Initial soil wetness and the vagaries of subsequent rainfall have an obvious effect on the pattern and migration of organics in the soil since these organics coinhabit the same network of pores and hence interact with the water initially present. There is a mutual interference between water, the wetting liquid, and organic liquids which are nonwetting. The latter are relegated to the interiors of the larger pores, whereas the smaller pores in the necks, between grains, as well as surfaces of the grains themselves, are occupied by the water which has a greater affinity to the mineral surfaces (water being a polar liquid).

Still another vexing issue is that of lateral heterogeneity. The field-scale, aerial, or spacial variability of basic soil properties, and of the transient conditions of the soil, strongly affects the pattern of migration of an introduced contaminant, depending on the particular time and place of its introduction. Among the properties of the invading material that affect the pattern of their movements and transformations within the soil are, of course, the following: volatility and vapor density, and solubility or miscibility in water. (Some materials are miscible in all proportions and others have a solubility product or limit so that only so much can be carried in the water stream.) In addition, of course, there are viscosity, density, and uptake by or toxicity to microbes or vegetation, that affect the pattern of movement, chemical reactivity, microbial degradability, adherence to mineral solids, and interactivity with soil organic matter. Most of these properties are dependent on temperature and moisture and hence on location and season.

So, altogether, there is a host of time- and space-variable factors which are often unpredictable in the field because they follow unpredictable patterns of weather and of soil boundary and initial conditions.

Another message I wish to convey is a word of caution on modeling of field-scale processes and its limitations. In modeling laboratory experiments under idealized conditions, one may get very nice correspondence or fit between predictions and measurements. In the field, however, one may get an entirely different kind of fit. I wish, therefore, to discuss some of the reasons for this discrepancy arising from the premises and inherent pitfalls of modeling.

The fundamental problem in our sciences is how to obtain knowledge of specific processes occurring in a complex dynamic system such as I've described, and then how to integrate that fragmented or partial knowledge in order to achieve an understanding of how the composite system as a whole operates. In principle, we can manage a system effectively only if we are able to define the major factors acting and interacting within it, and then to anticipate their combined and interactive effects.

The modeling approach allows us, in principle, to develop some statistical and simulation methods to handle several processes and effects simultaneously. The feeble human mind by itself cannot comprehend all that is happening in the field all at once, so we have to find some auxiliary way of processing and accounting for simultaneous and interactive processes. Fortunately, the computer is able to perform calculations quickly and in suffcient number that the human mind by itself cannot. Computer-based models appear capable of reducing a baffling system of seemingly hopeless complexity to orderly and manageable proportions. The very exercise of designing, operating, and testing such models has helped us to gain more insight into the workings of an actual system, and to develop criteria to predict its behavior under varying conditions. A model, however, is only a simplified representation of a reality that is too intricate to be formulated in total detail. Far from encompassing reality entirely, the model merely encapsulates selected aspects of it. The process of modeling consists of choosing the facts and factors deemed to be most relevant and essential to the solution of a perceived problem, and then formulating the relationships governing these factors. So there is something basically subjective in this exercise.

Recent efforts have extended beyond the initial approach to formulating mechanistic events in a deterministic fashion, and have gone toward recognition that some of the happenings in the field are not strictly deterministic, but are in fact subject to all sorts of vagaries. So attempts are now made to recognize the spatial and the temporal variation or stochasticism of field behavior, and then to devise a way of modeling a system which is partly deterministic and partly probabilistic, partly structured and partly random. That is a formidable challenge indeed to present-day modelers of field-scale processes.

The story of modeling the soil environment has produced impressive progress but is not, unfortunately, an uninterrupted parade of triumphs. The very success of modelers in devising ever more complex constructions seems to breed certain problems.

Modeling is a heady and exhilarating game with each new advance a tour de force of mathematical and programming virtuosity, but the proliferation of ever-new and more intricate models is not an unmixed blessing. Where once we had, or seemed to have, a plethora of disparate data without a redeeming theory, now we suffer from the opposite syndrome of having a myriad of still unproved theoretical models, and that can be just as baffling. As models become more complex, they require a greater number of parameters. More parameters are difficult to measure, and more complex models are harder to validate. Even the documentation required to understand such models has become too voluminous to publish and too elaborate, obscure, and tedious to study. Thus, more comprehensive models have become less comprehensible. What with the limitations of time, funds, and patience, as well as the difficulty of interdisciplinary cooperation and the compulsion to publish numerous papers quickly, precious few models have been tested or even documented adequately at the time of publication. To some, modeling has become an end in itself. Such is the seductive lure of the computer and the sense of power that it imparts to the modeler, that a critic is reminded of the proverbial Hindu metaphor, ''One who rides a tiger feels very powerful but cannot get off.'' And the question is whether the rider is captor or captive of his tiger.

Some models are practically inaccessible to persons other than the ones who create them, or at least not understandable to others; and even they, the creators, might lose familiarity with their models after a short time. Those who choose nevertheless to borrow unfamiliar models run the risk of using them out of context or beyond their seldom-specified limits of relevance. Consequently disappointment and doubt have set in. We now ask ourselves if our enthusiasm for modeling as a panacea has perhaps led us astray. Have we let modeling run away from reality? Has some of the effort invested in modeling been misallocated? In our zeal to solve problems quickly, have we fallen prey to an addictive fad, merely a higher version of the computer games played in amusement arcades? In seeking shortcuts to solve our problems, have we gone beyond the bounds of scientific constraint and have we been, in fact, creating fiction rather than truth? And when we rush to offer our incompletely tested models as readymade packages for industry or government agencies to use as management tools, are we acting responsibly or otherwise? Can we ignore the danger that some potential users of our models do not have the time or ability to evaluate the intricate codes which shroud today's models in the priestly garb of computer jargon, and may therefore take them uncritically to be literally true?

The crucial question is how good are our models, and that is answerable only by testing with independently obtained data. Without being pitted against systematic observations in the field, and the results of experiments again in the real world, a model is nothing but a tentative exercise in abstract logic; merely a construct of the mind, an artificial contrivance detached from reality. Moreover, a model expresses not only our conception of the system but also our misconception of it. So the model can be misleading and even do more harm than good.

Neither the modeling nor the experimental approach alone is self-correcting and likely to lead us toward the progressive acquisition of comprehensive knowledge and understanding. It is only by sallying back and forth between experimental data and theoretical models that we can advance, albeit in a tortuous and laborious way.

The ultimate test of a model, whether statistical, mechanistic, simulation, or stochastic, is the accuracy with which it predicts the behavior of the actual system in the future. Even here we must be wary of the human tendency to select the data which fit the model best. Modelers often develop a vested interest in the success of their creations and hence are in constant danger of losing their objectivity, like the mythic King Pygmalion who fell in love with his own Galatea. Models sometimes seem to acquire a life of their own and are accepted too easily at face value.

Model validation remains the most vexing, difficult, challenging, and necessary phase of our science. We have to remember that model validation is not a once-and-for-all deed. Having once proved that the model is valid for a given set of circumstances, we cannot rest on our laurels and become complacent. We must remain forever skeptical of our models and guard against accepting them as literal truths. Our models are always as vulnerable as we are. If we used a model with some degree of success in one year at one location, we cannot take it for granted that the same measure of success will be ours in another year or location. So we must constantly check model results against independent measurements. I think this warning needs to be sounded and repeated. Modeling is not a substitute for experimentation but possibly a more rational basis for experimentation. We need detailed, sound, comprehensive multidisciplinary experiments because the phenomena we are trying to understand and eventually control are the province of scientists in different disciplines working together. Although experiments in the field are costly and time-consuming, we cannot skip them and depend on our models alone. Models can, on the other hand, help to economize experimentation by guiding it to where it is needed most.

CHAPTER 8

Applicability of POSSM to Petroleum Product Spills

Walter J. Shields and Stuart M. Brown

Petroleum products are complex mixtures of hydrocarbons with widely varying physical-chemical properties. When spilled on soil, the environmental behavior of a petroleum product is controlled by a number of processes. A petroleum product can migrate through soil as a separate immiscible phase until it is completely redistributed in the vadose zone or until it reaches the water table. Redistribution is controlled by the volume of the petroleum product spill, the viscosity of the petroleum product, the characteristics of the soil, and the depth to the water table. During and following redistribution, water percolating through the soil can dissolve soluble components from the immiscible phase and subsequently leach these components to groundwater. The more volatile components of the petroleum product can also evaporate into the air-filled voids in the soil and subsequently migrate through vapor diffusion. Dissolution, degradation, and evaporation of components from the petroleum product cause the spill to weather. As a result, the chemical composition of the petroleum product changes with time.

The need to evaluate potential environmental and human health impacts has prompted risk analysts to use chemical transport and fate models to predict the potential environmental behavior of petroleum products in soil. In doing so, they find that there are relatively few models that consider all of the processes that control petroleum product behavior. Most of the available models simulate the transport and fate of chemicals in the soluble phase, rather than the immiscible phase. Most of the available models were also developed for individual chemicals, not mixtures of chemicals. Thus, risk analysts are currently faced with the problem of trying to apply single-component, solute transport models to petroleum product spills.

One model that potentially fills the gap that currently exists between the available modeling technology and needs of the risk analyst is the PCB Onsite Spill Model (POSSM). While POSSM was originally developed to simulate the behavior of PCB in soil following capacitor and transformer spills, it is also applicable to other organic compounds. This chapter describes the POSSM model and its two extensions, and then presents four potential strategies for using POSSM to evaluate petroleum product spills. The chapter focuses on limitations associated with each strategy.

POSSM MODEL

POSSM is a chemical transport and fate model for the vadose zone. The original version of POSSM was developed to simulate the environmental behavior of a single component. This version of POSSM was extended in two ways. First, POSSM was modified to simulate the behavior of a two-component mixture (i.e., a binary mixture). Second, POSSM was extended to allow the user to represent the potential variability of input parameter values with frequency distributions; a Monte Carlo technique is used to randomly select parameter values from the user-specified distributions. A description of the POSSM model and its two extensions follows.

Single-Component Version

The single-component version of POSSM is a transport and fate model capable of predicting (1) chemical losses from the land surface due to volatilization, runoff, and soil erosion, and (2) leaching to groundwater due to chemical movement in unsaturated soils.[1,2] POSSM is applicable to spill sites consisting entirely of soil, soil covered by an impervious surface (e.g., asphalt or concrete), or both. Figure 1 shows the specific hydrologic and physical/chemical mechanisms considered by POSSM. The upper portion of the figure shows the mechanisms considered for spills onto soils; the lower portion shows the mechanisms considered for impervious surfaces.

POSSM is a one-dimensional, compartmental model. It is one-dimensional in that water and chemical movement in unsaturated soils are represented in the vertical direction only. It is compartmental in that the soils underlying the spill site can be divided into a series of layers or compartments. Different model parameter values can be assigned to each horizon so that vertical variations in soil properties can be considered.

POSSM is also dynamic in that it operates on a daily time step. As a result, POSSM predicts chemical losses from a spill site on a daily basis, and daily

SOIL SURFACES

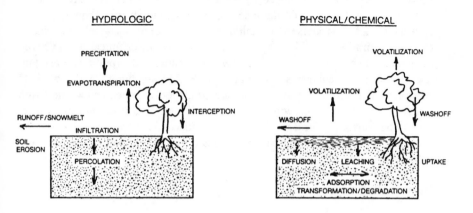

IMPERVIOUS SURFACES

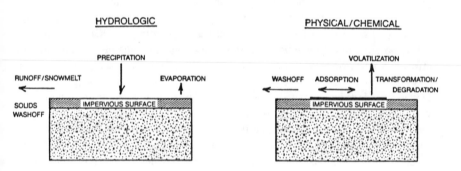

Figure 1.   Hydrologic and physical/chemical mechanisms considered by POSSM.

changes in onsite chemical concentrations. Daily changes in hydrologic conditions (e.g., soil moisture) and land surface conditions (e.g., vegetative cover) are also simulated. The latter are important because of their impact on runoff and soil erosion. The dynamic nature of the model requires that the user input daily changes in meteorological conditions (e.g., precipitation, temperature, and wind speed).

POSSM can consider multiple cleanup events and a range of cleanup practices. The user can specify cleanup levels for soil, the impervious surface, vegetation, or any combination of the three.

The basic framework for POSSM was taken from PRZM (Pesticide Root Zone Model).[3] PRZM was developed by the EPA Environmental Research Laboratory in Athens, Georgia. Of the mechanisms affecting chemical movement at a spill site, PRZM considers runoff, infiltration, soil erosion, snowmelt, percolation, evapotranspiration, chemical washoff from vegetation, chemical washoff from the land surface, leaching, transformation/degradation, adsorption, and plant uptake. The only mechanisms not included in PRZM are runoff, solids washoff, evaporation, adsorption, transformation/degradation and chemical washoff from impervious surfaces (e.g., concrete and asphalt), and volatilization from soil and impervious surfaces. These mechanisms were added to PRZM during the development of POSSM. Brown and Silvers[1] and Brown and Boutwell[2] provide a more complete description of POSSM.

Binary Mixture Version

The binary mixture version of POSSM accounts for the effect that one component will have on another when both are present together as an immiscible phase in soil.

Specifically, POSSM accounts for the affect that one component will have on the solubility in water and vapor pressure of the other component. The binary mixture version of POSSM was originally developed for transformer oil spills which consist largely of mineral oil and trace quantities of PCB.

It is possible to estimate how the presence of one nonaqueous component may affect the behavior of another, particularly with respect to leaching and volatilization. To estimate the effect on leaching, the concentration of each component in water needs to be estimated when the binary mixture is brought into equilibrium with water. Assuming that the solubilities of both components in water are relatively low, it can be assumed that there will be no interaction between the two components when they are dissolved in water. Thus, their activity coefficients together in the aqueous phase will be the same as when they are present alone. As a result, it is possible to invoke Raoult's Law and show that at equilibrium the concentration of one component in water, C_{w_1}, will be the product of its pure phase solubility, C_{s_1}, and its mole fraction in the binary mixture, x_1:

$$C_{w_1} = C_{s_1} \cdot x_1$$

A similar relationship can be derived to estimate the effect of one component on the volatility of the other. The partial pressure of one component, p_1, is the product of its pure phase vapor pressure, p_{s_1}, and its mole fraction in the binary mixture:

$$p_1 = p_{s_1} \cdot x_1$$

In POSSM, two complete sets of chemical transport equations, one for each component in the binary mixture, are solved each timestep. At the end of each timestep, mole fractions for each component are calculated and are subsequently used to adjust the pure phase solubility and vapor pressure of each component. Calculating the mole fraction on a daily basis makes it possible to account for the fact that the spilled material will weather and change composition with time. As the composition changes, so does the apparent solubility in water and vapor pressure of each component. Solubility and vapor pressure remain constant with time in the single-component version of POSSM.

Monte Carlo Version

MCPOSSM, the Monte Carlo version of POSSM, is an expansion of the binary mixture version. MCPOSSM allows the user to acknowledge that there is some degree of uncertainty or variability associated with the estimation of model input parameters. Rather than forcing the risk analyst to qualitatively determine how this uncertainty may impact model results, MCPOSSM provides a quantitative estimate. This is accomplished by using frequency distributions to represent the potential variability of model input parameters. A Monte Carlo procedure is used to randomly select a value for each parameter from its respective distribution, prior to executing a POSSM run (see Figure 2). A correlation matrix is used to ensure that combinations of selected parameter values are physically meaningful. For instance, the correlation matrix ensures that parameter values selected for all soil properties (e.g., bulk density, field capacity, and wilting point) are internally consistent. The Monte Carlo parameter selection procedure is conducted a large number of times (usually 300 to 500), thus producing a large number of model simulations. MCPOSSM statistically summarizes the model results in terms of means and standard deviations, and of cumulative frequency distributions. Shields et al.[4] provide a complete description of MCPOSSM.

The current version of MCPOSSM allows the user to select two different frequency distributions to represent the variability of model parameters: normal and lognormal. Input parameters that can be varied include:

- spill area
- spill volume
- soil bulk density
- field capacity
- wilting point
- organic carbon content
- solubility in water
- organic carbon partition coefficient
- chemical degradation rate
- initial concentration of one component in the binary mixture

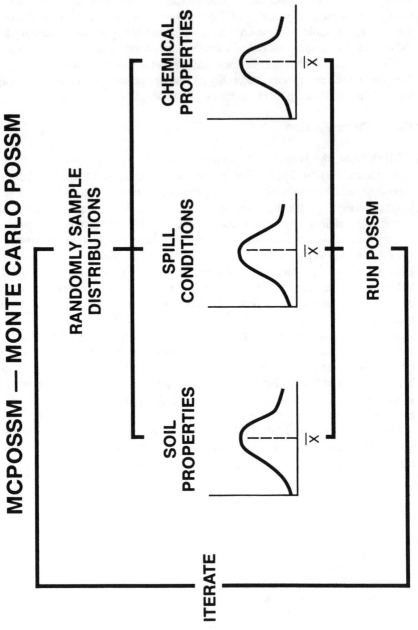

Figure 2.   MCPOSSM simulation process.

POTENTIAL SIMULATION STRATEGIES

There are at least four potential strategies for using POSSM, and its extensions, to simulate the behavior of a petroleum product spill.

1. Treat the petroleum product as a single component with bulk average properties representative of the entire mixture.
2. Subdivide the petroleum product into groups of components (e.g., compound classes) with similar physical-chemical properties and independently simulate the behavior of each component group.
3. Subdivide the petroleum product into groups of components with similar physical-chemical properties and independently simulate the behavior of each component group in the presence of the remainder of the petroleum product mixture.
4. Subdivide the petroleum product into soluble and insoluble fractions and independently simulate behavior of soluble fraction components in the presence of the insoluble fraction.

Strategy 1. Bulk Average Properties Strategy

This simulation strategy would involve treating the petroleum product as a single component with bulk average properties rather than as a mixture of components with widely varying properties. Bulk average physical-chemical properties (e.g., solubility in water and vapor pressure) would be calculated based on the composition of the mixture.

This strategy is consistent with the structure of the single-component version of POSSM. It is also consistent with the structure of most other chemical transport and fate models for soil and groundwater.

The advantage of this strategy is that the behavior of the petroleum product can be evaluated with a single model simulation.

Strategy 2. Individual Component Strategy

This strategy would involve subdividing the petroleum product into component groups (e.g., aromatics, n-alkanes, iso-alkanes, and polar compounds), each group having similar physical-chemical properties. This is a special case of the bulk average properties strategy in that with average properties would be estimated for each component group. Another approach would be to use the properties for the most representative chemical in each component group. A model run would be conducted for each component group assuming the remainder of the mixture was not present. The results of each simulation would then be combined to determine the behavior of the overall petroleum product mixture.

This strategy is also consistent with the structure of the single-component version of POSSM.

More effort would be involved with this strategy because the petroleum product would have to be subdivided into component groups, with bulk average physical-chemical properties estimated and a model run performed for each group.

Strategy 3. Individual Component-Mixture Strategy

This strategy is an extension of the individual component strategy. The primary difference is that the petroleum product would be treated as a binary mixture. One component would be the component group of interest (e.g., aromatics); the second component would be the remainder of the petroleum product mixture. A model run would be conducted for each component group using bulk average properties for both components. The overall behavior of the petroleum product would be determined by combining the model simulation results for each component group.

This strategy is consistent with the structure of the binary mixture version of POSSM. Additional effort would be involved in using this strategy because the bulk average physical-chemical properties of the component group of interest and the remainder of the petroleum product mixture would need to be estimated for each model simulation.

Strategy 4. Soluble-Insoluble Fraction Strategy

This strategy would involve dividing the petroleum product into two fractions: soluble and insoluble. The soluble fraction would be further subdivided into component groups having similar physical-chemical properties. Model runs would be conducted for each component group in the presence of the insoluble fraction. The overall behavior of the petroleum product mixture would be determined by combining the simulated behavior of each component group.

The soluble-insoluble fraction strategy is consistent with the structure of the binary mixture version of POSSM. One component of the mixture would be the insoluble fraction; the other would be a component group selected from the soluble fraction.

Another approach to this strategy would be to use frequency distributions to represent the variability of physical-chemical properties of those compounds in the soluble fraction. Model runs would then be made by selecting parameter values for each physical-chemical property from user-specified distributions. This strategy is consistent with the structure and assumptions of MCPOSSM.

POTENTIAL LIMITATIONS OF EACH SIMULATION STRATEGY

To provide a basis for identifying potential limitations associated with each simulation strategy, POSSM was used to simulate the behavior of a diesel oil spill over a period of one year. Soil, meteorological, and spill conditions for this model application are summarized in Table 1 and are described more fully in the POSSM manual.[2] The spill volume is assumed to be equally distributed throughout the 0–10-cm soil layer.

Table 1.   Diesel Oil Spill and Environmental Conditions.

Spill Conditions	
Spill area	0.004 ha (430 ft²)
Spill volume	50 L of diesel oil
Spill depth	10 cm
Environmental Conditions	
Meteorology	O'Hare Airport meteorological record for August 1, 1979, through July 31, 1980
Soil type	Dana silt loam
Soil bulk density	1.4 g/cm³
Soil field capacity	0.23 cm/cm
Soil wilting capacity	0.10 cm/cm
Soil organic carbon content	1.77%
Depth to groundwater	155 cm

Physical-chemical properties required for input to POSSM are the degradation rate, soil-water partition coefficient, vapor pressure, diffusivity in air, molecular weight, solubility in water, and density. Property values are required for both components in the binary mixture version. Frequency distributions of values are required in the Monte Carlo version.

Data from Mackay et al.[5] were used to estimate values for all properties except diffusivity in air. These researchers developed a model that estimates and correlates the amounts, properties, and composition of diesel fuel as a function of time of exposure to evaporation, dissolution, and biodegradation.

Mackay et al. divided diesel fuel into five compound classes based on structural groups, and estimated their composition in unweathered diesel oil:

- n-alkanes (degrade somewhat rapidly)—37.6% of fresh diesel oil
- iso- and cyclo-alkanes (degrade somewhat slowly)—37.6%
- isoprenoids (e.g., pristane and phytane, highly branched iso-alkanes very resistant to degradation)—3.5%
- aromatics (water-soluble hydrocarbons, predominated by parent and alkylated benzenes, napthalenes, hydrindenes, phenanthrenes, and fluorenes)—20.2%
- polars (water-soluble sulfur, nitrogen, and oxygen compounds)—1.1%

Their model estimated physical-chemical properties and percent composition for C_6 through C_{20} compounds for each compound class.

To obtain input values for POSSM, weighted means for molecular weight and density for each compound class were calculated (see Table 2). Assuming that Raoult's Law applies, the mole fraction was multiplied times the solubility for each C_n compound; the products were summed to get the solubility for a com-

Table 2. Physical and Chemical Property Input Data for Each Strategy.

Strategy	Component Group	Component Concentration in Binary Mixture (g/100 g oil)	Degradation Rate (/day)[a]	Soil-Water Partition Coeff.[b] (mL/g)	Vapor Pressure (mm Hg)[a]	Diffusivity in Air (cm²/day)[c]	Molecular Weight[a]	Solubility (mg/L)[a]	Density (g/cm³)[a]
Bulk average properties	Bulk average	NA	0.010	1.1×10^3	0.03	4,000	202	0.2	0.80
Individual component and individual component-mixture[d]	N-alkanes	37.6	0.014	1.5×10^6	0.04	4,000	204	0.00001	0.74
	Iso-alkanes	37.6	0.005	3.8×10^4	0.04	4,000	218	0.0019	0.77
	Isoprenes	3.5	0.002	6.4×10^7	0.000006	4,000	261	0.000000009	0.79
	Aromatics	20.2	0.012	3.4×10^2	0.01	4,000	162	1.0	0.97
	Polar	1.1	0.014	1.4×10^3	0.00007	4,000	156	0.16	1.07
Soluble-insoluble fraction	Soluble[e]	20.2	0.012	3.4×10^2	0.01	4,000	162	1.0	0.97
	Insoluble[f]	NA	0.010	2.1×10^4	0.04	4,000	212	0.004	0.76

[a]Derived from data of Mackay et al.[5].
[b]Calculated from solubility.
[c]General value estimated from Thibodeaux[7].
[d]Parameters for Individual Component Mixture Strategy are the weighted means of the parameters for the remaining components.
[e]Aromatic component group.
[f]Remaining four component groups (i.e., n-alkanes, iso-alkanes, isoprenes, and polar compounds).
NA = Not applicable.

pound class. The same procedure was used to calculate vapor pressure. For the bulk average simulation, weighted means were calculated from the compound class means, solubilities, and vapor pressures. Soil-water partition coefficient was estimated from solubility by the relationships given in Verschueren.[6] A general diffusivity in air value was obtained from Thibodeaux.[7]

For all four strategies, simulations were conducted for a period of one year. The vertical distribution of each component group and total diesel fuel in the soil after one year were used as a basis for identifying limitations associated with each strategy.

Strategy 1. Bulk Average Properties Strategy

Figure 3 shows the vertical distribution of diesel oil in the soil column simulated by POSSM after a period of one year using the bulk average properties strategy. The results show that the peak diesel oil concentration occurs at a depth of 10 cm with a gradual decrease in concentration to about 35 cm. Using bulk average property values, POSSM estimated that about 19% of the oil would remain after one year. This is less than the 30% and 34% residual estimated for Strategies 2 and 3, respectively (see Table 3).

The results shown in Figure 3 illustrate one of the major limitations associated with the bulk average properties strategy. This strategy provides no information on the vertical distribution of individual components or compound groups. Risks associated with groundwater contamination by more mobile compounds, such as benzene, cannot be evaluated with this strategy.

A second limitation is that the use of bulk average properties does not adequately represent the expected behavior of a petroleum product. In using this strategy, leaching and volatile losses of the more immiscible components and higher molecular weight compounds will be overestimated. Conversely, the leaching and volatile loss of the lighter and more miscible compounds will be underestimated.

Finally, this strategy does not account for changes in the composition of the pretroleum product mixture and the resulting impact on the behavior of individual

Table 3. Oil Composition—Percent of Initial Mass Remaining After One Year.

Compound Class	Strategy 1 Bulk Average Properties	Strategy 2 Individual Component	Strategy 3 Individual Component-Mixture
N-alkanes	NA	38	43
Iso-alkanes	NA	36	40
Isoprenoids	NA	86	87
Aromatics	NA	5	19
Polars	NA	0.4	6
TOTAL	19	30	38

NA = not applicable.

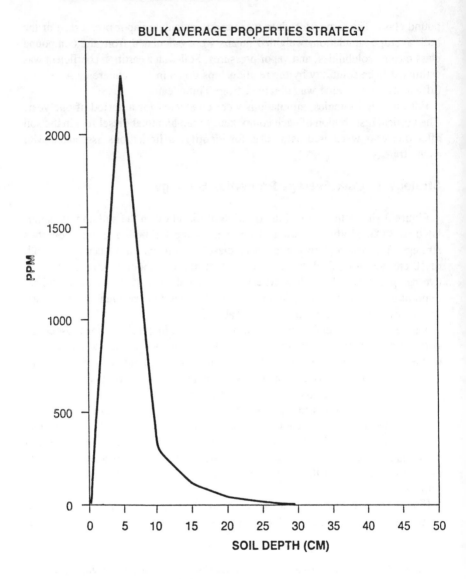

Figure 3. Simulated depth distribution of diesel fuel one year after spill using bulk average properties strategy.

components. Over time, the water-soluble fraction will be depleted from the spill, thus changing the relative composition of the mixture. The rate of depletion of the water-soluble fraction depends on initial composition, spill size, and volume of infiltrating water. The mixture will also weather, due to degradation reactions. Petroleum components exhibit a wide range of susceptibilities to microbial attack.[8]

Strategy 2. Individual Component Strategy

To demonstrate this strategy, the transport and fate of the five compound classes identified by Mackay et al.[5] were simulated individually, using the single-component version of POSSM. The 50 L of diesel oil were proportioned among the five components based on their mole fraction in the original mixture.

Figure 4 shows the estimated vertical distribution of each component and the vertical distribution of the overall mixture, assuming it is reasonable to simply combine the simulation results for each of the five component groups. The simulation results show that most of the components were lost from the upper 5 cm of soil, mainly through volatilization and degradation. After one year, only the isoprene group remained stable in the upper 5 cm because of the low vapor pressure and solubility of compounds in this group. Limited leaching of any of the components occurred below the original spill depth of 10 cm (see Figure 4).

This strategy overcomes several of the limitations associated with the bulk average mixture strategy. Information on the vertical distribution of individual components or component groups is provided and the expected behavior of the overall mixture is more accurately represented.

However, the individual component strategy does not account for changes in the composition of the petroleum product mixture (i.e., weathering). In addition, this strategy neglects any effects that the bulk mixture would have on any one compound or component group. In simulating the independent behavior of each component group, it is inherently assumed that the component group is present in the soil alone, and that the rest of the mixture has no impact on its behavior. This is probably not a reasonable assumption because the leaching and volatilization of each component group will be affected by the bulk mixture.

Strategy 3. Individual Component-Mixture Strategy

To obtain results for comparison with Strategy 2, all five compound classes were also run using the binary version of POSSM. One component of the binary mixture was assumed to be the compound class of interest. The second component was considered to be the remaining four compound classes. Thus, the behavior of each compound class in the presence of the rest of this mixture was simulated.

The simulated vertical distribution of each compound class and the total diesel oil mixture (see Figure 5) is similar to the distribution obtained for Strategy 2 (see Figure 4). The amount remaining in the soil column one year after the spill was slightly greater for the individual component-mixture strategy than for the individual component strategy. As the results in Table 3 indicate, a higher percentage of most of the component groups remained after one year, particularly for the more mobile component groups (i.e., aromatics and polars). The single-component version estimated that 5% of the initial amount of aromatics would remain, whereas the binary version estimated 19%. This difference illustrates

INDIVIDUAL COMPONENT STRATEGY

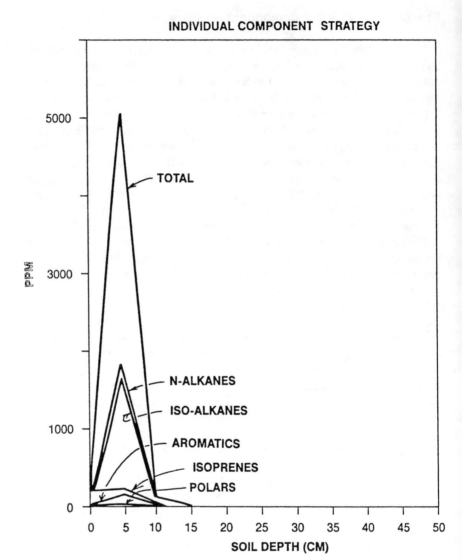

Figure 4. Simulated depth distribution of diesel fuel one year after spill using individual component strategy.

the effect of considering the presence of the remainder of the mixture when simulating the behavior of a given component group. The apparent solubility in water and vapor pressure of the component group will be reduced, leading to lower leaching and volatilization losses. Thus, a higher percentage of each compound will remain in soil.

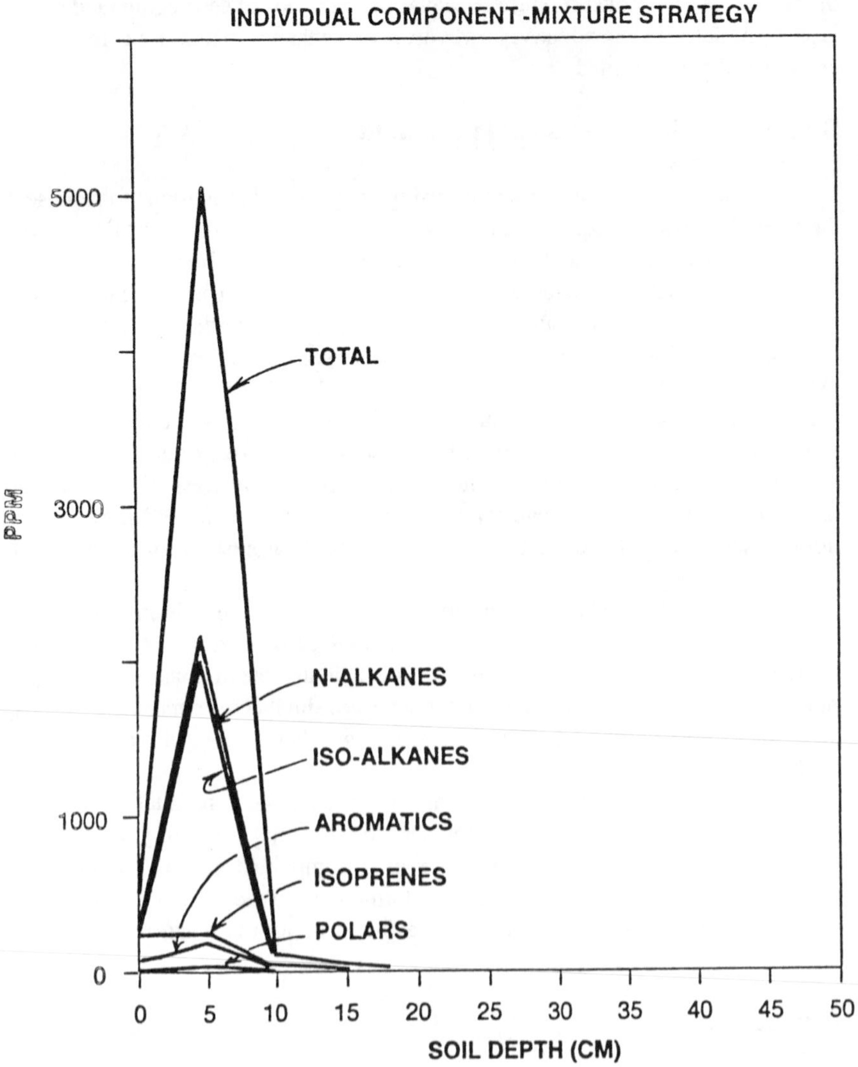

INDIVIDUAL COMPONENT-MIXTURE STRATEGY

Figure 5. Simulated depth distribution of diesel fuel one year after spill using individual component-mixture strategy.

This strategy partially overcomes two of the major limitations associated with the individual component strategy. This strategy does represent the effects that the overall mixture will have on the behavior of any one compound or component group. Changes in mixture composition are also accounted for by simulating the combined behavior of the selected compound group and the remainder of the petroleum product mixture. Because the behavior of each compound group is still simulated independently, however, actual changes in mixture composition are not fully represented.

Strategy 4. Soluble-Insoluble Fraction Strategy

This strategy was demonstrated by using frequency distributions to represent the potential variability of solubility in water and organic-carbon partition coefficient of compounds in the soluble fraction of a diesel oil. Means and standard deviations for each parameter (see Table 2) were estimated using data in Mackay et al.[5] The aromatic compound class was selected to represent the soluble fraction; the remaining four compound classes were selected to represent the insoluble fraction.

MCPOSSM was used to simulate the behavior of the aromatic compound class in the presence of the remainder of the diesel oil mixture. Figure 6 shows the simulated vertical distribution of the aromatic compound class. The solid line in Figure 6 represents the mean soil concentration based on 400 POSSM simulations, each with a different set of physical-chemical properties for the aromatic component group.

As with the individual component-mixture strategy, the soluble-insoluble fraction strategy overcomes many of the limitations associated with the other two strategies. It is still somewhat limited, however, in that the dissolution and evaporation of individual compounds or components are simulated independently. Thus, this strategy does not fully represent the overall weathering of a petroleum product spill.

One of the benefits of this strategy, as compared to the individual component-mixture strategy, is the ability to represent the potential variability of physical-chemical properties of those compounds in a component group. The effect of physical-chemical property variability within the aromatic compound class is illustrated by the band of one standard deviation about the mean in Figure 6.

CONCLUSIONS

The POSSM model may fill a gap currently facing risk analysts needing to evaluate potential impacts associated with petroleum product spills. An initial evaluation of potential simulation strategies suggests that either the binary mixture or Monte Carlo versions of POSSM are the most applicable.

Of the simulation strategies that were identified, the individual component-mixture and soluble-insoluble fraction strategies appear to be the most applicable. Field or laboratory data are needed to test the applicability of each version of POSSM and each simulation strategy, and to further assess potential limitations.

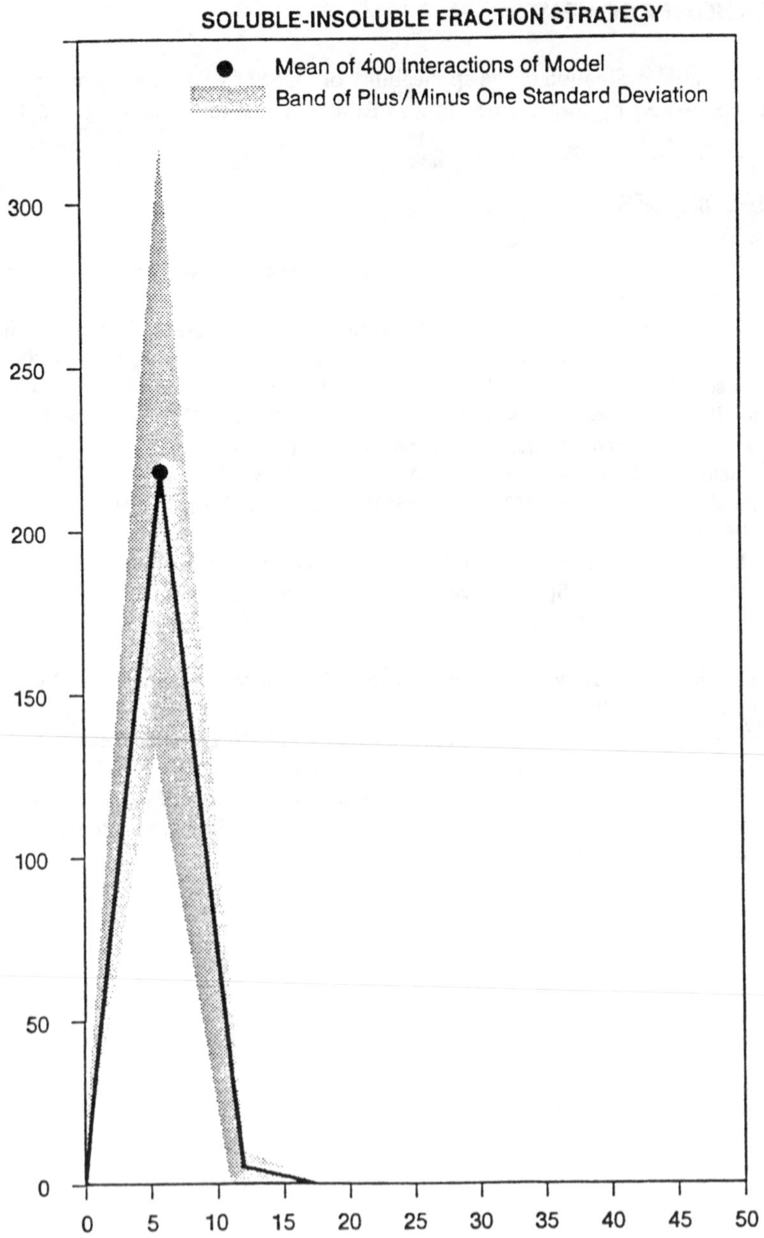

Figure 6. Simulated depth distribution of the soluble fraction using soluble-insoluble fraction strategy and MCPOSSM.

ACKNOWLEDGMENTS

This initial evaluation of the applicability of POSSM to petroleum product spills was sponsored by the Electric Power Research Institute under RP 2634-1.

REFERENCES

1. Brown, S. B., and A. Silvers. "Chemical Spill Exposure Assessment," Risk Analysis. 6(3) (1986).
2. Brown, S. B., and S. H. Boutwell. "Chemical Spill Exposure Assessment Methodology." RP 2634-1. Electric Power Research Institute, Palo Alto, CA (1988).
3. Carsel, R. F., C. N. Smith, L. A. Mulkey, J. D. Dean, and P. Jowise. User's Manual for Pesticide Root Zone Model (PRZM), Release I. U.S. Environmental Protection Agency Research Laboratory. Athens, Georgia (1984).
4. Shields, W. J., E. W. Strecker, J. D. Dean, and S. M. Brown. "Chemical Spill Uncertainty Analysis." RP 2634-1. Electric Power Research Institute, Palo Alto, California (1987).
5. Mackay, D., W. Y. Shiu, A. Chau, J. Southwood, and C. I. Johnson. "Environmental Fate of Diesel Fuel Spills on Land," Report for Association of American Railroads. Department of Chemical Engineering and Applied Chemistry, University of Toronto (1985).
6. Verschueren, K. *Handbook of Environmental Data on Organic Chemicals* (New York: Van Nostrand Rheinhold Co., 1983).
7. Thibodeaux, L. J. *Chemodynamics: Environmental Movement of Chemicals in Air, Water, and Soil* (New York: John Wiley & Sons, 1979).
8. Abriola, G. F., and L. M. Pinder. "Migration of Petroleum Products Accidentally Introduced into the Subsurface Through Spills," Water Resources Program. Princeton University (1981).

CHAPTER 9

The Utility of Environmental Fate Models to Regulatory Programs

Michael A. Callahan*

The environmental fate of pollutants has been an important issue at least as early as the 1960s, when environmentalists began to raise public consciousness that chemicals may ultimately travel far from their original sources.[1,2] The U.S. Environmental Protection Agency (EPA) itself was founded at least in part to answer the public concern that the environmental fate of chemicals be understood and chemicals be controlled, so that people or the ecosystem would not be unexpectedly exposed to toxic chemicals.

During the 1970s and 1980s, the EPA and other organizations have supported research designed to better understand, and better predict, the fate of chemicals in the environment. Computational tools, usually computer-based programs or models, were developed to help in predictions. Early models predicted the transport or dispersion of the medium itself (i.e., air, water), and used these phenomena as a first approximation of where the pollutant would go. Later, properties of the pollutant, how these properties affected the pollutant's interaction with the environment, and degradation of the pollutant were taken into account. Still, by the late 1970s, most such models were "single-medium" models; that is, they were air models, groundwater models, surface water models, etc., but didn't follow pollutants from one medium to another. In the late 1970s, "multimedia"

---

*The opinions expressed in this chapter are the author's and do not necessarily reflect the policy of the U.S. Environmental Protection Agency.

models began to be developed, either by connecting several single-medium models together, or by designing more sophisticated integrated multimedia models.

Throughout the years of development of environmental fate models, their primary utility in the regulatory process of EPA has remained the same: fate models were, and are, useful tools to help decisionmakers understand *some* of the factors that need to be weighed in a regulatory decision. In order to understand how environmental fate models fit into the regulatory scheme, it is necessary to show the relationships among risk management, risk assessment, exposure assessment, and environmental fate.

RISK ASSESSMENT AND RISK MANAGEMENT

The National Research Council (NRC) in 1983[3] noted that regulatory actions taken by government agencies such as EPA are based on two separate and distinct, albeit related, processes. These are *risk assessment* and *risk management.* According to the NRC, risk assessment is "the use of the factual base to define the health effects of exposure of individuals or populations to hazardous materials and situations." Risk management, on the other hand, is "the process of weighing policy alternatives and selecting the most appropriate regulatory action, integrating the results of risk assessment with engineering data and with social, economic, and political concerns to reach a decision."

The NRC report further describes risk assessment as containing the following four steps: *hazard identification* (the determination of whether a particular chemical is or is not causally linked to particular health effects), *dose-response assessment* (the determination of the relation between the magnitude of exposure and the probability of occurrence of the health effects in question), *exposure assessment* (the determination of the extent of human exposure before or after application of regulatory controls), and *risk characterization* (the description of the nature and often the magnitude of human risk, including attendant uncertainty).

In considering these definitions, some interesting points arise. First, regulatory decisions are not made only on the basis of risk assessment; that is, only on the basis of the factual science of the extent and nature of risk. Rather, regulatory decisions use the scientific analysis (risk assessment) as only one of several inputs needed. "Engineering data," or the practicability of a regulatory control being considered, is a major factor in regulatory decisions, as are "economic concerns," or the cost of implementing the control being considered. Second, exposure assessment is a component of risk assessment, and as such includes a responsibility to determine exposure both before and after regulatory controls. At the time a regulatory decisionmaker is considering alternatives for controlling risks, the "after control" exposures cannot be measured, since they do not yet exist. Exposure assessment must then be used as a predictive tool in estimating potential future exposures. It is this predictive nature of exposure assessment which relies heavily on models, including environmental fate models.

APPROACHES TO EXPOSURE ASSESSMENT

The EPA Guidelines for Estimating Exposure[4] define exposure as the *contact* with a chemical or physical agent. The magnitude of this contact is determined by measuring or estimating the amount of an agent available at the exchange boundaries (i.e., lungs, gut, skin) during some specific time. Once the agent is absorbed through these boundaries, the absorbed amount becomes a dose. Exposure assessment is the qualitative or quantitative determination/estimation of the magnitude, frequency, duration, and route of exposure, and often also describes the resultant absorbed dose.

Figure 1 shows a schematic of environmental exposure, dose, and effects. On the left side of the diagram, there is a source of an environmental pollutant ("environmental situation"), from which chemicals are released into the environment. These chemicals may be degraded or transported, or both, in the environment, leading to various environmental concentrations in environmental media (air, groundwater, etc.). Most environmental fate models simulate the *pathways* chemicals take in the environment and the resultant media concentrations. Note, however, that the pollutant concentrations alone do not constitute exposure. Since exposure is *contact*, there must be an individual or population contacted. The activities that bring the population into contact with the chemical are termed "population habits." It is at the point where an individual or population contacts the environmental pollutant that exposure occurs.

After the exposure (following the top line in Figure 1 to the right), some of the pollutant may be absorbed, after which there may be changes in levels of the pollutant in body tissues or fluids, or changes in such things as enzyme production. These are measurable changes often referred to as biomarkers. Subsequent health effects may or may not occur.

A regulatory agency such as EPA is especially interested in the point of exposure, since it is at this point that risk reduction makes the biggest impact. Since toxicity is an intrinsic property of a chemical, seldom if ever can the toxic nature of a pollutant be changed. Risk reduction then becomes a matter of reducing exposure, either by reducing the concentrations of the pollutant in the environmental media or by changing population habits so that individuals will no longer come in contact with the pollutant.

The lower line in Figure 1 illustrates three approaches that exposure assessors have used to determine or estimate exposures. First, there is direct measurement. It is possible to measure the magnitude of contact directly as it happens. The best-known example of this is the radiation dosimeter, a small badge-like device worn in areas where exposure to radiation is possible. The dosimeter effectively measures exposures to radiation while it is taking place, then indicates when a preset level has been exceeded. Another example of direct measurement of pollutant exposure is provided by the Total Exposure Assessment Methodology (TEAM) studies conducted by EPA.[5] In the TEAM studies, a small pump with a collector and absorbent is attached to a person's clothing and measures the exposures to

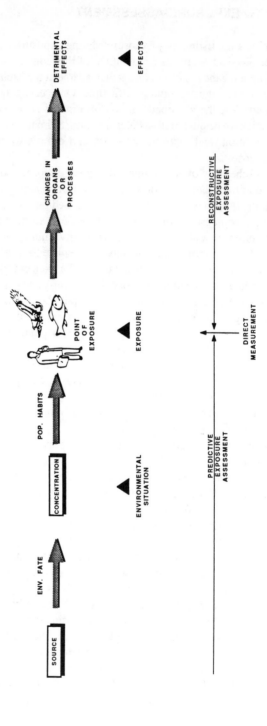

Figure 1. A schematic of environmental exposure, dose, and effects.

airborne solvents or other pollutants while the exposure takes place. The absorbent cartridges are then analyzed for a variety of chemicals. The key to direct measurement techniques is that they must be at the interface between the person and the environment and measure the exposure while it is taking place.

A second approach to assessing exposure is by using tissue levels or biomarkers along with pharmacokinetics to reconstruct what absorbed dose must have been at some time in the past. The key to using this approach is an understanding of the relationship between exposure and the observable change within the body. This will many times require some knowledge of pharmacokinetics of specific chemicals, data often lacking at present.

The third approach is the predictive approach, where one tries to predict, or estimate, exposure based on knowledge of sources, environmental pathways/fate, monitoring results of concentrations in environmental media, modeling results, and knowledge about population habits. This is an approach widely used for exposure assessment, and often it makes extensive use of environmental fate models.

Each of these approaches has strengths and weaknesses. The direct measurement approach, for example, can give us the most accurate measurement of what exposure is actually taking place. On the other hand, it is difficult to use this approach for predicting future exposures, since it involves monitoring present activities and exposures. For the same reason, it is difficult to extrapolate the results from a study of one group in one geographic location to another group elsewhere. The sources of the pollutant are not always clear in a direct measurement study, which is a disadvantage for an agency trying to control sources. Finally, this approach tends to be costly, and methods are not available for studying all chemicals.

The reconstructive approach's strengths are twofold; it can provide a positive indication that pollutants have actually crossed the exchange boundaries after exposure has occurred, and it may provide a good indication of past exposure levels. In order to do this, however, data are needed to link exposure levels to the levels found in the body, and except for a relatively few environmental pollutants, these data are unavailable. This approach does not predict future exposures, and it will not work for all chemicals (e.g., metabolites of other chemicals may cause interferences).

The predictive approach has a major strength in that it can be used to predict present or future exposures or estimate past exposures. This makes it a powerful tool for risk managers who are evaluating alternatives for possible regulation. Also, it tends to be less costly to use, since it uses whatever data are available. However, it has one glaring weakness: without valid models or (sometimes unavailable) environmental data, or data on population habits, it may give results that bear little resemblance to reality. Many of the models considered for use in predictive exposure assessment may be difficult to validate. Although an uncertainty assessment is necessary for exposure assessments in general, for detailed predictive assessments an uncertainty analysis is critical.

Tables 1–3 summarize the advantages and disadvantages of each approach relative to the others. The best exposure assessments will normally use a combination of these approaches to reduce uncertainty and add credibility to the assessment.

Table 1.   Direct Measurement of Exposure.

Description:	Direct, real-time measurements of the contact of a chemical or substance with an organism
Examples:	Radiation Dosimeters
	TEAM study measurements with solvents
Advantage:	Can be best indication of actual exposures in sampled population
Disadvantages:	Sources not always clear
	Methods not well developed for all chemicals
	May be costly
	Results at one location may not apply elsewhere

Table 2.   Reconstructive Exposure Assessment Approach.

Description:	Measurement of chemical or other indications of changes in body tissues, fluids, etc., and relating these measurements back to exposure
Examples:	Biomarkers
	Calculation of absorbed dose via use of body burden and pharmacokinetics
Advantages:	Can provide direct evidence that chemical has crossed exchange boundaries after exposure
	Can be a good indication of past exposures
Disadvantages:	Cannot predict future exposures
	Will not work for all chemicals
	Sources not always clear
	Research/data base not well developed

Table 3.   Predictive Exposure Assessment Approach.

Description:	Estimation of contact intensity, frequency, duration, and route by estimation of concentration in media and/or estimation of the habits/activities of individuals or populations that bring them into contact with the chemical
Examples:	Estimation of exposure via source estimation, monitoring data, fate models, use of exposure scenarios, etc.
Advantages:	Can be used to estimate past, present, or future exposures
	Usually less costly
	Powerful risk management tool to evaluate options
Disadvantages:	Can have limited accuracy, may be misleading
	May be difficult to validate
	Relies on data that may not be available

THE ROLE OF ENVIRONMENTAL FATE MODELS IN EXPOSURE ASSESSMENT

As the above discussion indicates, if exposure assessment is to be the determination of exposure, both before and after the application of regulatory controls, it becomes necessary to at least partly use the predictive approach to exposure assessment. An environmental fate model usually simulates what happens to a pollutant from the time it is released from the source into the environment until it has reached some future point in time where it is distributed as a concentration in one or more media.

For example, EPA's Office of Toxic Substances must evaluate potential risk for new chemicals submitted under the Premanufacturing Notice (PMN) program. Since these chemicals have not yet been manufactured, there are no monitoring data, populations currently exposed, or body burden levels for most of these new chemicals. A predictive assessment, usually involving some form of environmental fate model, is a necessary tool for estimating potential risks from new chemicals.[6-12] Decisionmakers can also use the results from such predictive analyses to estimate the reduction in potential exposure if certain controls are imposed upon the use of the new chemical.

Environmental fate models are often used for existing chemical assessment also, especially where collection of data on ambient concentrations would be prohibitively time-consuming, expensive, or impracticable. EPA's Office of Water uses fate models in Waste Load Allocation studies,[13] where models are useful to calculate the effluent quality required to meet ambient water quality criteria in a receiving stream. EPA's Office of Air and Radiation uses fate models to estimate pollutant dispersion and degradation in evaluation of New Source reviews and State Implementation Plans, to calculate concentrations of pollutants, and to compare these predictions with the National Ambient Air Quality Standards. EPA's Office of Underground Storage Tanks used fate models to help evaluate the environmental benefit of several alternative strategies before proposing a rule for underground storage tank requirements. All EPA's program offices use environmental fate models at one time or another as part of the process of evaluating risk or potential risk.

This does not mean, however, that a fate model by itself provides regulatory decisions. A fate model is merely a tool used to assess exposure, either before or after a regulatory control being contemplated. As was discussed above, the risk assessment is only one of the inputs for making regulatory decisions.

SUMMARY

In summary, the role of an environmental fate model in a regulatory agency such as EPA is as a predictive tool useful in estimating both present and potential exposures. Risk management decisions in EPA depend not only on risk assessment, but also on other factors such as cost, practicability, etc. Environmental fate models can be powerful tools in helping assess both current situations and predicting the results of control alternatives being considered, but fate models themselves do not direct decisions. Models are an important part of a much larger process of risk assessment and risk management, and it is this process that the agency decisionmakers use to help them make regulatory decisions.

REFERENCES

1. Carson, R. *Silent Spring* (New York: Houghton Mifflin Co., 1962).
2. Graham, F., Jr. *Since Silent Spring* (New York: Houghton Mifflin Co., 1970).
3. "Risk Assessment in the Federal Government: Managing the Process," National Research Council, National Academy Press, Washington, DC, 1983.
4. "The Risk Assessment Guidelines of 1986," U.S. EPA Report-600/8-87/045 (1987a).
5. "The Total Exposure Assessment Methodology (TEAM) Study" (3 vol.), U.S. EPA Report-600/6-87/022a,b,c (1987b).
6. "Methods for Assessing Exposure to Chemical Substances, Vol. 1, Introduction," U.S. EPA Report-560/5-85/001 (1985a).
7. "Methods for Assessing Exposure to Chemical Substances, Vol. 2, Methods for Assessing Exposure to Chemical Substances in the Ambient Environment," U.S. EPA Report-560/5-85/002 (1985a).
8. "Methods for Assessing Exposure to Chemical Substances, Vol. 3, Methods for Assessing Exposure from Disposal of Chemical Substances," U.S. EPA Report-560/5-85/003 (1985c).
9. "Methods for Assessing Exposure to Chemical Substances, Vol. 4, Methods for Enumerating and Characterizing Populations Exposed to Chemical Substances," U.S. EPA Report-560/5-85/004 (1985d).
10. "Methods for Assessing Exposure to Chemical Substances, Vol. 5, Methods for Assessing Exposure to Chemical Substances in Drinking Water," U.S. EPA Report-560/5-85/005 (1985a).
11. "Methods for Assessing Exposure to Chemical Substances, Vol. 6, Methods for Assessing Occupational Exposure to Chemical Substances," U.S. EPA Report-560/5-85/006 (1985f).
12. "Methods for Assessing Exposure to Chemical Substances, Vol. 8, Methods for Assessing Environmental Pathways of Food Contamination," U.S. EPA Report-560/5-85/008 (1986).
13. "Technical Support Document for Water-Quality Based Toxics Control," U.S. EPA Report-440/4-85/032 (1985g).

PART III

Remedial Options

Available Remedial Technologies for Petroleum Contaminated Soils

Lynne Preslo, Michael Miller, Wendell Suyama,
Mary McLearn, Paul Kostecki, and Edwin Fleischer

INTRODUCTION

The electric utility industry owns and operates many underground and above-ground storage tanks, as well as other facilities for using, storing, or transferring petroleum products, primarily motor and heating fuels. The prevention, detection, and correction of leakage of these products from underground storage tanks (UST) has gained high priority in the utility industry and within the regulatory agencies. The 1984 amendments to the Resource Conservation and Recovery Act (RCRA) required the U.S. Environmental Protection Agency to develop new federal regulations for reducing and controlling environmental damage from underground storage tank leakage by 1987. In addition, many states and localities have developed and are implementing strict regulations governing underground storage tanks and remedial action for releases from tanks to soil and groundwater.

This chapter summarizes the results of a jointly funded study by the Electric Power Research Institute (EPRI) and the Utility Solid Waste Activities Group (USWAG).[1] The study describes and evaluates available technologies for remediating soil and groundwater containing petroleum products released from an underground storage tank leak or other discharges, leaks, or spills.

The intent of the EPRI/USWAG study was to provide a general introduction to the state-of-the-art cleanup technology, including a listing of feasible methods, a description of their basic elements, and some discussion of the factors to be considered in their selection and implementation for a remedial program. There was not intention to develop a remediation design and implementation manual. All soil and groundwater remediation problems require site-specific considerations in choosing, designing, and implementing the most appropriate remedial programs.

The study results provide sufficient information to enable the users to become familiar with the methods which may be most applicable to their specific problems, and perhaps eliminate from further consideration those methods which are clearly inappropriate.

This chapter summarizes the advantages and limitations of 13 possible remedial technologies. In addition, short discussions are provided on the environmental fate of hydrocarbon constituents, relative costs of each technology, and current state cleanup practices permitted or used.

APPLICABLE REMEDIAL TECHNOLOGIES

The study summarized information on the technical, implementational, environmental, and economic feasibility of 13 remedial technologies for soils and/or groundwaters impacted by leaking underground storage tanks. The remedial technologies were divided into two categories: in situ treatment and non-in situ treatment. In situ treatment refers to treatment of soil or groundwater in place. Non-in situ refers to treatment of the contaminated soil or groundwater at another location. The remedial technologies were further subdivided within these categories as follows:

In situ technologies:

    volatilization
    biodegradation
    leaching and chemical reaction
    vitrification
    passive remediation
    isolation/containment

Non-in situ technologies:

    land treatment
    thermal treatment
    asphalt incorporation
    solidification/stabilization
    groundwater extraction and treatment

chemical extraction
excavation

Table 1 summarizes information for each remedial technology with regard to potential exposure pathways, application of the technology to specific petroleum products, advantages and limitations of each technology, and costs.

Table 1. Summary of Remedial Technologies.

Technology	Exposure Pathways[c]	Applicable Petroleum Products[b]	Advantages	Limitations	Relative Costs[c]
In Situ					
Volatilization	1–7	1, 2, 4	Can remove some compounds resistant to biodegradation	VOCs only	Low
Biodegradation	1–7	1, 2, 4	Effective on some non-volatile compounds	Long-term timeframe	Moderate
Leaching	1–7	1, 2, 4	Could be applicable to wide variety of compounds	Not commonly practiced	Moderate
Vitrification	1–7	1, 2, 3, 4		Developing technology	High
Passive	1–7	1, 2, 3, 4	Lowest cost and simplest to implement	Varying degrees of removal	Low
Isolation/ containment	1–7	1, 2, 3, 4	Physically prevents or impedes migration	Compounds not destroyed	Low to moderate
Non-In Situ					
Land treatment	1–7	1, 2, 3	Uses natural degradation processes	Some residuals remain	Moderate
Thermal treatment	1–6	1, 2, 3, 4	Complete destruction possible	Usually requires special facilities	High
Asphalt incorporation	1–6	1, 2	Use of existing facilities	Incomplete removal of heavier compounds	Moderate
Solidification	1–6	1, 2, 3, 4	Immobilizes compounds	Not commonly practiced for soils	Moderate

Table 1.   (Continued)

Technology	Exposure Pathways[c]	Applicable Petroleum Products[b]	Advantages	Limitations	Relative Costs[c]
Groundwater extraction and treatment	1–6	1, 2, 4	Product recovery, groundwater restoration		Moderate
Chemical extraction	1–6	1, 2, 3, 4		Not commonly practiced	High
Excavation	1–6	1, 2, 3, 4	Removal of soils from site	Long-term liability	Moderate

[a]Exposure pathways: 1 = vapor inhalation; 2 = dust inhalation; 3 = soil ingestion; 4 = skin contact; 5 = groundwater; 6 = surface water; and 7 = plant uptake.
[b]Applicable petroleum products: 1 = gasolines; 2 = fuel oils (#2, diesel, kerosenes); 3 = coal tar residues; and 4 = chlorinated solvents.
[c]Costs are highly dependent on site conditions. For additional information on costs, refer to Table 5.
Source: *Remedial Technologies for Leaking Underground Storage Tanks,* EPRI CS–5261, (Palo Alto, CA: Electric Power Research Institute).

ENVIRONMENTAL FATE OF HYDROCARBON CONSTITUENTS

Petroleum products (gasoline, fuel oil, etc.) are complex mixtures of hydrocarbons. From 100 to 150 compounds can be identified in a typical gasoline, although many more are known to be present. Each constituent in a mixture has different physical and chemical characteristics that control the behavior of petroleum products in a soil system. Alternatively, insights that may apply to the study of complex fuel mixtures can be gained through the study of a limited number of individual hydrocarbon compounds. Research efforts by Fleischer et al.[2] have centered on 13 specific compounds because of their presence in petroleum products, their tendency to be released to the subsurface environment, and their potential toxicity. These compounds are listed in Table 2.

Table 2.   Common Constituents of Petroleum Products.

Gasoline and Fuel Oils	Heavy Oils and Waste Oils
Benzene	Benz (a) Anthracene
Ethylbenzene	Benzo (a) Pyrene
(n) Heptane	Naphthalene
Pentane	Phenanthrene
(n) Hexane	
1-Pentene	
(o) Xylene	
Toluene	
Phenol	

Source: Fleischer, E. J., P. R. Noss, P. T. Kostecki, and E. J. Calabrese. "Evaluating the Subsurface Fate of Organic Chemicals of Concern Using the SESOIL Environmental Fate Model." In *Proceedings of the Third Eastern Regional Groundwater Conference,* National Water Well Association, Springfield, MA, 29–31 July 1986.

The potential environmental fate of these organic compounds was examined by conducting a computer simulation using the unsaturated zone environmental fate model SESOIL. The Seasonal Soil Compartment Model (SESOIL) was developed by Bonazountas and Wagner at Little, Inc., for the U.S. EPA Office of Toxic Substances.[3] The theoretical soil column approach utilized by SESOIL is schematically shown in Figure 1. SESOIL has been used by Bonazountas and Wagner to evaluate the potential environmental effects of two land treatment sites[4] and the potential fate of six buried solvents.[5] A testing program conducted by Anderson-Nichols, Inc. for the EPA concluded that the SESOIL model could be a useful model for chemical mobility and fate screening.[6]

Results of the SESOIL simulations are presented in Table 3. Based on these results, the organic compounds that were evaluated can be divided into four groups: (1) those that preferentially adsorb onto soil particles; (2) those that volatilize rapidly; (3) those that pose an immediate threat to groundwater supplies; and (4) those for which no one compound migration pathway dominates. The results imply that the lighter hydrocarbons associated with gasoline are more likely to volatilize, while heavier constituents can be bound tightly to soil particles. Table 4 presents these compound groupings.

Table 3.   Relative Environmental Partitioning of Petroleum Constituents Based on SESOIL Results.

Petroleum Compound	Adsorption onto Soil Particles (%)	Volatilization (%)	Soluble Portion in Groundwater and Soil Moisture (%)
Benzene	3	62	35
Ethylbenzene	21	59	20
(n) Heptane	0.1	99.8	0.1
(n) Hexane	0.1	99.8	0.1
(n) Pentane	0.1	99.8	0.1
Benz (a) Anthracene	100	0	0
Benzo (a) Pyrene	100	0	0
Naphthalene	61	8	31
Phenanthrene	88	2	10
1-Pentene	0.1	99.8	0.1
Phenol	9	0.01	91
Toluene	3	77	20
(o) Xylene	15	54	31

Source: Fleischer, E. J., P. R. Noss, P. T. Kostecki, and E. J. Calabrese. "Evaluating the Subsurface Fate of Organic Chemicals of Concern Using the SESOIL Environmental Fate Model." In Proceedings of the Third Eastern Regional Groundwater Conference. National Water Well Association, Springfield, MA, 29–31 July 1986.

These results have implications regarding both human exposure pathways and the potential effectiveness of various remedial actions. For those compounds that tend to volatilize rapidly, the primary exposure pathway can be expected to be vapor inhalation. An effective remedial action for soils containing these

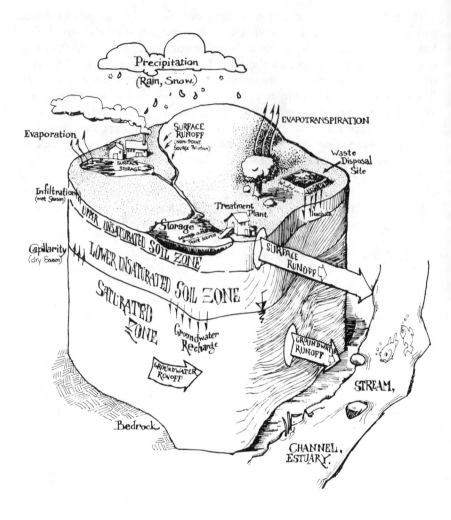

Reprinted from
Proceedings of the 1984 Speciality Conference on Environmental Engineering
ASCE/Los Angeles, CA, June 25-27, 1984

Figure 1.   SESOIL: Schematic presentation of the soil compartment (cell).

Table 4. Categories of Migration Pathways.

Adsorb to Soil Particles	Volatilize in Air	Solubilize in Groundwater	Multiple Pathways
Benzo (a) Pyrene	(n) Hexene	Phenol	Benzene
Phenanthrene	(n) Heptane		Ethylbenzene
Benz (a) Anthracene	(n) Pentane		Napthalene
	1-Pentene		Toluene
			(o) Xylene

Source: Fleischer, E. J., P. R. Noss, P. T. Kostecki, and E. J. Calabrese. "Evaluating the Subsurface Fate of Organic Chemicals of Concern Using the SESOIL Environmental Fate Model." In *Proceedings of the Third Eastern Regional Groundwater Conference,* National Water Well Association, Springfield, MA, 29–31 July 1986.

compounds might take advantage of their volatile nature. In situ volatilization is one such process. For compounds that adsorb tightly to soil, skin contact or soil ingestion are important exposure pathways. In situ volatilization would not be an effective remedial action in such a case; however, the use of a soil cover system might provide a solution to the problem. Multiple exposure and migration pathways are possible for many of the compounds contained in petroleum products. Thus, the minimization of health risks associated with exposure to soils containing petroleum products as well as the maximization of benefits obtained from remedial action efforts requires serious consideration.

RELATIVE COSTS FOR APPLICABLE TECHNOLOGIES

It is important to note that site-specific conditions will dictate the actual design decisions which, in turn, will dramatically affect the final cost for each technology. In order to develop relative cost comparisons for these technologies, cost ranges were derived for each technology by assuming a hypothetical site of moderate size. The size of this site, as well as the relative costs and the hypothetical quantities of impacted soil and groundwater, are presented in Table 5. Key design decisions, such as offsite versus onsite operation and leased versus purchased equipment, are noted.

STATE PRACTICES FOR REMEDIAL ACTIONS

Table 6 provides a state-by-state summary of remedial technology practices that have been permitted or used.[7] Information for Table 6 was acquired through a national survey, using both mail and follow-up telephone surveys.

The issues covered in these surveys were:

- type of remedial options that are currently allowed
- state regulations that trigger site investigations and existing levels for cleanup

- analytical techniques used for sample analysis
- human health effects

Table 5.   Relative Cost Comparisons for Remedial Technologies.

Remedial Technologies	Relative Total Cost[a]	Design Assumptions[b]
In Situ		
Volatilization	Low	7.63 meter (25 ft) centers for eight venting pipes; no treatment for effluent gases
Biodegradation	Moderate	Three extraction wells with infiltration galleries for injection; flow rate of 0.002572 m³/sec (40 gpm) through reactor
Leaching	Moderate	Same assumptions as biodegradation except treatment unit differs
Vitrification	High	Unit costs are based on a larger site (Pacific Northwest Lab., 1986)
Passive remediation	Low	Monitoring costs only; four monitoring wells with quarterly sampling of aromatic volatile hydrocarbon indicator compounds
Isolation/ containment	Low	Cap composed of liner, soil, and bentonite; no slurry wall
	Moderate	Same cap with slurry wall
Non-In Situ		
Land treatment	Moderate	Purchased (not leased) equipment; onsite operation
Thermal treatment	High	Leased mobile unit; onsite operation
Asphalt incorporation	Moderate	Offsite operation; shipping costs are additional
Solidification	Moderate	Leased equipment; 30% portland cement, 2% sodium silicate
Groundwater treatment	Low to Moderate	Moderately-sized carbon unit or air stripper without effluent treatment
Chemical extraction	High	Leased mobile unit; onsite operation
Excavation and landfill	Moderate to High	Leased equipment; costs relative to landfill disposal fees and transporation costs

[a]Unit costs: Low = Less than $13.00/m³ ($10/yd³) of soil or 3,780 L (1,000 gal) of water.

Moderate = $13 to $130/m³ ($10–$100/yd³) of soil or 3,780 L (1,000 gal) of water.

High = Greater than $130/m³ ($100/yd³) of soil or 3,780 L (1,000 gal) of water.

[b]Site with dimensions of 30.5 m x 15.25 m x 6.1 m (100 ft x 50 ft x 20 ft) depth, a volume of 2,837.3 m³ (3,700 yd³) weighing 3,636.4 metric tons (4,000 tons) and 37,800,000 L (10,000,000 gal) of impacted groundwater. Depth to the water table is 6.1 m (20 ft).

Source: *Remedial Technologies for Leaking Underground Storage Tanks,* EPRI CS-5261, (Palo Alto, CA: Electric Power Research Institute).

Table 6.   State Survey of Remedial Technology Practices.

State	Land Treatment	Air Stripping of Groundwater	Approved State	Hazardous Waste	As Cover Material	Road Use	Aeration of Soil	In Situ Biodegradation	Asphalt Batching	Land Spreading	General Road Use	Leave in Place	Moderate	Soil Venting	Cement Kiln	Heat in Rotary Kiln	Soil Shredding
Alabama	X		X	X													
Alaska	X						X										
Arizona		X		X			X										
Arkansas	X	X	X				X										
California	X			X	X		X										
Colorado			X														
Connecticut	X	X															
Delaware	X			X	X		X										X
Florida				X	X		X	X						X		X	
Georgia					X	X											
Idaho			P										X				
Illinois			X														
Indiana	X		X				X										
Iowa	X	X															
Kansas	X	X			X												
Kentucky	X		X	X													
Louisiana	X	X	X	X	X						X			X			
Maine	P	X	X														
Maryland	X		X											X			
Massachusetts		X	X	X					X	X							
Michigan	X		X														
Minnesota	X	X		X	X				X	X	X						
Mississippi	X		X	X													
Missouri	X		X														
Montana			X	X											X	X	
Nebraska	X	X				X						X					
Nevada			X	X								X					
New Hampshire			X	X						X							
New Jersey	X	X	X														
New Mexico	X																
New York	O	X	X	X													
North Carolina			X														
North Dakota			X								X						
Ohio			X	X					X		X	X					
Oklahoma			X	X													
Oregon	X		X														
Pennsylvania	X	X	X		X												
Rhode Island			X							X							
South Carolina			X	X													
South Dakota	X	X					X										
Tennessee	X													X			
Texas	X		X														
Utah				X										X			

Table 6.   (Continued)

State	Land Treatment	Air Stripping of Groundwater	Approved State	Hazardous Waste	As Cover Material	Road Use	Aeration of Soil	In Situ Biodegradation	Asphalt Batching	Land Spreading	General Road Use	Leave in Place	Moderate	Soil Venting	Cement Kiln	Heat in Rotary Kiln	Soil Shredding
Vermont				X				X									
Virginia	X	X															
Washington	X	X							X								
West Virginia	X			X													
Wisconsin			X		X	X									X		
Wyoming	X			X													
Total	29	19	31	17	8	1	7	7	5	3	4	3	3	2	1	1	1

P = Proposed, not tabulated in totals
O = Attempted one time
Note: This table presents the remedial technologies that have been accepted and implemented within each state. Hawaii reported no remedial technologies.
Source: Kostecki, P. T., E. J. Calabrese, and E. Garnick. "Regulatory Policies for Petroleum Contaminated Soils: How States have Traditionally Dealt with the Problem." In *Proceedings of the Conference on the Environmental and Public Health Effects of Soils Contaminated with Petroleum Products,* University of Massachusetts, Amherst, MA, 30–31 October 1985 (in press).

It should be noted that data in the table are incomplete, and in some cases out-of-date. Nonetheless, the results are useful in demonstrating in a qualitative sense which remedial technologies have been most popularly accepted and most widely applied in recent years. The most commonly used methods are land treatment, air stripping, and landfilling.

SUMMARY

A variety of options are available to clean up contaminated soils and groundwater from leaking underground storage tank sites. The most viable option will depend on the site-specific conditions involved. Costs, risk factors, exposure pathways, and type and quantity of leak are some of the factors that need to be considered in selecting the most appropriate remedial action.

REFERENCES

1. *Remedial Technologies for Leaking Underground Storage Tanks,* EPRI CS-5261 (Palo Alto, CA: Electric Power Research Institute).
2. Fleischer, E. J., P. R. Noss, P. T. Kostecki, and E. J. Calabrese. "Evaluating the Subsurface Fate of Organic Chemicals of Concern Using the SESOIL Environmental Fate Model." In *Proceedings of the Third Eastern Regional Groundwater Conference,* National Water Well Association, Springfield, MA, 29–31 July 1986.
3. Bonazountas, M., and J. Wagner. *SESOIL: A Seasonal Soil Compartment Model* (Cambridge, MA: A. D. Little, Inc., for the U.S. Environmental Protection Agency, May 1984).
4. Bonazountas, M., J. Wagner, and B. Goodwin. *Evaluation of Seasonal Soil/Groundwater Pollutant Pathways via SESOIL* (Cambridge, MA: A. D. Little, Inc., July 1981).
5. M. Bonazountas, J. Wagner, and M. Alsterburg. *Potential Fate of Buried Halogenated Solvents via SESOIL* (Cambridge, MA: A. D. Little, Inc., January 1983).
6. Watson, D. B., and S. M. Brown. *Testing and Evaluation of the SESOIL Model* (Anderson-Nichols, Inc., for the U.S. Environmental Protection Agency, September 1984).
7. Kostecki, P. T., E. J. Calabrese, and E. Garnick. "Regulatory Policies for Petroleum Contaminated Soils: How States have Traditionally Dealt with the Problem." In *Proceedings of the Conference on the Environmental and Public Health Effects of Soils Contaminated with Petroleum Products,* University of Massachusetts, Amherst, MA, 30–31 October 1985 (in press).

CHAPTER 11

An Evaluation of Organic Materials That Interfere with Stabilization/Solidification Processes

M. John Cullinane, Jr., and R. Mark Bricka

INTRODUCTION

Background

The Environmental Protection Agency (EPA) is responsible for evaluating the suitability of hazardous waste and materials for land disposal. Chemical stabilization/solidification (S/S) is one technique that has been proposed as a means for controlling the release of contaminants from landfilled wastes to surface and ground waters. Indeed, S/S of hazardous wastes is recognized in regulations implementing both the Superfund Amendments and Reauthorization Act of 1986 (SARA) and the Hazardous and Solid Waste Act Amendments of 1984 (RCRA).

A variety of S/S technologies have been proposed for treating hazardous wastes. The most commonly applied technologies use portland cement, pozzolan, or portland cement-pozzolan combinations as the primary means of contaminant immobilization.[1,2] A potential problem with using S/S technology involves chemical interferences with the hydration reactions typical of the portland cement and pozzolan processes. Experience in the construction industry has demonstrated that small amounts of some chemicals can significantly affect the setting and strength development characteristics of concrete. Consequently, the concrete industry has developed fairly stringent criteria for the quality of cement, aggregate, water, and additives (accelerators or retarders) that are allowed in concrete.[3]

Of particular concern to S/S technology is the effect of organic compounds on the strength and contaminant immobilization characteristics of the final product. It is well documented that small concentrations of organic compounds,[4,5] sugars,[6] formaldehydes,[7] and various chemical contaminants typically found in hazardous waste affect the setting mechanisms of pozzolanic cements and lime/flyash pozzolans. Roberts[8] and Smith[9] reported on the effects of methanol, xylene, benzene, adipic acid, and an oil and grease mixture on the strength and leaching characteristics of a typical lime/flyash S/S formulation. Smith[9] concludes that there was a good correlation between the effects of organic compounds on lime/flyash pozzolanic systems and the reported effects on the hydration of portland cement. More recently, Chalasani et al.[10] and Walsh et al.,[11] using X-ray diffraction and scanning electron microscopy techniques, reported on the effects of ethylene glycol and p-bromophenol on the microstructure of portland cement hydration products. Ethylene glycol was found to produce significant changes in the microstructure even after a year of curing time.

Purpose and Scope

The purpose of the research described in this chapter is to develop data on the compatibility of organic waste constituents with three binding agents: portland cement, portland cement/flyash, and lime/flyash. Only the results of the unconfined compressive strength (UCS) test are presented at this time. The remainder of the data will be presented in a comprehensive report scheduled for publication later in 1988.

MATERIALS AND METHODS

The study reported herein was conducted in three phases: (1) preparation of a synthetic wastewater and sludge; (2) addition of a binder and interfering material to the sludge; and (3) UCS testing of cured specimens containing sludge, binder, and interference chemicals.

Synthetic Wastewater and Sludge Production

Initial laboratory tests revealed that a synthetic wastewater containing nitrate salts of cadmium, chromium, and mercury at 600 times the EPA extraction procedure limit, and nickel at 600 times the California limit, could be treated with calcium hydroxide to produce a hydroxide sludge with typical metal concentrations of 86.2, 84.1, 18.8, and 0.137 mg/g (dry weight basis) of nickel, chromium, cadmium, and mercury, respectively.

Typically, the raw sludge contained 8% solids (by weight) and was very fluid. The sludge was dewatered to approximately 30% solids using a rotary drum vacuum filter. A constant moisture content between sludge batches was maintained by adjusting the solids content of the dewatered sludge to 25%, using the supernatant liquid from the sludge production process as a dilution liquid.

Specimen Preparation

The 25% solids content sludge was divided into three 150-gal samples and binder material was added to each at the following ratios.

Binder/Sludge Ratio

Binder	Ratio
Portland cement (Type 1)	0.3:1 cement:sludge
Portland cement (Type 1)/ flyash (Type F)	0.2:1 cement:sludge 0.5:1 flyash:sludge
Lime/flyash (Type C)	0.3:1 lime:sludge 0.5:1 flyash:sludge

After mixing the sludge with the binder, each binder/sludge sample was subdivided into four equal parts. A single organic interfering substance was added to each of the subsamples at ratios of 0.0, 0.02, 0.05, and 0.08 (by weight) interference chemical to binder/sludge material. The subsample to which no interference chemical was added was used as a control specimen. To account for variability between batches, control specimens were prepared each time an interference chemical was processed. This allows for the comparison of UCS results between batches.

Interference/binder/sludge mixtures (I/B/S) were then molded into 2 in. cubes in accordance with ASTM Method C-109-77/86.[12] Because the I/B/S mixture was usually viscous and could not be tamped into the molds, the ASTM method was modified to include vibration of the I/B/S mixture to remove any air pockets that developed during the molding process.

The specimens were cured in the molds at 23°C and 98% relative humidity for a minimum of 24 hr and removed from the molds whenever they developed sufficient strength to be freestanding. After removal from the molds, the specimens were cured under the same conditions for periods of 4, 11, and 28 days. At the end of each curing period, the UCS of the specimens was determined in accordance with ASTM C 109-77/86.[12] A minimum of four replicates were performed for each interference/binder/sludge mixture.

DISCUSSION OF RESULTS

Space limitations do not allow for presentation of all study results, but typical results for selected interference/binder/sludge mixtures are discussed below.

Table 1 presents the results, reported as the percent increase or decrease in 28-day UCS from the control specimen. Figures 1 through 7 provide a graphical representation of selected study results. Figures 1 through 3 present the UCS versus curing time for the phenol-portland cement, oil-portland cement, and grease-portland cement interference/binder/sludge combinations. Each curve shown on these figures represents the strength development curve for one interference material concentration. Figure 4 presents the 28-day UCS versus interference concentration for each of the three binders for the phenol interference. Figure 4 illustrates that increasing the concentrations of the interference chemical does not necessarily affect the different binders to the same degree. Figures 5, 6, and 7 present a graphical representation of the relative effects of interference concentration on the 28-day UCS of each of the three binders.

The data for the 28-day UCS are an indicator of the UCS trends observed at the earlier curing periods. This is clearly illustrated in Figure 1, which is a plot of cure time versus UCS for the phenol/portland cement/sludge material. Although the slope of the curve varies between binder and interference treatments, in most cases the lines of constant interference concentration do not cross.

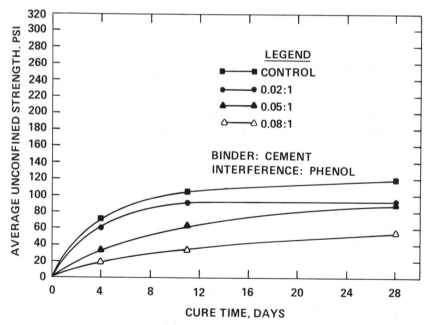

Figure 1.   Unconfined compressive strength as a function of curing time and interference material concentration for the phenol-portland cement interference-binder combination.

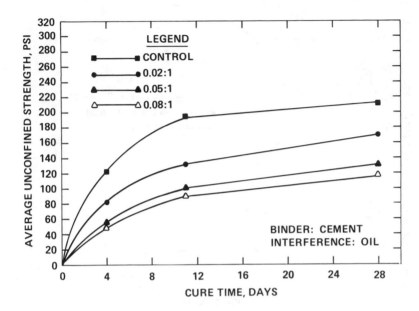

Figure 2. Unconfined compressive strength as a function of curing time and interference material concentration for the oil-portland cement interference-binder combination.

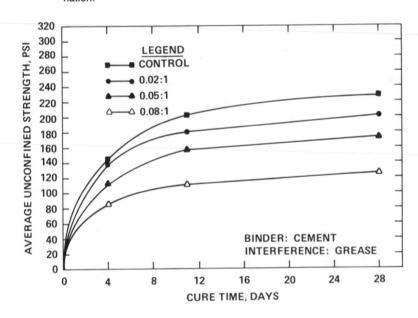

Figure 3. Unconfined compressive strength as a function of curing time and interference material concentration for the grease-portland cement interference-binder combination.

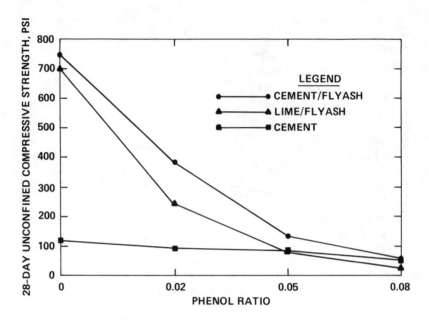

Figure 4.   28-day unconfined compressive strength for three binders as a time and interference material concentration.

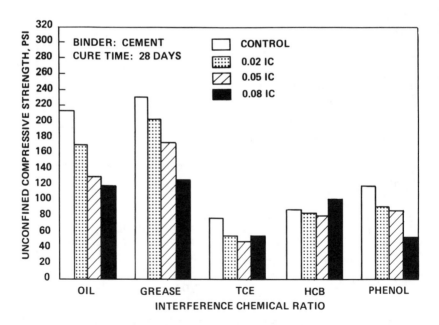

Figure 5.   Effect of interference material concentration on the 28-day UCS for the Type I portland cement binder.

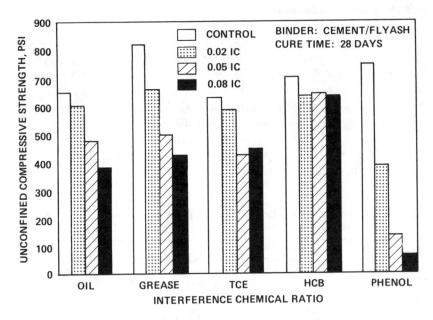

Figure 6. Effect of interference material concentration on the 28-day UCS for the Type I portland cement/flyash binder.

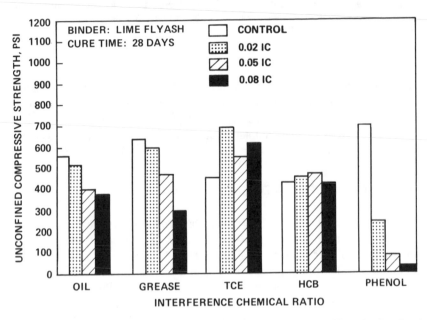

Figure 7. Effect of interference material concentration on the 28-day UCS for the lime/flyash binder.

Portland Cement Binder

The data presented in Table 1 show that the interference effects may be positive or negative, depending on the interfering material and concentration of the interfering material. For portland cement, the UCS generally declined with increasing organic interference concentration. The addition of oil, grease, and phenol resulted in a consistent decrease in the UCS with increasing concentration of interfering material. A 0.08 phenol ratio resulted in a 54% decrease in the UCS of the specimen. The addition of hexachlorobenzene was shown to have only a marginal effect on UCS, with a 15% increase in UCS at the 0.08 ratio. This unexpected result remains unexplained.

Table 1. 28-Day Unconfined Compressive Strength as a Percent of Control Specimen.[a]

Interference Chemical	Portland Cement Binder I/B/S Ratio			Cement/Flyash I/B/S Ratio			Lime/Flyash I/B/S Ratio		
	0.02	0.05	0.08	0.02	0.05	0.08	0.02	0.05	0.08
Oil	−20	−38	−44	−8	−28	−42	−7	−27	−32
Grease	−12	−25	−45	−20	−40	−48	−7	−27	−54
Trichloroethylene	−28	−36	−27	−7	−33	−29	+51	+20	+34
Hexachlorobenzene	−5	−6	+15	−10	−9	−10	+6	+9	−1
Phenol	−22	−26	−54	−49	−82	−92	−65	−88	−96

[a]All results reported as percent increase (+) or decrease (−) from the control specimen rounded to nearest whole percent. Average of four replicates.
[b]Interference to binder/sludge ratio.

By comparing the UCS results for the control specimens of different binders, it is evident that the portland cement specimens developed less strength than the lime/flyash or the cement/flyash specimens. Upon closer examination, however, it can also be observed that the total binder/sludge ratio for the portland cement binder is lower than for the other binders.

Portland Cement/Flyash Binder

The portland cement/flyash interference/binder/waste mixture showed a consistent decrease in UCS for all interference materials. In general, the greater the concentration of interfering material, the greater the impact on 28-day UCS. The addition of a 0.08 ratio of phenol resulted in a decrease of over 90% in 28-day UCS. The effect of hexachlorobenzene was not concentration-dependent and resulted in a consistent 9% decrease in 28-day UCS regardless of concentration. The addition of 0.08 ratio of oil resulted in a 42% reduction in UCS.

Lime/Flyash Binder

With the exception of trichloroethlyne and hexachlorobenzene, the addition of organic interfering materials had a consistently negative impact on the 28-day UCS of the lime/flyash interference/binder/sludge mixture. The addition of a 0.08 ratio of phenol resulted in an 80% decrease in 28-day UCS. The addition of oil or grease at a 0.08 ratio resulted in a 32% and 54% decrease, respectively, in 28-day UCS.

The addition of trichloroethylene appeared to result in a gain in 28-day UCS; 51%, 20%, and 34% at the 0.02, 0.05, and 0.08 interference ratios, respectively. The addition of hexachlorobenzene resulted in a slight increase in 28-day UCS for the 0.02 and 0.05 interference ratios, 6% and 9%, respectively. However, a 1% reduction in UCS occurred at the 0.08 interference ratio.

CONCLUSIONS

Several conclusions can be drawn that characterize the effects of the interference materials investigated in this project on the UCS of stabilized/solidified waste materials.

1. The interference chemicals tested had a measurable effect on the setting and strength development properties of the stabilized/solidified waste. The magnitude of the effect depended on the type of binder, the curing time, and the type and concentration of the interfering compound.
2. Stabilized/solidified waste showed decreased UCS development with increasing oil or grease concentrations.
3. Although the waste stabilized with portland cement resulted in lower 28-day strength development, it appears that the concentration of interference material had less effect on the UCS development properties for the portland cement binder than for the portland cement/flyash or lime/flyash binders.
4. Phenol concentrations above 5% resulted in marked decreases in 28-day UCS development for all the binders tested.
5. The chlorinated hydrocarbons evaluated in this study had less effect on the UCS development properties than the other interference materials investigated.

ACKNOWLEDGMENTS

The tests described and the resulting data presented, unless otherwise noted, were obtained from research conducted by the U.S. Army Engineer Waterways Experiment Station and were sponsored by the U.S. Environmental Protection

Agency, Hazardous Waste Engineering Research Laboratory, Cincinnati, Ohio, under Interagency Agreement DW96930146-01. Mr. Carlton Wiles, Hazardous Waste Engineering Research Laboratory, was the EPA project officer. Permission to publish this information was granted by the Chief of Engineers and the U.S. Environmental Protection Agency.

REFERENCES

1. "Guide to the Disposal of Chemically Stabilized and Solidified Waste," U. S. EPA Report-SW-872, Office of Research and Development, Municipal Environmental Research Laboratory, Cincinnati, OH (1980).
2. Cullinane, M. J., L. W. Jones, and P. G. Malone. "Handbook for Stabilization/Solidification of Hazardous Waste," U. S. EPA Report-540/2-86/001, Hazardous Waste Engineering Research Laboratory, Cincinnati, OH (1986).
3. Jones, J. N., M. R. Bricka, T. E. Myers, and D. W. Thompson. "Factors Affecting Stabilization/Solidification of Hazardous Wastes," Proceedings: International Conference on New Frontiers for Hazardous Waste Management, U.S. EPA Report-600/9-85-025, Hazardous Waste Engineering Research Laboratory, Cincinnati, OH (1985).
4. Young, J. F. "A Review of the Mechanisms of Set-Retardation of Cement Pastes Containing Organic Admixtures," Cement and Concrete Research 2(4) (1972).
5. Young, J. F., R. L. Berger, and F. V. Lawrence. "Studies on the Hydration of Tricalcium Silicate Pastes. III. Influences of Admixtures on Hydration and Strength Development," Cement and Concrete Research 3(6) (1973).
6. Ashworth, R. "Some Investigations Into the Use of Sugar as an Admixture to Concrete," Proceedings of the Institute of Civil Engineering, London, England (1965).
7. Rosskopf, P. A., F. J. Linton, and R. B. Peppler. "Effect of Various Accelerating Chemical Admixtures on Setting and Strength Development of Concrete," Journal of Testing and Evaluation 3(4) (1975).
8. Roberts, B. K. "The Effect of Volatile Organics on Strength Development in Lime Stabilized Fly Ash Compositions," Master's Thesis, University of Pennsylvania, Philadelphia, PA (1978).
9. Smith, R. L. "The Effect of Organic Compounds on Pozzolanic Reactions," I. U. Conversion Systems, Report No. 57, Project No. 0145 (1979).
10. Chalasani, D., F. K. Cartledge, H. C. Eaton, M. E. Tittlebaum, and M. B. Walsh. "The Effects of Ethylene Glycol on a Cement-Based Solidification Process," Hazardous Wastes and Hazardous Materials 3(2) (1986).
11. Walsh, M. B., H. C. Eaton, M. E. Tittlebaum, F. K. Cartledge, and D. Chalasani. "The Effect of Two Organic Compounds on a Portland Cement-Based Stabilization Matrix," Hazardous Wastes and Hazardous Materials 3(1) (1986).
12. Annual Book of ASTM Standards: Construction, Volume 04.01, Cement; Lime; Gypsum, American Society for Testing and Materials, Philadelphia, PA (1986).

CHAPTER 12

In Situ Vitrification Processing of Soils Contaminated with Hazardous Wastes

Craig L. Timmerman, James L. Buelt, and Vincent F. FitzPatrick

INTRODUCTION

As management of hazardous materials gains increased attention in the United States, new, more effective technologies are being sought to immobilize and/or destroy the wastes, either in situ for previously disposed wastes, or at the waste generation site. The new Resource Conservation and Recovery Act (RCRA) and Comprehensive Environmental Response, Compensation and Liability Act (CERCLA) legislation, combined with proposed rulemaking by the Environmental Protection Agency (EPA), is making landfill disposal very costly and is moving in a direction that will severely limit future land disposal. Further, the extended liability associated with future environmental impairment provides a significant corporate incentive to dispose and delist wastes within the plant or waste site boundary.

Pacific Northwest Laboratory (PNL) is developing a remedial action process for contaminated soils that is potentially significant in its application to these concerns. Although the process was initially developed to demonstrate a potential technology for disposal of transuranic waste contaminated soil sites, recent tests have shown that many hazardous chemical wastes are also destroyed or immobilized as a result of the treatment. The process, in situ vitrification, was originally developed for the U.S. Department of Energy and is more recently being adapted for selected commercial clients for hazardous wastes.

137

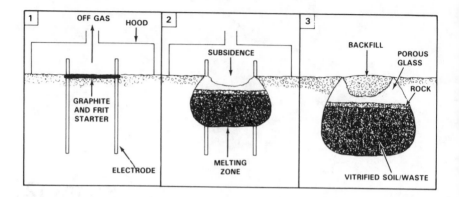

Figure 1. The in situ vitrification process.

In situ vitrification (ISV) is a thermal treatment process that converts contaminated soil into a chemically inert and stable glass and crystalline product. Figure 1 illustrates how the process operates. A square array of four electrodes is inserted into the ground to the desired treatment depth. Because the soil is not electrically conductive once the moisture has been driven off, a conductive mixture of flaked graphite and glass frit is placed among the electrodes to act as the starter path. An electrical potential is applied to the electrodes, which establishes an electrical current in the starter path. The resultant power heats the starter path and surrounding soil up to 3600°F (2000°C), well above the initial melting temperature or fusion temperature of soils. The normal fusion temperature of soil ranges between 2000 and 2500°F (1110 and 1400°C). The graphite starter path is eventually consumed by oxidation, and the current is transferred to the molten soil, which is now electrically conductive. As the vitrified zone grows, it incorporates nonvolatile elements and destroys organic components by pyrolysis. The pyrolyzed byproducts migrate to the surface of the vitrified zone, where they combust in the presence of oxygen. A hood placed over the processing area provides confinement for the combustion gases, and the gases are drawn into the off-gas treatment system.

Pacific Northwest Laboratory began developing ISV technology in late 1980 under the support of the U.S. Department of Energy. Since then, numerous experimental tests with varying conditions and waste types have been conducted.[1-4] Table 1 describes the different scales of test units that PNL used in developing the technology. The successful results of the 46 bench-, engineering-, and pilot-scale tests have proved the feasibility of the process. Also, economic studies have indicated that tremendous economies of scale are attainable with the ISV process.[1] These factors led to the commitment to design, fabricate, and test a large-scale prototype. Its successful testing has demonstrated the field utility of the large-scale unit, and has proved the initial economic projections.

Table 1. ISV Test System Characteristics.

System, Scale	Power, kW	Electrode Spacing, ft	Vitrified Mass per Setting	No. of Tests
Bench	10	0.36	2 to 5 lb	5
Engineering	30	0.75 to 1.2	0.05 to 1.0 ton	25
Pilot	500	3.0 to 5.0	10 to 50 ton	16
Large	3,750	11.5 to 18.0	400 to 800 ton	5

This chapter describes the large-scale ISV system, discusses its capabilities, and summarizes the results of testing to date. Pacific Northwest Laboratory recognizes that ISV is not the solution to all hazardous waste management problems; but judiciously applied, ISV can offer technical and economic improvements to state-of-the-art remedial action technology. With an understanding of the process design and functions, the waste manager can make sound judgments about the applicability of ISV to site-specific disposal problems.

PROCESS AND OPERATION DESCRIPTION

In situ vitrification uses the surface starting technique to establish melting of the soil. As the melt grows downward and outward, power is maintained at levels sufficient to overcome the heat losses from the surface and to the surrounding soil. Generally, the melt grows outward for a width of about 50% beyond the spacing of the electrodes. Therefore, if the electrode spacing is 18 ft (5.5 m), a melt width of about 28 ft (8.5 m) would be observed under nominal conditions. The molten zone is roughly a square with slightly rounded corners, a shape that reflects the higher power density around the electrodes.

As the melt grows in size, the resistance of the melt decreases, making it necessary to periodically adjust the ratio between the voltage and the current to maintain operation at constant power. This is done by adjusting the tap position on the primary power supply to the electrodes. There are 14 effective taps that permit adjusting the voltage from a maximum of 4000 V to a minimum of 400 V per phase, and the current from a minimum of 400 A to a maximum of 4000 A per phase. Operations follow the power equation $P = I^2 \times R$, where P is power, I is current, and R is resistance.

The large-scale process equipment for in situ vitrification is depicted in Figure 2. The process immobilizes contaminated soil and isolates it from the surrounding environment. Controlled electrical power is distributed to the electrodes, and special equipment contains and treats the gaseous effluents. The process equipment required to perform these functions is divided into five major subsystems: (1) electrical power supply, (2) off-gas hood, (3) off-gas treatment, (4) off-gas support, and (5) process control.

Except for the off-gas hood, all components are contained in three transportable trailers (Figure 3): a support trailer, a process control trailer, and an off-gas trailer.

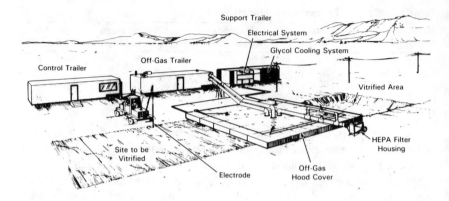

Figure. 2. Setup for in situ vitrification.

All three trailers are mounted on wheels sufficient for a move to any site over a compacted ground surface. The off-gas hood and off-gas line, which are installed on the site to collect the gaseous effluents, are dismantled and placed on a flatbed trailer for transport between the sites to be treated. The effluents exhausted from the hood are cooled and treated in the off-gas treatment system. The entire process is monitored and controlled from the process control trailer.

The off-gas trailer is the most complex and expensive of the three trailers. The off-gas treatment system cools, scrubs, and filters the gaseous effluents exhausted from the hood. The primary components include: a gas cooler, two wet scrubber systems (tandem nozzle scrubbers and quenchers), two heat exchangers, two process scrub tanks, two scrub solution pumps, a condenser, three mist eliminators (vane separators), a heater, a particulate and charcoal filter assembly, and a blower system.

A major element of the off-gas support system is the glycol cooling system, which is mounted on the support trailer. This system interfaces with the scrub solution and extracts the thermal energy that builds up in the off-gas treatment system from cooling the combustion gases from the hood. The heat is rejected to the atmosphere in a fin tube, air-cooled heat exchanger. This makes the entire process system self-sufficient in terms of site services, except for electrical supply. In cases where electrical supply is remote and costly to bring in, diesel generators can be used to supply the required electricity. Details of the large-scale process equipment and the process capabilities are found in Buelt and Carter.[3]

The normal processing rate for the large-scale system is 3 to 5 tons/hr or nominally 3 to 5 yards/hr, a rate competitive with many other remediation technologies. The average processing operation lasts about 150 to 200 hr, depending upon the depth and electrode spacing, although for processing to depths of 50 ft, single processing operations can range from 300 to 400 hr. The production rate will remain constant at 3 to 5 tons during the entire time period, resulting in a vitrified mass greater than 1,000 tons.

For routine operations on a site, all three trailers are coupled together and moved from one processing position to another by pulling them as a unit. The hood is moved from one position to another with a crane. The crane is also used to assist in coupling and uncoupling the off-gas lines. Moving from one processing position to another takes about 16 hr; thus, a relatively high operating efficiency can be achieved. This 16-hr interim for movement also provides time for performing routine maintenance.

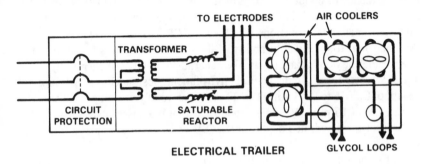

ELECTRICAL TRAILER

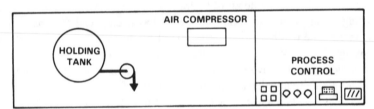

PROCESS CONTROL TRAILER

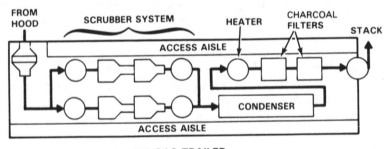

OFF-GAS TRAILER

Figure 3. Process trailers for the large-scale in situ vitrification unit.

PRODUCT CHARACTERISTICS

The ability of the waste form to retain the encapsulated or incorporated heavy metals is of prime importance to the usefulness of the ISV process.

The vitrified waste form has been subjected to a variety of leach tests, including the EPA's Extraction Procedure Toxicity Test (EP Tox) and Toxic Characteristics Leach Test (TCLP). All of these tests show a uniformly low leach rate for heavy metals of about 1×10^{-4} lb/ft²/day (5×10^{-5} g/cm²/day) or lower. Based on limited tests, it is reasonable to assume that the vitrified material can be delisted under the provisions of either the EP Tox or the TCLP.

Another indication of the durability of the ISV waste form is found in a study of the weathering of obsidian, a glasslike material physically and chemically similar to the ISV waste form.[5] In the natural environment, obsidian has a hydration rate constant of 0.0016 to 0.03 mil² (1 to 20 μm²) per 1000 yr.[6] Using a linear hydration rate of 0.016 mil² (10 μm²) produces a highly conservative estimate of a $\ll 40$ mil (1 mm) hydrated depth for the ISV waste form over a 10,000-yr time span.

Data for the release of sodium from vitrified Hanford soil during a leach test at 90°C are available for durations of 7, 14, and 28 days.[1] Because the sodium is soluble in the leachate, its normalized release is a measure of the extent of hydration of the glass and, in particular, its normalized release divided by the density of the glass is the depth of hydration. If the glass is assumed to hydrate according to the same parabolic rate law as has been found for obsidian, then the square of the depth of hydration divided by the duration of the test should be constant. Using the data in Buelt et al.,[1] the result of this calculation increases between the 7- and 14-day data, but then is constant between the 14- and 28-day data. Taking this final value and using the density of the glass, the hydration rate at 194°F (90°C) appears to be about 0.0032 mil² (2 μm²/yr). In the literature on field studies of obsidian hydration, the rate is found to obey an Arrhenius relation with an activation energy of 20 kcal/mole.[6] Applying this to the ISV glass hydration, we can predict rates of 0.008 mil² (5 μm²)/1000 yr at 77°F (25°C) (e.g., for glass exposed to the air) and 0.0016 mil² (1 μ²)/1000 yr at 50°F (10°C) (e.g., for glass buried underground). These values are comparable to those found for obsidian hydration rates in the field for similar average weathering temperatures.

The long-term stability of obsidian in nature is controlled by three mechanisms:[5] alteration (weathering), devitrification (recrystallization), and hydration (water absorption).[6] Review of the literature indicates that the usual controlling mechanism is devitrification. Studies of the mean age of natural glasses indicate that obsidian has a mean life of about 18 million yr.[5] Considering the similarity of the ISV waste form to obsidian, it is reasonable to postulate that the mean life of the vitrified material would be on the order of 1 million yr.

The ISV waste form is a glass with the atomic structure that is random, rather than the highly structured nature of a crystalline material. This leads to another benefit: the fracture mechanism is conchoidal, which means that the waste form is not subject to significant damage by freeze/thaw mechanisms that can accelerate natural degradation. Accelerated fracturing by alternately freezing and thawing would increase the surface area and the amount of material that could be leached into groundwater.

ECONOMIC ANALYSIS

The economics of the process have been examined under various conditions typical of those that might be encountered throughout the United States. The methodology for the ISV cost estimates has been developed and reported for a large-scale system.[1] The approach and estimates have been validated by actual design, fabrication, and operational testing of the large-scale system. Key features of the economic projections were: (1) that the system could be operated with two people per shift, and (2) that the system could be moved from one processing area to another processing area in less than 24 hr. Both of these assumptions have proved correct.[2]

Highlights of the cost estimate technique used for the cost projections are summarized here; details can be obtained from the original reference.[1] The cost estimate is divided into five main categories: (1) site costs; (2) equipment cost, or capital recovery; (3) operations and labor; (4) electrode costs; and (5) electrical costs. The factors that most significantly affect total cost are the amount of moisture in the soil, labor rate costs, and the cost of electricity. The amount of moisture directly affects operational time and, therefore, has a direct bearing on the labor and operations costs. The electrical energy equivalent of the heat of vaporization for the moisture in the soil must be supplied, and the water boiled off, before vitrification can proceed. The cost of labor and electrical power also has a direct effect on the operational costs. Equipment or capital recovery costs and electrode costs are significant; however, they are treated as constants. Site costs are based on nominal amounts of civil work that must be performed, which include acquiring and placing clean backfill in the subsidence zone.

Equipment costs or capital recovery include the costs of the ISV system and the necessary support equipment such as a front loader and crane for earth-moving operations and for moving the hood and trailers, respectively. There is also a nominal allowance for extending existing power lines and installing a substation. All equipment is assumed to have a 10-yr life. The sum of the equipment costs is multiplied by a 20% capital recovery factor and added in as a unitized cost factor.

Operations and labor costs consist of the labor and materials for those activities that must be performed to support normal operations. These activities include labor time for the two operators required to operate the system. System operation is calculated on the basis of 24-hr continuous operations. Other support activities include digging the holes and placing the electrodes, moving the trailers and hood from one processing position to another, and performing routine maintenance operations when the system is being moved. Also included are costs for placing the starter material, and connecting, disconnecting, and testing the electrodes. The operational cost also includes an allowance for secondary waste disposal; i.e., treating and/or disposing of the scrub solution once per week.

Electrode costs represent the purchase of electrode materials, which are used only once, assuming that the electrodes are left in the melt. For operations in a chemically hazardous environment, the electrodes can be retrieved and sold for salvage value, which is about 20% of the original cost. For operations with a process sludge where significant decomposition of the sludge occurs, there is a potential for electrode reuse. However, estimating electrode recovery values that are higher than the salvage value is not conservative planning at this time.

These cost factors have been calculated and plotted in graphical form, as shown in Figure 4. A maximum cost for the various soil and moisture cases in Figure 4 occurs at $.0825/kWh. This maximum cost for electricity will fluctuate with the price of diesel fuel; however, the indicated value is realistic in today's economy. At local electrical rates that are higher than this value, the use of portable diesel generator power is recommended. Whether the units are rented or purchased depends on the length of the remediation operation and business decisions regarding future operations. The flat rate can also be used in planning for sites where local electrical power is unavailable and bringing in a power line would be very expensive. The case for wet industrial sludges was included to cover those situations where consolidation and immobilization at an active plant site was the primary objective. This may be important to the typical owner/generator who has sludge surface impoundments that must be closed and the owner/generator does not want to incur the long-term liability and costs associated with shipping to a secure landfill.

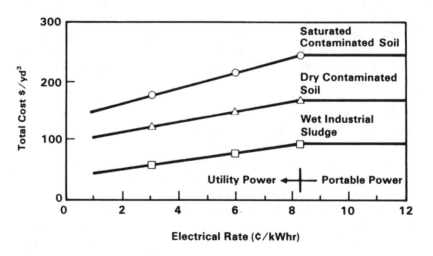

Figure 4.   Cost of ISV applications.

PROCESS PERFORMANCE

The discussion of process performance is divided into two sections: (1) experience with hazardous wastes, and (2) applications considerations. This discussion will focus on the overall treatment efficiency; that is, the retention and/or destruction in the melt and the capture and removal of the material released from the melt by the off-gas system. The sum of the two functions represents the overall system Destruction Removal Efficiency (DRE).

Hazardous Waste Experience

The results of the process performance testing with hazardous materials are subdivided into two categories: metals and organics.

Metals

During the processing operations with ISV, metals are either dissolved in the glass or incorporated in the vitreous matrix. The three factors that have the largest effect on retention are burial depth, solubility, and vapor pressure. Burial depth has a direct function on retention, increasing the amount of retention with

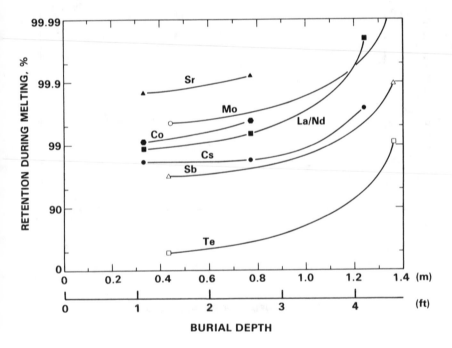

Figure 5. Element retention versus burial depth during pilot-scale tests.

increased burial depth. Metals are retained in the melt as a direct function of the solubility and are inversely proportional to their vapor pressure.

The measure of the material retained in the melt is the retention factor, defined as:

Retention Factor $= RF = [A]i/[A]e$

where $[A]i$ is the concentration of the element A initially present and $[A]e$ is the concentration of element A exiting the component. This terminology is used for the retention in the melt as well as the retention in the off-gas treatment system. The retention factor is the inverse of the quantity [1-DRE].

The effect of burial depth on retention of elements is charted in Figure 5. In these pilot-scale tests, the metals were in common chemical forms such as nitrates, fluorides, and oxides. Figure 6 shows the results of the engineering-scale tests with lead and cadmium. The lead shows a constantly higher retention in

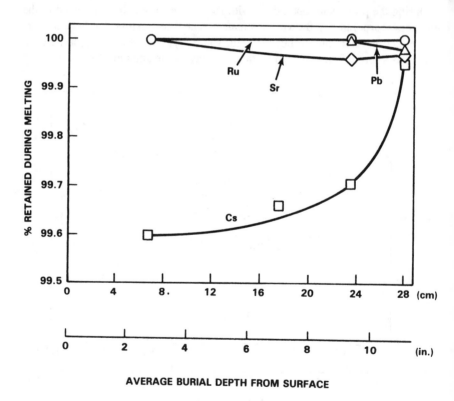

Figure 6.   Element retention versus burial depth during engineering-scale tests: (a) shows ruthenium, stronthium, lead, and cesium; (b) shows cesium and cadmium.

the melt with increasing depth, most likely reflecting the high solubility of lead in glass. Cadmium, which is less soluble in glass, also shows a depth-dependent retention. When estimating retention factors for metals in the melt or off-gas system, the solubility of the metal in the glass and also the likely oxidation state should be considered. The melt is reducing in nature, so the most likely form of most metals is either the pure state or the lowest oxidation state that will accommodate a stable oxide. Data from other glass processes, such as melters, is not considered a reliable source in estimating retention factors, because of the difference in reducing conditions in the glass. Ruthenium in an oxidizing environment can form RuO_4 which sublimes at about 1830°F (1000°C) and can cause serious problems in melter off-gas treatment systems. In the pilot-scale radioactive test, ruthenium exhibited a retention of 99.82%, or a retention factor of 550. The enhanced retention because of oxidization state and solubility is very important when assessing potential applications to traditionally volatile metals.

Retention factors measured in pilot-scale testing are shown in Table 2 for both the melt and the off-gas treatment system. These data are sufficient to infer the retention of other hazardous metals. Results for large-scale testing show that retention continues to increase with increasing depth below 3.3 ft (1 m). An order of magnitude increase in retention was observed when the depth was increased from 3 to 16 ft (1 to 5 m).[2] The presence of combustibles can provide a path to the surface by entraining the metals in the combustion product or pyrolytic gases, increasing the retention fraction. The closer to the surface, the less likely the entrained material will not be scrubbed out by the molten glass and recaptured. Even the decomposition of nitrates can provide an elution path if the reaction occurs near the surface.

Table 2.   Retention Factors of Metals.

Type of Metal	Soil	Off-Gas	Combination
Particulates Sr, Pu, U, La, Nd	10^4	10^5	10^9
Semi-Volatiles Co, Cs, Sb, Te, Mo	10^2	10^4	10^6
Volatiles Cd, Pb	2–10	10^4	10^5

Enhanced releases associated with combustion events are shown in Figure 7. These results are also from pilot-scale tests where the metals were deliberately placed with combustibles. At each combustion event, the concentration of the metals in the scrub solution increases measurably with a clear change in the slope of the concentration curve.

These data are important for estimating the amount of secondary waste that will be generated by contaminating the scrub solution with hazardous chemicals, necessitating potential treatment of the scrub solution before its disposal. With a meter of clean overburden, retention factors of at least 100 can be expected

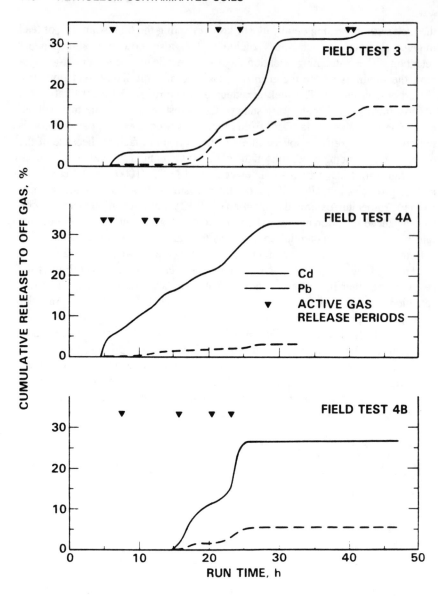

Figure 7. Cadmium and lead release as a function of run time.

even for the volatile elements. Retention factors of 1000 and greater can be expected for the semi- and nonvolatile metals, assuming that there is no significant quantity of combustibles.

Organics

During processing, organics that are contacted by the vitrified material are destroyed by pyrolysis. The pyrolytic gases move upward through the melt and combust when contacted by the oxygen atmosphere in the hood. The data base for processing organics has been limited to selected organics in containers, PCBs, dioxins, and organics associated with electroplating sites. As the ISV process gains additional acceptance as a remediation tool, the data base will grow through feasibility tests, field demonstrations, and actual remediation applications.

Combustible testing with organics has included up to 110 lb (50 kg) of solid combustibles and 50 lb (23 kg) of liquid organics in a single pilot-scale experiment. The materials were packaged in a container such as what might be found in a solid-waste burial ground. Chromatographic, sample bomb, and mass spectrometric analyses of the effluent from both the hood and the exhaust stack showed less than 5×10^{-3} vol% release for light hydrocarbons during peak combustion periods. This level of release indicates nearly complete pyrolysis and combustion.[7] A limited number of experiments have been conducted to define the pressure rise and rate of release associated with organics in sealed containers. Theoretical calculations predicted that the internal pressures would be several hundred psi, and that the pyrolized material would produce a transient pressure wave that would move through the melt in a few seconds. The tests were performed using the engineering-scale system; the sealed containers were equipped with pressure sensors. The maximum pressure observed was 32 psig. The pressure attenuated over a brief period of about 1 min; however, the organic contents in the containers were slowly released over a period of a few minutes (<10) to up to 90 min.[8] These data indicate that the metal softened as the vitrified zone approached the sealed container, and that the intrusion of the glass was slower than the theoretical maximum rate. This could be a scaling effect, related to the relatively high viscosity of the glass and smaller size of the container. Intrusion into a buried 55-gal drum is expected to be more rapid; however, effective destruction is expected by the extensive molten soil of a large-scale melt.

An engineering-scale test was conducted using soils contaminated with 500 ppm of PCBs. The data from the test showed that the process destruction was slightly greater than 99.95%.[9] The small amount of material released to the off-gas system was effectively removed, yielding an overall system DRE of >99.9999%. Analysis of the vitrified block showed that there were no residual PCBs; considering the processing temperature, the data are reasonable. The soil adjacent to the vitrified area was examined for PCBs; limited quantities were detected (0.7 ppm of PCBs). These data indicate that the vitrification rate is greater than the PCB diffusion rate and that migration away from the vitrification zone during processing is not a significant concern. The overall results were sufficiently promising that a pilot-scale test has been authorized. The scope of the test will use 1000 ppm PCB-contaminated soil and be performed in accordance with a Toxic Substance Control Act (TSCA) permit issued by EPA headquarters.

A similar bench-scale test[10] was performed on soils contaminated with dioxin (2,3,7,8 TCDD). The results gave a destruction efficiency for the ISV process alone of greater than 99.995% based on analytical detection limits. Additional off-gas removal with a conventional off-gas treatment system that utilizes a carbon filter would result in an overall DRE of >99.99999% for dioxin. No dioxins or furans were detected in the off-gas emissions during the test. These data confirm and support the PCB results and indicate that ISV is a high potential process for treatment of chlorinated organic contaminated soils.

Engineering-scale tests on electroplating wastes have shown that the destruction efficiency for contaminated soils is >97% for the process, even when the contaminated soil is not covered with a layer of clean soil. Other tests have shown that an uncontaminated layer of soil increases the efficiency of the process to greater than 99.99%. Additional removal can be obtained by the use of charcoal filters in the off-gas treatment system, thus improving the overall system DRE.

The observed system DREs indicate that the process has a potential to be a very valuable tool for the remediation of sites that contain both organic and metallic hazardous wastes. While the results are promising, feasibility testing to confirm applicability is strongly recommended prior to any commitment to deploy the process on a site that contains significant quantities of organics that are unconfined in the soil column.

APPLICATIONS CONSIDERATIONS

Before exploring various hazardous waste application scenarios, the operational capabilities and limitations of the large-scale system will be reviewed. The capabilities of the large-scale system to treat various soil characteristics and inclusions can logically be divided into two categories: (1) capabilities of the power supply system, and (2) capabilities of the off-gas system to maintain a negative pressure during transient events. The capabilities of the electrical system in terms of electrode width, depth, and shape have been reported in References 1 and 2.

The two factors that can influence the ability of the power supply system are the presence of groundwater and buried metals. As a general rule, soils with low permeabilities do not inhibit the ISV process even in the water table, because the rate of recharge is not significant in terms of the processing rate. The melt proceeds at a rate of about 3 to 6 in. per hr. Thus, soils with permeabilities in the range of 10^{-5} to 10^{-9} cm/sec are considered vitrifiable even in the presence of groundwater or in the water table. Soils with permeabilities in the range of 10^{-5} to 10^{-4} cm/sec are considered marginal. Soils with permeabilities higher than 10^{-4} cm/sec are difficult to vitrify in the water table unless additional steps are taken, such as drawing the local water table down by pumping, and installing underground barriers. This is not considered a significant impediment because it is highly unlikely that any single method would be used to completely remediate a site. For example, ISV might be used to destroy/immobilize the source and standard pump and treat methods used to clean up plumes that have migrated from the source.

The presence of buried metals can result in a conduction path that would lead to electrical shorting between the electrodes; however, the processing margins are quite generous. Buried metals that occupy up to 90% of the linear distance between the electrodes can be accommodated without suppressing the voltage between the electrodes. Also, once melted, the impact of the metal is less significant. Miscellaneous buried metal, such as drums, should have little or no effect on the ability to process a candidate site. Metal limits are currently 5 wt% of the melt. This is a large fraction when considering drums of waste. In fact, drums containing hazardous and/or classified wastes can be placed in an array that will take advantage of the melt configuration. Such an array is shown in Figure 8. Here, the metal content of the 273 drums is 1.5% of the melt weight, leaving considerable capacity for miscellaneous metal contained within the drums.

Capacity of the off-gas system to maintain a negative pressure during processing, thus preventing the spread of contamination or resulting in fugitive emissions,

ACCEPTABLE CONDITIONS: 90% LINEAR DISTANCE & 5 wt%

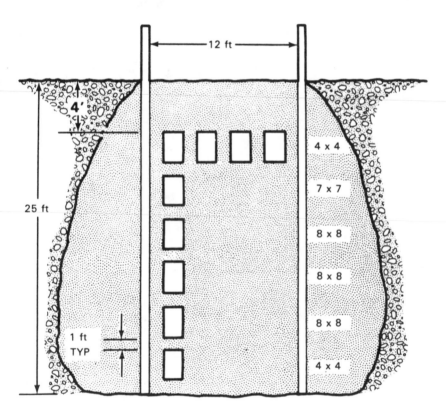

METAL wt = 1.5% OF MELT MASS

Figure 8. Use of in situ vitrification for buried metals.

is a function of the gas generation rate within the processing area. Gas generation resulting from the decomposition of humus and other natural chemicals within the soil is considered insignificant. Generic gas generating situations are shown in Figure 9. These represent the intrusions of the molten glass into void spaces.

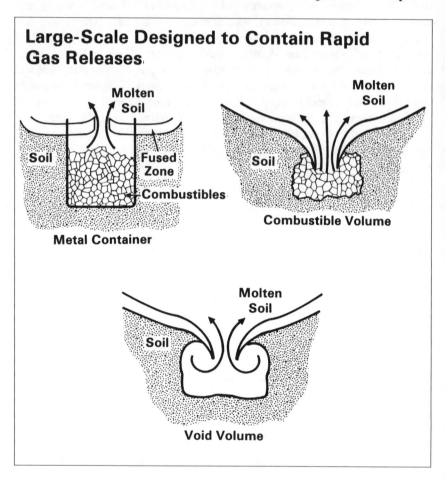

Figure 9.   Gas-generating configurations.

Such intrusions result in release of the entrapped air, penetration of a drum that contains combustible materials, and intrusion into soil inclusions that contain combustible materials, either solids or liquids. Schematically, the capacity of the off-gas system to contain the gas resulting from the processing event is shown in Figure 10. The varying capacities represent what might be encountered in a solid-waste burial ground. The release of the gas is a transient event, ending in about 1 min. Therefore, once the transient event has passed, the system still has the capacity to handle another transient event. The events are time-order limits, not cumulative capacities.

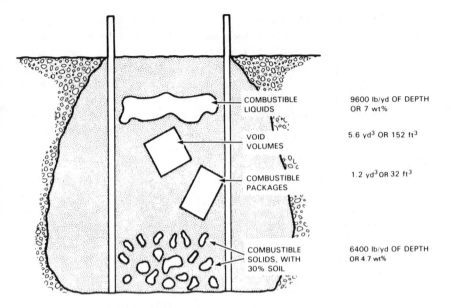

COMBUSTIBLE LIQUIDS	9600 lb/yd OF DEPTH OR 7 wt%
VOID VOLUMES	5.6 yd^3 OR 152 ft^3
COMBUSTIBLE PACKAGES	1.2 yd^3 OR 32 ft^3
COMBUSTIBLE SOLIDS, WITH 30% SOIL	6400 lb/yd OF DEPTH OR 4.7 wt%

Figure 10. Combustible limits for in situ vitrification processing.

The ISV process is particularly well suited to in-place disposal of hazardous waste. The toxic heavy metals are encapsulated or incorporated into the glass, and the organics in containers are destroyed. Certain inorganic compounds such as nitrates are also destroyed by reducing the compound to the diatomic gases by the temperature and reducing conditions of the melt. Sulfates are partially decomposed; the remainder can easily be removed by the off-gas treatment system. Fluorides are dissolved into the glass to 98% for source terms of several hundred ppm. Chlorides are dissolved to the limits of solubility, which are much less than those for fluorides, but up to 1 wt% of the melt, which can be a large quantity. The fluorides and chlorides not dissolved in the glass can be scrubbed out by the off-gas treatment system, using a caustic scrub solution.

There are five general areas where the ISV process might be applied to mixed hazard waste: (1) contaminated soil sites, (2) burial grounds, (3) tanks that contain a hazardous heel in the form of either a sludge or salt cake, (4) classified waste that is already containerized or amenable to containerization, and (5) process sludges and tailings piles. The application of the ISV process to contaminated soil sites and burial grounds is similar to the previously discussed application to general soil and burial ground sites, with the same processing limits for metal and combustibles.

The use of ISV to destroy the hazardous heel in tanks has been tested on the engineering scale with chemical salts. The results of the feasibility study showed that the release was within acceptable limits for the off-gas system and that a vitreous mass was formed. The original tests were performed using additions of glass formers during the processing to achieve a vitreous waste form. The data could be extended to a scenario to dispose of the residual heel and the tank, and to immobilize contaminated soil in the immediate vicinity of the tank. By adding soil and/or rock backfill, the tank could be filled with glass-forming materials prior to processing. This technique could eliminate the concerns of tank dome and/or wall collapse identified during the original testing. Techniques for filling to the peak of the tank dome have been developed. Electrodes would be inserted into the tank through existing openings. The vitreous area would grow downward and outward, encompassing the tank, the contents, and a portion of the surrounding soil. Estimates of the maximum size tank have not been completed, but tanks from 100,000 to 300,000 gal could be permanently disposed of by this technique. The metal content of the tank structure should not impose a processing limit.

Process sludges and tailing piles that contain natural radioactive materials and hazardous chemicals can be disposed of using the ISV process. Applications that involve natural radioactive elements that result in relatively high radon fluxes at the surface are considered potential candidates for remediation by ISV. Tests with zirconia-lime sludges[11] showed that the material was able to be vitrified, and the radon emanation level was reduced by a factor of 10^4 to 10^5 after processing. This is a practical solution where the radon emanation levels are high, the wastes also contain hazardous chemicals that could be leached into the groundwater, and the local infiltration rate is high. In contrast, for large piles in remote areas and where the infiltration rate is very low, barriers over the pile have been shown to be quite effective in preventing the release of hazardous chemicals. Each potential application must be examined on its own merit.

Valuable land that is contaminated can be reclaimed by the use of ISV processing, converting a corporate liability to a capital asset. Old transformer areas and capacitor storage and repair areas that are now in the business district, but contaminated with PCBs, are examples of this concept. The ISV process has been shown to be effective on PCB-contaminated soils, achieving a system destruction removal efficiency of 99.9999%.

Other applications considered, but not yet developed, include shaft sealing, foundations and erosion barriers for remote locations, and the generation of impermeable barrier walls to prevent groundwater seepage into a site. Barrier generation is considered an interim solution that would mitigate an existing hazardous situation until a final solution could be implemented.

SUMMARY OF PROCESS STATUS AND BENEFITS OF APPLICATION

The ISV process has been demonstrated at field-scale conditions, thus eliminating uncertainties of scale-up. This is significant because scale-up is the major risk area in the development and deployment of a new technology. Technology adaptation is a much smaller investment risk than technology development. The applicability of the ISV process to a particular waste can be determined with existing ISV equipment for a few thousand dollars. Thus, feasibility testing is relatively inexpensive. The focus of the feasibility testing is on the performance requirements for the off-gas treatment system and the type and quantity of secondary waste generated. It has already been shown that almost all of the soils encountered can be vitrified, so this is not a consideration during most feasibility testing.

The experience to date indicates that the process is ready for deployment for soil sites contaminated with heavy metals and inorganics. Experience with low boiling point organics that are uncontaminated in the soil column is very limited, and feasibility testing with actual site samples prior to application is strongly recommended. The experience with PCBs, process sludges, and plating wastes is very encouraging. It is anticipated that the ISV process will be used for a broader application of waste management problems. It is also recognized that no single treatment process is applicable to all waste management needs. Within this context, ISV is a new and powerful tool that should be considered and evaluated for radioactive, mixed hazardous, and hazardous chemical applications that fall within the treatment capabilities of the process.

Specific benefits inherent in the ISV process are:

- safety for workers and public
- long-term durability of the waste form (>1 million yr)
- applicability to a variety of soils
- cost-effectiveness ($100–$250/ton for soils)
- volume reduction for sludges (<$100/ton)
- efficient processing rates (3–5 tons/hr)

In summary, the ISV process is cost-effective, applicable to a wide variety of soils and wastes, and responsive to the regulatory changes that emphasize the need for onsite treatment and remediation.

ACKNOWLEDGMENT

The Pacific Northwest Laboratory is operated by Battelle Memorial Institute for the Department of Energy under contract DE-AC06-76RLO 1830.

The authors gratefully acknowledge the contributions of Ken Oma, Gary Carter, and Mike Longaker to much of the data presented in this chapter.

REFERENCES

1. Buelt, J. L., C. L. Timmerman, K. H. Oma, V. F. Fitzpatrick, and J. G. Carter. Pacific Northwest Laboratory, Richland, WA. "In Situ Vitrification of Transuranic Wastes: An Updated Systems Evaluation and Applications Assessment," PNL-4800 Supplement 1 (1987).
2. Buelt, J. L. and J. G. Carter. Pacific Northwest Laboratory, Richland, WA. "In Situ Vitrification Large-Scale Operational Acceptance Test Analysis," PNL-5828 (1986).

3. Buelt, J. L. and J. G. Carter. Pacific Northwest Laboratory, Richland, WA. "Description and Capabilities of the Large-Scale In Situ Vitrification Process," PNL-5738 (1986).
4. Timmerman, C. L. and K. H. Oma. Pacific Northwest Laboratory, Richland, WA. "An In Situ Vitrification Pilot-Scale Radioactive Test," PNL-5240 (1984).
5. Ewing, R. C. and R. F. Hocker. Pacific Northwest Laboratory, Richland, WA. "Naturally Occurring Glasses: Analogues for Radioactive Waste Forms," PNL-2776 (1979).
6. Larsen, T. and W. A. Langford. "Hydration of Obsidian," *Nature* 276(9):153–156 (1978).
7. FitzPatrick, V. F., et al. Pacific Northwest Laboratory, Richland, WA. "In Situ Vitrification—A Potential Remedial Action Technique for Hazardous Waste," PNL-SA-12316 (1984).
8. Koegler, S. S. Pacific Northwest Laboratory, Richland, WA. "Disposal of Hazardous Waste by In Situ Vitrification," PNL−6281 (1987).
9. Timmerman, C. L. Electric Power Research Institute, Palo Alto, CA. "In Situ Vitrification of PCB-Contaminated Soils," EPRI.CS-4839 (1986).
10. Mitchell, S. J. American Fuel and Power Corporation, Panama City, FL. "In Situ Vitrification of Dioxin-Contaminated Soils," Contract 2311211874 (1987).
11. Buelt, J. L. and S. T. Freim. Teledyne Wah Chang, Albany, OR. "Demonstration of In Situ Vitrification for Volume Reduction of Zirconia-Lime Sludges," Contract 2311205327 (1986).

CHAPTER 13

Field Studies of in Situ Soil Washing

James H. Nash and Richard P. Traver

INTRODUCTION

Surface and near-surface contamination often serve as the source for ground-water contamination. Percolation of rainwater through spill sites quickly carries soluble and semisoluble contaminants away from the point of origin. Contaminants considered "insoluble" above parts per million nevertheless migrate more slowly. Gross contaminant sources supply pure product that, over many years, flows deeply through unsaturated soils.

Part of the Environmental Protection Agency's (EPA's) Superfund site cleanup research has been directed at washing such contaminated soil with the aid of aqueous surfactant solutions. The research takes two directions. The first is to excavate the soil and mix it in a wash solution. The second research objective concentrates on the application or injection of a surfactant solution into undisturbed soil in situ. A segment of this in situ research is the subject of this project summary.

This demonstration effort grew out of mutual need between the EPA and the U.S. Air Force. From 1982 to 1985, the EPA researched soil washing technology, using surfactants in laboratory studies. Recompacted soils were used in these studies to simulate in situ conditions. Truly undisturbed contaminated soil was not tested up to that time. The U.S. Air Force, as part of its Installation Restoration Program, was seeking processes to clean up 128 fire training pits at Air Force installations. The Air Force selected the Air National Guard base in Camp

Douglas, Wisconsin as a candidate site for the EPA to test either excavated or in situ soil washing. The EPA and the Air Force representatives chose in situ washing after further consideration.

THE LABORATORY STUDY

Previous laboratory work identified a 50:50 blend of two commercially available surfactants that work well in removing contaminants from soil. They are Adsee 799 and Hyonic PE-90, sold by Witco Chemical and Diamond Shamrock, respectively. To determine if this same blend would work at the Volk Field fire training pit, contaminated soil samples were collected.

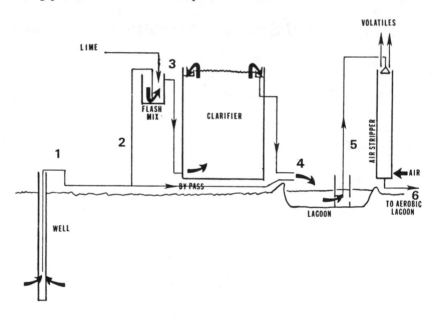

Figure 1. Volk Field pilot treatment for water.

Five physical tests characterized the soil. They were: grain size, TOC, cation exchange capacity (CEC), mineralogy by X-ray diffraction, and permeability. The grain size of the contaminated soil was 98% sand. By X-ray diffraction, alpha-quartz comprised the major portion of the soil with a minor amount of feldspar being present. TOC was as high as 14,900 $\mu g/g$. The cation exchange capacity of 5 mq/100 g was not significant to the contamination levels; however, it did support the X-ray diffraction mineralogic findings. The permeability of the fire permeability of the fire pit soil, at 10^{-3} and 10^{-4} cm/sec, was one to two orders of magnitude less than adjacent uncontaminated soil.

Chlorinated hydrocarbons were part of the volatile contamination. Dichloromethane, chloroform, 1, 1, 1-trichloroethane (TCA), and trichloroethylene (TCE) at concentrations up to 3 μg/g and total chlorinated solvents up to 3.5 μg/g were determined by the volatile organic analysis (VOA) procedure. Other hydrocarbons are aliphatic, aromatic, and polar constituents. The level of hydrocarbon contamination is in the hundreds of μg/g based on the laboratory analysis.

Contaminated groundwater from the aquifer below the fire training pit, a significant problem, was also characterized in the lab study. VOA, TOC, and ultraviolet spectroscopy (UV) were used. The investigations determined that the groundwater contains chlorinated and nonchlorinated hydrocarbons in excess of 300 μg/L.

The soil adsorption constant (K) is a measure of a pollutant's tendency to adsorb and stay on soil. A value of 2,000 for PCBs indicates a two-hundredfold greater adsorption (holding power) than benzene at $K=10$. Benzo(a)pyrene, a toxic substance, and oil have similar values—$K=30,000-40,000$. Grouping contaminants according to a K value and evaluating removal efficiencies (RE) gives order to an otherwise complex collection of chemical classes. This is a report of the EPA's and the U.S. Air Force's field evaluation of in situ soil washing of compounds having K values between 10^1 and 10^6.

THE FIELD STUDY

The field study was conducted by laying out ten 60-cm$\times$60-cm$\times$30-cm pits, dug into the contaminated surface of the fire training area, which served as reservoirs that held various surfactant solutions. Field technicians applied wash solutions into the holes at the maximum rate of 77 L/m^2 per day. The daily dosage was applied in four increments. Since each hole percolated the solutions at different rates, the time interval between doses varied from hole to hole. Testing in three of the pits stopped when the time intervals for the next application approached 10 hours, indicating unacceptable permeabilities being created. Following seven days of washing, the pits received rinses with local, potable well water.

A combination of infrared spectroscopy (IR) and gravimetric determinations of soil extracts was used to evaluate "before and after" contaminant concentrations. To determine contaminant concentration, soil samples were taken after the rinse process, extracted with carbon tetrachloride, and analyzed by IR spectrophotometer for spectral absorbance by the carbon hydrogen bond. The extracts were then air dried and weighed to determine gravimetrically the contaminant concentration (nonvolatile).

The contaminant concentration before soil washing was based on the extracts of soil samples taken adjacent to the test holes. No samples were taken directly from the test holes before washing in order not to bias permeation rates.

Based on both the gravimetric and IR determinations of contaminant concentrations, there was no measurable decrease in contaminants following as many as 14 pore volumes of soil washing in the field tests.

In addition to the soil washing, the field crew conducted a bench-scale groundwater treatment study. From that study a treatment system was assembled and operated which successfully reduced TOC, VOA, and BOD_5 by 50, 99, and 50%, respectively. At these effluent levels, discharge to the local aerobic sewage lagoon was below the Wisconsin Department of Natural Resources' permit limits. A total of 320,000 L of contaminated groundwater was treated at rates of 15,000 to 45,000 L/day.

The bench-scale study investigated the use of lime, alum, ferric sulfate, hydrogen peroxide, polymeric electrolytes, and mineral acids. The application of these chemicals was guided by conventions appropriate to wastewater treatment plants. The resulting water treatment process (shown in Figure 1) was based on the addition of lime at 2 g/L. The lime created a flocculation of iron oxides and organics. The contaminant plume contained up to 52 mg/L iron. Particulate sedimentation in a clarifier, followed by additional residence time in a holding lagoon, reduced the TOC, BOD_5, and VOA to acceptable discharge levels. A final polishing of the volatiles in an air stripper was the final step in the process. Table 1 is a summary of the analytical data.

Table 1.   Analytical Tests and Sampling Points Table for the Water Treatment Process.

Pt. No.	Description	Tests Performed	Approximate Values, Average or Range
1	Individual well head	volatile organic total organic chemical oxygen demand oil and grease pH	10–20 mg/L 60–760 mg/L 6–500 mg/L 0.2–46 mg/L 5.1–6.2
2	Well field effluent	volatile organic total organic iron pH chemical oxygen demand flow rate	10–20 mg/L 250 ± 14% mg/L 32 mg/L 6.0 ± 0.2 41 mg/L .25 – 2 L/sec
3	Flash mixer effluent	total organic (dissolved) suspended solids pH flow rate	160 mg/L 350 mg/L 6.8–9.7 .5–2 l/sec
4	Clarifier effluent	total organic suspended solids pH flow rate	205 ± 7% mg/L 13.6–104 mg/L 7.6 .5–2 L/sec

Table 1.   (Continued)

Pt. No.	Description	Tests Performed	Approximate Values, Average or Range
5	Air stripper feed	volatile organic	3.5–7.0 mg/L
		total organic	151 ± 13% mg/L
		temperature	6–15°C
		flow rate (water)	.95–1.26 L/sec
		oil and grease	3.6 mg/L
6	Air stripper effluent	volatile organic	0.3–0.5 μg/L
		total organic	146 mg/L
		flow rate (air)	101 L/sec
		oil and grease	3.6 mg/L
		biochemical oxygen demand	2.5 mg/L
		chemical oxygen demand	180 mg/L
7	Clarifier	suspended solids	4.4 mg/L
8	Clarifier bottom	suspended solids	2331 mg/L
9	Soil	oil and grease	800–16,000 mg/kg

CONCLUSIONS

1. In situ soil washing of the Volk Field fire training pit with aqueous surfactant solutions was not measurably effective. It is likely that this same ineffectiveness would occur at other chronic spill sites that have contaminants with high soil-sorption values $(K > 10^3)$.

2. In situ soil washing requires groundwater treatment. Groundwater treatment at this site was very successful with the simple addition of lime. Air stripping effectively removed the volatile organics. Advantages at this site were its remoteness for workable air emission limits that facilitated groundwater treatment operations and a local sewage treatment system owned by the responsible party. TOC levels of the recovered groundwater were reduced to one-half the initial values by precipitation with lime, which allowed for direct discharge to the aerobic treatment lagoons. Obviously, not all waste sites have these favorable conditions.

CHAPTER 14

Land Treatment of
Hydrocarbon Contaminated Soils

John Lynch and Benjamin R. Genes

Land treatment has been used as a waste treatment and disposal technology by U.S. petroleum refineries for more than 25 years. In addition, the technology is being applied by the exploration and production sector of the petroleum industry as a cost-competitive alternative in Superfund cleanup studies. This chapter presents information and operational data for a specific Superfund site involving land treatment as the final cleanup option. The waste material consists of contaminated soils from a wood-treating plant. The constituents of concern are hydrocarbons found in petroleum industry waste. Operational data and removal rates are quantified for gross hydrocarbons and specific polynuclear aromatic hydrocarbons.

Land treatment uses the assimilative capacity of the soil to decompose and contain the applied waste in the surface soil layer (usually the top 15–30 cm or 6–12 in.). The upper soil layer is the *zone of incorporation* (ZOI). The incorporation zone, in conjunction with the underlying soils where additional treatment and immobilization of the applied waste constituents occur, is referred to as the *treatment zone*. The *treatment zone* in the soil may be as much as 1.5 m or 5 ft. Soil conditions below this depth generally are not conducive to oxidation of the applied waste constituents. The transformations, biological oxidations, and immobilization will occur primarily in the zone of incorporation.

BACKGROUND

Wastewaters from a creosote wood-preserving operation have been sent to a shallow, unlined surface impoundment for disposal since the 1930s. The discharge of wastewater to the disposal pond generated a sludge which is a listed hazardous waste under the Resource Conservation and Recovery Act (RCRA). Due to groundwater contamination of the shallow aquifer at the site by polynuclear aromatic hydrocarbons (PNAs), the State of Minnesota nominated the site for listing on the Superfund National Priorities List in 1982. Since 1982, numerous remedial investigation activities have been undertaken to determine the nature and extent of contamination at the site. Based on the results of these studies and extensive negotiations, the Minnesota Pollution Control Agency (MPCA), the U.S. Environmental Protection Agency (EPA), and the owner of the facility signed a Consent Order in March 1985 specifying actions to be taken at the site.

In general terms, the remedial actions selected by the MPCA and EPA involve a combination of offsite control measures and source control measures. The offsite controls involve a series of gradient control wells to capture contaminated groundwater. The source control measures include onsite biological treatment of the sludges and contaminated soils, and capping of residual contaminants located at depths greater than 5 ft. Costs for onsite treatment and capping were estimated to be $59/ton.

PILOT-SCALE STUDIES

Before the onsite treatment alternative was implemented, bench-scale and pilot-scale studies were conducted to define operating and design parameters for the full-scale facility. Several performance, operating, and design parameters were evaluated in the land treatment studies. These included:

- soil characteristics
- climate
- treatment supplements
- reduction of gross organics and PAH compounds
- toxicity reduction
- effect of initial loading rate
- effect of reapplication

Three different loading rates were evaluated in the test plot studies: 2%, 5%, and 10% benzene extractable (BE) hydrocarbons. The soils used in the pilot study consisted of a fine sand which was collected from the upper 2 ft of the RCRA impoundment. The soil was contaminated with creosote constituents consisting primarily of PNA compounds. Total PNAs in the soil ranged from 1000 to 10,000 ppm, and BE hydrocarbons in the contaminated soil ranged from approximately 2% to 10% by weight.

Because the natural soils are fine sands and extremely permeable, it was decided that the full-scale system would include a liner and leachate collection system to prevent possible leachate breakthrough. To simulate the proposed full-scale conditions, the pilot studies consisted of five lined, 50-ft square test plots with leachate collection. The studies were designed to maintain soil conditions which promote the degradation of hydrocarbons. These conditions included:

- maintain a pH of 6.0 to 7.0 in the soil treatment zone
- maintain soil carbon-to-nitrogen ratios between 50:1 and 25:1
- maintain soil moisture near field capacity

Hydrocarbon losses in the test plots were measured using benzene as the extraction solvent. The analysis of BE hydrocarbons provides a general parameter which is well suited to wastes containing high molecular weight aromatics such as creosote wastes. Reductions of BE hydrocarbons were fairly similar between all the field plots. Average removals for all field plots over four months were approximately 40%, with a corresponding first-order kinetic constant (k) of 0.004/day.

The reduction of PNA constituents was monitored by measuring decreases in 16 PNA compounds. The following compounds were monitored in the test plots:

2 Rings	3 Rings	4, 5, and 6 Rings
Naphthalene	Fluorene	Fluoranthene
Acenaphthylene	Phenanthrene	Pyrene
Acenaphthene	Anthracene	Benzo(a)anthracene
		Chrysene
		Benzo(j)fluoranthene
		Benzo(k)fluoranthene
		Benzo(a)pyrene
Dibenzo(a,h)anthracene		
		Benzo(g,h,i)perylene
Indeno(1,2,3,c,d)pyrene		

Greater than 62% removals of PNAs were achieved in all the test plots and laboratory reactors over a four-month period. PNA removals for each ring class are shown below:

- 2-ring PNA: 80–90%
- 3-ring PNA: 82–93%
- 4+-ring PNA: 21–60%
- Total PNA: 62–80%

Table 1 summarizes first-order rate constants and half-life data for BE hydrocarbons and PNA compounds for the 5% and 10% BE hydrocarbon test plots. With the exception of the 4- and 5-ring PNAs, the table shows that the kinetic values are approximately equal for the 5% and 10% loading rates. In the case of the 4- and 5-ring compounds, the 5% loading rate resulted in higher kinetic rates for these compounds as compared to the 10% loading rate. This difference may have been due to more 2-ring and 3-ring compounds being available to soil bacteria at the 10% loading rate. These compounds may be preferentially degraded by soil bacteria.

Table 1.   Comparison of Pilot-Scale Kinetic Data at Two Initial Loading Rates.

	First Order Rate Constant (day⁻¹)		Half-Life (days)	
	5% Plot	10% Plot	5% Plot	10% Plot
Benzene extractable	0.003	0.003	231	231
2-Ring PAH	0.023	0.023	30	30
3-Ring PAH	0.016	0.016	43	43
4-Ring PAH	0.004	0.001	173	693
TOTAL PNAs	0.009	0.008	77	87

OPERATING AND DESIGN CRITERIA

The pilot-scale studies were successful in developing operating and design criteria for a full-scale system. These criteria are summarized below:

- treatment period can be extended through October
- soil moisture should be maintained near field capacity
- soil pH should be maintained between 6.0 and 7.0
- soil carbon: nitrogen ratios should be maintained between 25:1 and 50:1
- fertilizer applications should be completed in small frequent doses
- initial benzene extractable hydrocarbon contents of 5% to 10% are feasible
- waste reapplication should occur after initial soil concentrations have been effectively degraded
- waste reapplication rates of 2–3 lb of benzene extractables per cubic foot of soil per 3 degradation months can be effectively degraded

The studies suggest that all the loading rates tested are feasible. First-order rate constants were fairly similar between all the test plots, although the intermediate loading rate (5% benzene extractable hydrocarbons) may demonstrate a slightly higher removal of high molecular weight PNA compounds. The higher loading rates, however, showed the greatest mass removals. The selection of an initial loading rate should balance additional land area requirements against time requirements for completing the treatment process. Moderate loading rates (5%)

will result in a faster detoxification, whereas higher loading rates will decrease land area requirements.

CONSTRUCTION AND START-UP OF FULL-SCALE SYSTEM

Construction of the full-scale system involved preparation of a treatment area within the confines of the existing RCRA impoundment (Figure 1). The treatment area was constructed on top of the impoundment to avoid permitting a new RCRA facility. If the facility were located outside the impoundment, then a Part B permit would have to be obtained before the treatment facility could be constructed. By locating the treatment area within the confines of the impoundment, the treatment system was considered part of closure of the impoundment. This enabled us to fast track the cleanup and avoid the delays associated with permitting a new RCRA unit.

The principal construction activities at the site involved:

- preparation of a lined waste pile for temporary storage of the sludge and contaminated soil
- removal of all standing water in the impoundment
- excavation and segregation of the sludges for subsequent free oil recovery
- excavation of approximately 3–5 ft of "visibly" contaminated soil from the impoundment and subsequent storage in the lined waste pile
- stabilization of the bottom of the impoundment as a base for the treatment area
- construction of the treatment area, including installation of a 100-mL high density polyethylene (HDPE) liner, a leachate collection system, and 4 ft of clean backfill
- installation of a sump for collection of the stormwater and leachate
- installation of a center pivot irrigation system

As previously discussed, a lined treatment area was constructed because the natural soils at the site are highly permeable. A cap also was needed for the residual contaminants left in place below the liner. Therefore, the treatment area liner serves two functions at the site. The first function is to provide a barrier to leachate from the treatment area. The second is to provide a cap over the residual contaminants that were left in place.

The treatment area was constructed on top of the existing wastewater disposal pond after all contaminated materials were removed. The surface area for treatment is approximately 125,000 ft². Containment berms with 3:1 slopes enclose the treatment area and prevent surface runoff from leaving the site.

The treatment area is lined with a 100-mil HDPE membrane. The base of the liner slopes 0.5% to the south and west. A sump with a 50,000-gal capacity is located in the southwest corner of the treatment area. A layer of silty sand ballast

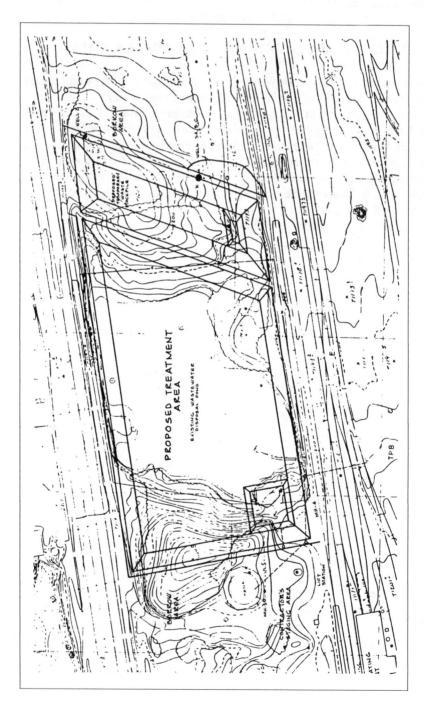

Figure 1. Site plan for onsite treatment system.

18 in. thick was placed on top of the treatment area liner. A 6 in. gravel layer was placed on top of the ballast. This layer serves as a leachate collection system and as a marking layer for land treatment operations.

The leachate collection system includes 2-ft wide leachate collection drains at 100-ft centers. The drains are filled with gravel and perforated pipe to carry leachate from the collection system to the sump. The drains were wrapped in filter fabric to prevent clogging. A 2-ft layer of uncontaminated sand was placed above the leachate collection system. This sand serves as an initial mixing layer for the contaminated soils, and is the treatment zone for the full-scale system.

Water in the leachate collection sump is discharged by gravity flow to a manhole, and is automatically pumped via a lift station to a 117,000-gal storage tank. Water in the storage tank is recycled back to the treatment area via a spray irrigation system. Water in excess of irrigation requirements is discharged to the municipal wastewater treatment plant.

Construction of the waste pile and treatment area was completed in October 1985. In late April 1986, a center pivot irrigation system was installed and 120

Table 2. Summary of Start-Up Data (5/23/86).

Parameter	Average
Benzene extractables, %	5300
TOC, ppm	29710
TKN, ppm	1367
Ammonia, ppm	2.37
Total phosphorus, ppm	522
Total potassium, ppm	502
pH	7.66
Polynuclear Aromatic Hydrocarbons (PAH), ppm:	
Naphthalene	1148
Acenaphthylene	21
Acenaphthene	1082
Total 2-ring PAH	2251
Fluorene	1885
Phenanthrene	4190
Anthracene	3483
Total 3-ring PAH	9558
Fluoranthene	1575
Pyrene	958
Benzo(a)anthracene and chrysene	837
Total 4-ring PAH	3370
Benzofluoranthenes	368
Benzopyrenes	294
Indeno(123cd)pyrene	111
Dibenzo(ah)anthracene	100
Benzo(ghi)perylene	106
Total 5-ring PAHs	979
Total PAHs	16159

tons of manure were spread in the treatment area. Manure loading rates were based on achieving a carbon:nitrogen ratio of 50:1. In addition to nitrogen, the manure provides organic matter which enhances absorption of the hazardous waste constituents.

In May 1986, a 3-in. lift of contaminated soil was applied to the treatment area. The target loading rate for start-up was a BE hydrocarbon concentration of 5%. The soil was mixed (rototilled) with 3 in. of native soil to achieve a treatment depth of 6 in. This application involved approximately 1200 yd³ of sludge and contaminated soil. Table 2 summarizes start-up data for the full-scale facility.

The treatment area is irrigated almost daily due to dry weather during the summer months. Irrigation needs are determined from soil tensiometer readings, soil moisture analyses, and precipitation and evaporation records. Typical irrigation rates range from 1/4 in. to 3/8 in. per application. This application rate keeps the soils in the cultivation zone moist without saturating soils in the lower treatment zone. Maintaining soil moisture near field capacity was determined to be a key operating parameter in the pilot-scale studies.

PERFORMANCE OF THE FULL-SCALE FACILITY

Benzene extractable (BE) hydrocarbons and 16 polynuclear aromatic (PNA) compounds are being monitored to evaluate the performance of the facility. Figure 2 shows the BE hydrocarbon concentrations measured in the zone of incorporation

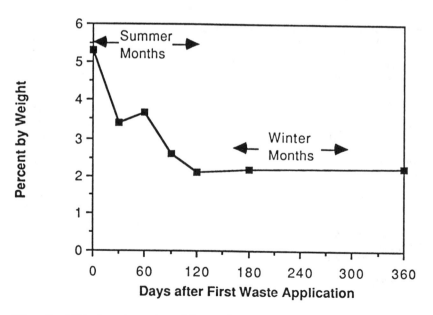

Figure 2.   BE hydrocarbons degradation vs time.

(ZOI) during the first year of treatment. BE hydrocarbon concentrations decreased approximately 60% over the first year of operation. Most of the decrease occurred during the first 120 days (May through September). Little decrease in BE hydrocarbon concentrations was observed during the fall and winter months.

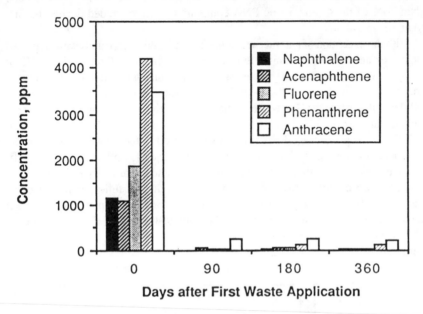

Figure 3. 2-ring and 3-ring PNA degradation vs time.

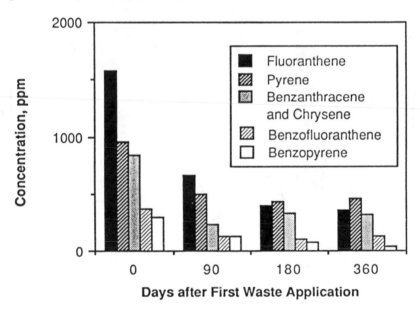

Figure 4. 4-ring and 5-ring PNA degradation vs time.

Figures 3 and 4 show PNA concentrations measured in the treatment facility during the first year of treatment. Figure 3 summarizes data for 2- and 3-ring PNAs. Figure 4 summarizes data for the 4- and 5-ring compounds. Greater than 95% reductions in concentration were obtained for the 2- and 3-ring PNAs. Greater than 70% of the 4- and 5-ring PNA compounds were degraded during the first year of operation.

With the exception of anthracene, all the 2- and 3-ring compounds were degraded below or near detection limits after 90 days of treatment. Greater than 92% of the anthracene present in the waste was degraded during the first 90 days of treatment. Similarly, most of the 4- and 5-ring removals occurred during the first 90 days of treatment. This was expected because the warmest weather occurred during this period.

Table 3 shows average PNA removals measured in the pilot-scale studies and compares them with the full-scale removal efficiencies. Full-scale removal efficiencies were higher than test plot removal efficiencies for every PNA ring class and BE hydrocarbons. However, it must be noted that the full-scale facility operated for 360 days, compared to only 126 days for the test plot units. Table 3 also presents average half-life data for both the test plots and the full-scale unit. Full-scale half-lives were consistently in the low end of the range of half-lives reported for the test plot units.

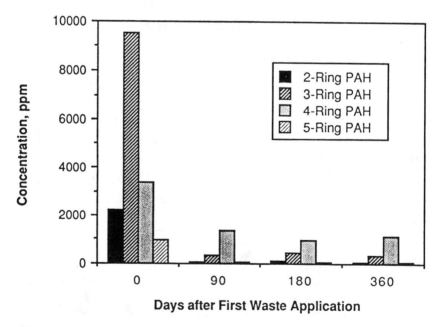

Figure 5. PNA degradation by ring class.

Table 3.   Comparison of Full-Scale and Test Plot Removals.

	Avg. Percent Removal		Avg. Half-Life (Days)	
Parameter	Full Scale[a]	Test Plots[b]	Full-Scale	Test Plots
2-Ring PAHs	95	93–95	<45	29–33
3-Ring PAHs	95	83–85	45	46–49
4- and 5-Ring PAHs	72	32–60	115	95–226
Total PAHs	90	65–76	65	61–83
BE Hydrocarbons	60	35–56	150	106–202

[a]Removal efficiency calculated after 193 days of treatment.
[b]Removal efficiency calculated after 126 days of treatment.

In summary, the rate and amount of PNA degradation is proportional to the number of rings contained by the PNA compounds (Figure 5). The 2- and 3-ring PNAs degraded most rapidly. The 4- and 5-ring PNAs degraded at slower rates; however, these compounds are strongly adsorbed to soils and are immobilized in the treatment zone of the facility. Table 4 summarizes water quality data for the leachate collection system of the facility. Only acenaphthene and fluoranthene were detected in the drain tile water samples. Concentrations for these two compounds were near analytical detection limits.

Table 4.   Drain Tile Water Quality.

	Concentration, ppb		
Compound	June 1986	August 1986	October 1986
Naphthalene	<1	<1	<1
1-Methylnaphthalene	<1	<1	<1
2-Methylnaphthalene	<1	<1	<1
Acenaphthylene	<1	<1	<1
Acenaphthene	<1	3.7	2.7
Fluorene	<1	<1	<1
Phenanthrene	<1	<1	<1
Anthracene	<1	<1	<1
Fluoranthene	<1	2.1	1.4
Pyrene	<1	<1	<1
Benzo(a)anthracene	<1	<1	<1
Chrysene	<1	<1	<1
Benzofluoranthenes	<5	<1	<1
Benzopyrenes	<5	<1	<1
Indeno(123cd)pyrene	<5	<1	<1
Dibenzo(ah)anthracene	<5	<1	<1
Benzo(ghi)perylene	<5	<1	<1

CONCLUSION

The data developed during this project has shown that onsite treatment of creosote contaminated soils is feasible. Based on the data developed in pilot-scale studies, a conservative design for a full-scale system was developed and constructed. The full-scale unit has matched or surpassed the performance of the pilot-scale unit in degrading creosote organics. The advantages of onsite treatment are that it reduces the source of contaminants at the site in a very cost-effective manner. In addition, it satisfies the developing philosophical approach that EPA has to onsite remedies, and it reduces the liability of the owner/operator due to offsite disposal.

CHAPTER 15

Asphalt Batching of Petroleum Contaminated Soils as a Viable Remedial Option

Paul T. Kostecki, Edward J. Calabrese, and Edwin J. Fleischer

TECHNICAL CONSIDERATIONS

Theory

One recently proposed remedial option for soils containing petroleum hydrocarbons and associated hydrocarbons involves the incorporation of these soils into hot asphalt mixes as a partial substitute for stone aggregate. The primary mechanisms of remediation in this instance are volatilization and thermal destruction by incineration. During the asphalt batching process, aggregate is passed through a dryer where average temperatures range from 500°F to 800°F.[1] Retention times are approximately five minutes. When the aggregate leaves the dryer it is at approximately 300°F. The purpose of the dryer is to remove moisture from the aggregate, but it can also drive volatile compounds from oily soils.

Large quantities of the lighter petroleum hydrocarbons contained in gasolines, kerosenes, and fuel oils are incompatible with asphalt. These compounds can act as solvents to soften the final asphalt product. Heavier fractions are similar in nature to asphalt and, therefore, may not be as damaging to the asphalt mix. Thus, it is essential, from a product quality standpoint, that the lighter hydrocarbons be removed within the dryer. Solidification or encapsulation within the asphalt/aggregate matrix is a secondary mechanism of soil remediation which serves to contain heavier hydrocarbon fractions that might not be removed during the drying process.

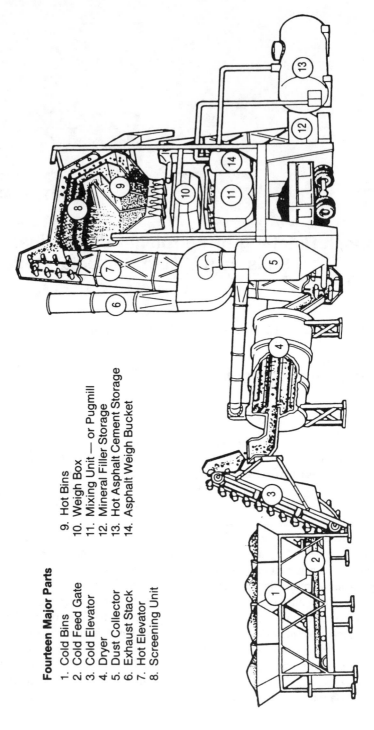

Fourteen Major Parts

1. Cold Bins
2. Cold Feed Gate
3. Cold Elevator
4. Dryer
5. Dust Collector
6. Exhaust Stack
7. Hot Elevator
8. Screening Unit
9. Hot Bins
10. Weigh Box
11. Mixing Unit — or Pugmill
12. Mineral Filler Storage
13. Hot Asphalt Cement Storage
14. Asphalt Weigh Bucket

Figure 1. Typical asphalt batching plant. (*Source:* Asphalt Institute, 1983).

The Hot Mix Asphalt Process

Figure 1 illustrates a typical hot mix asphalt batching operation.[1] Unheated aggregates stored in cold bins [1] are proportioned by cold feed gates [2] onto a conveyer system [3], which delivers the material to a dryer [4], where it is heated and dried. The dust collector [5] and the exhaust system [6] remove exhaust gases from the dryer. A more modern plant than the one pictured would also include a baghouse for air quality control. The aggregates are then delivered by hot elevator [7] to a screening unit [8], which separates the material into different fractions by particle size and deposits them into hot storage bins [9]. The aggregates are measured into the weigh box [10] and are then dumped into the mixing chamber [11], along with any mineral filler that may be needed [12]. Hot asphalt which is pumped from storage [13], is weighed [14] and delivered to the mixing chamber, where it is completely mixed with the aggregates. The final product is then dumped in a truck or conveyed to heated storage.

When petroleum contaminated soils are utilized in a hot mix asphalt batching plant, the soils are simply added to the aggregate feed stream as a small percentage of the total feed. In practice, the soil feed must be limited to less than 5% of the total aggregate feed at any one time.[2] A restriction is required in order to maintain final product quality and to minimize the danger of air emissions caused by volatilization of the compounds.

Massachusetts Experience

The incorporation of soils containing petroleum products into asphalt mixes is not widely practiced. A review of the current literature on asphalt mixes and additives found no mention of this process. Two asphalt industry representatives who were interviewed revealed that it is not considered to be common practice.[3,4] However, a national survey of state agencies has found that five states (Massachusetts, Minnesota, Rhode Island, New Hampshire, and Vermont) have had experience with the disposal of petroleum-laden soils by this method.[5] The following paragraphs will describe in detail the Massachusetts experience.

In 1984, the Massachusetts Department of Environmental Quality Engineering (DEQE) approved an application from the Henley-Lundgren Company in Shrewsbury, Massachusetts, which allows them to use petroleum-bearing soils as part of their mineral aggregate feed. The soil is incorporated at less than 5% by weight of the total feed. Even at this small percentage, this plant has the capacity to use 8000 tons of soil annually.

The approval letter regarding this application specifies strict management guidelines which must be followed by the plant. Construction of a sloped concrete pad capable of holding 1000 cubic yards of soil in dead storage was required. The pad was surrounded by a dike lined with a coal tar emulsion sealer to prevent possible leaching of petroleum products into surrounding soils. The pad and storage pile must be covered when the plant is not in operation. Any runoff from the pad is collected and removed periodically to a hazardous waste treatment facility.

In addition, the plant may only accept soils that contain virgin fuel products. Used and waste oils could contain unknown constituents that would pose a health threat to workers and occupants of the surrounding area. Only waste material generated in Massachusetts can be accepted at the site. During the winter months, when the plant is closed, no hydrocarbon-bearing material can be kept in storage at the plant site. Even though the soil is not considered to be a hazardous waste under the terms of the application approval, the plant must also comply with a strict manifest and recordkeeping system.

This method of disposal for petroleum-laden soils is now considered to be the most favored option in the state of Massachusetts. State officials plan to encourage at least one operator in each of the state's four regions to retrofit their plants in a manner that is similar to that described above.

As a postscript to this case study, it should be noted that the Shrewsbury plant has stopped taking oily soils as of the summer of 1986, because they cannot comply with Massachusetts Department of Public Works specifications for asphalt mixes if ungraded soils are included in the mix. This issue will be discussed further in the following section.

IMPLEMENTATION CONSIDERATIONS

Design Considerations

In theory, the high temperatures encountered by petroleum-bearing soils in the aggregate dryer should completely destroy any hydrocarbons that might be present. To date, however, there has not been any research conducted to support this theory. Proponents of this disposal option point out that secondary containment is provided by the asphalt mix, since the soil is ultimately locked into the asphalt/aggregate matrix. This may not be true. Asphalt industry representatives and geotechnical engineers have pointed out that incomplete destruction of the fuel hydrocarbons, as well as characteristics of the soil itself, could have deleterious consequences for the final asphalt product that must be considered in the design and operation of such a system.[3,4,6]

Any residual hydrocarbons that are not removed in the dryer could soften the mix, as when asphalt is "cut" with fuel oil, kerosene, or gasoline to create slow, medium, or rapid curing asphalts, respectively. Soil requiring remediation that is stored in bulk may contain all three of these substances, so it is possible that curing times for different portions of the same mix may vary unpredictably.

The strength and durability of asphalt paving mixes is dependent on the size, type, and amounts of aggregates used. Normally, the amount of fine material which passes the No. 200 sieve is limited to 2–10% of the mixture. As the size of a particle decreases, it tends to absorb a larger percentage of the hot asphalt, due to increasing surface area. Too many fines could lead to coating problems or a mixture that is too dry. In an extreme case, this imbalance could lead to an inferior product which exhibits cracking or instability. Also, clays and organic matter should not be included in asphalt cements. Soils from mixed sources which contain petroleum products are highly variable in quality, and could easily contain components which could damage the final asphalt mixture. Unfractured stones within the mix also decrease product stability. In the Massachusetts example cited above, the final product could not meet specifications set by the Massachusetts Department of Public Works for use on state highways due to the amount of unfractured material in the mix. However, the mix could be used for all other paving purposes. This disadvantage could possibly be solved by screening the soil to remove any stone-size particles.

Equipment Requirements

Additional equipment is needed when petroleum-laden soils are utilized in an asphalt batching plant. Specifically, an additional cold storage bin or hopper is needed as well as a system for conveying the soil to the rotary dryer. The conveyer system must provide some means of metering the amount of soil that is sent to the dryer. Facilities for storage of the material must also be constructed if not available. This equipment is simple to design; however, it may require a large capital expense depending on the size and configuration of the asphalt plant that is being retrofitted. The storage hopper and conveyers are of the same type that are normally used to handle the other aggregate feed. The concrete storage pad described in the case study may not be a requirement in all states. Any existing plant would already own any heavy equipment needed to move the soil from place to place at the plant site.

Treatment Requirements

No pre- or post-treatment is needed for soils that are incorporated into asphalt mixes. One might be concerned that volatilized hydrocarbon gases could be released to the atmosphere as soil is passed through the dryer. However, most hot mix asphalt batching plants already have extensive facilities to control air emissions from the plant. The Massachusetts plant described above was limited to a 5% soil feed because officials felt that any more than that might compromise the neighborhood air quality, but no data are available to determine how much or to what extent this is true.

Disposal Issues

There are two main disposal issues which must be addressed when one is considering the reuse of soils which contain petroleum hydrocarbons via the asphalt batching process. The first involves uncertainties concerning the actual extent to which hydrocarbon compounds are destroyed via incineration or are volatilized to the atmosphere.

The second disposal issue was discussed in detail earlier, and concerns effects that petroleum-laden soils and ungraded soils in general could have on the integrity of the final asphalt mix. Both issues impose practical limits on the quantity of soil that can be included in the aggregate feed.

Monitoring Requirements

A hot mix asphalt batching operation is a complex process that requires constant monitoring. A plant which incorporates petroleum-bearing soils into its mix may require additional monitoring. Plant operators must make sure that the soil does not adversely affect the product mix or the plant air emissions quality. In addition, regulators may require that soils stored on site be sampled periodically to ensure compliance with the virgin product rule.

Permitting Requirements

Permitting requirements for this disposal option will vary on a state-by-state basis. The asphalt industry is already highly regulated by transportation regulators and air quality regulators. Asphalt plant operators must comply with the permitting requirements of these agencies and may face additional permitting requirements if their plant is considered to be a soil treatment facility. The imposition of additional regulatory requirements on asphalt plant operators may have considerable effect on their willingness to accept petroleum contaminated soils for processing.

Requirements for the Henley-Lundgren plant in Shrewsbury, Massachusetts were discussed earlier. A copy of the approval letter outlining these requirements is presented as an appendix to this chapter.

ENVIRONMENTAL CONSIDERATIONS

Exposure Pathways

Asphalt incorporation of soils containing petroleum products offers the advantage of soil remediation via incineration and volatilization of most of the hydrocarbons contained in those soils. As was mentioned in the case study, incomplete combustion of volatile organic compounds could compromise air quality in the vicinity of the plant. The extensive air pollution control equipment normally required for hot mix asphalt batching plants and the restrictions on the allowable percentage of soil contained in the feed minimizes this concern.

A number of exposure pathways for asphalt incorporation of petroleum contaminated soils have been identified.[7] Three primary exposure pathways are skin contact, vapor inhalation, and particle ingestion. Workers involved in the excavation and transportation of soils from the field to the plant site face the highest level of exposure to the compounds contained in those soils.

A secondary pathway is water ingestion. This could result if the soil is stock-piled in a manner that allows water to leach through the material and eventually reach a potable water supply. A concrete storage pad and provisions for cover-ing the storage pile during the hours when the plant is closed can help to prevent the generation of runoff from the storage pile which might contain hydrocarbon constituents.

Finally, if the material used at the plant is restricted to soils containing virgin fuels only, individuals associated with the process will be protected from exposure to unexpected compounds that can sometimes be found in used or waste oils.

Effectiveness

The time required to dispose of hydrocarbon-laden material through the asphalt batching process is limited only by the size of the batching plant. Material may be excavated and stored until it can be used.

As stated previously, acceptable material should be limited to that containing only virgin fuel oils. This is a quality control and safety measure, required because the material often arrives at the site in small quantities from various sources.

At present, no data on hydrocarbon removal efficiency are available. Data could easily be collected by testing the aggregate for hydrocarbon content after it exits the dryer. A removal efficiency for the dryer could then be calculated. Present-ly, it is only assumed that the high temperatures encountered in the dryer and the secondary encapsulation of the soil within the asphalt mix provide adequate remediation measures.

ECONOMIC CONSIDERATIONS

Capital Costs

The cost of retrofitting an asphalt batching plant to accept soils containing petroleum products must be incorporated into the general plant costs by the plant operator. This cost is considered to be proprietary by the one operator interviewed; however, the cost is likely to be on the order of tens of thousands of dollars. This cost is offset by the fees that are charged to the waste soil generators. Ex-perience has shown that the soil disposal business can be a profitable sideline for an asphalt batching firm. Capital costs include the cost of storage facilities, an additional feed hopper, a conveyer system, and a metering system.

Installation Costs

Installation costs include the cost of assembling the above-mentioned compo-nents. This cost is also borne by the operator, who usually maintains a crew that is capable of performing customized modifications to the existing plant structure. Other workers may need to be hired to design and build structures, such as the concrete storage pad and runoff collection system.

Operation and Maintenance

Maintenance of the equipment is an additional expense; however, the operations involved are the same as those required for the other plant equipment. Record-keeping requirements can be substantial. These are a function of state regulations, and will differ on a state-by-state basis.

Qualitative Cost Analysis

Fees charged to generators of petroleum-laden soils are relatively low compared to other forms of offsite disposal. The average charge is around $60 per ton of material (1985 costs). Excavation costs are the same as for other offsite disposal methods; however, transportation costs may be dramatically reduced if plants are available on a regional basis.

While not widely practiced at this time, the disposal of soils via the asphalt batching plant option offers potential disposal sites within short driving distances of many locations throughout the country. Successful reuse of soils via this process is dependent on the existence of an interested and diligent asphalt contractor and also on the aid of a supportive regulatory agency. Future experience will eventually lead to a complete evaluation of this technology.

AN ALTERNATIVE PROPOSAL

Modified Asphalt Plants

The questions cited above regarding asphalt integrity could be avoided altogether if one were to utilize only the rotary dryer portion of the asphalt plant to volatilize hydrocarbons from the soil. This approach has been suggested by a consultant from Florida who has done a feasibility study on the process.[8] Figure 2 details the process. Soil is processed through the dryer, stockpiled, and then backfilled. Loading rates would be much higher (100% oily feed vs 5%) than the system described earlier in this report. Provisions could be made to include a secondary combustion chamber that would be used to combust the off-gases from the dryer. Figure 3 is a flow diagram of such a system. Two such systems are currently being built.[8] It may be possible to use an existing asphalt plant in a similar manner. Specifically, oily soil could be processed for a short period of time at the beginning or end of the asphalt producing season. Costs would likely be higher than those currently charged because the asphalt producer would have no product sales during this period. The feasibility of such a system is completely unknown,

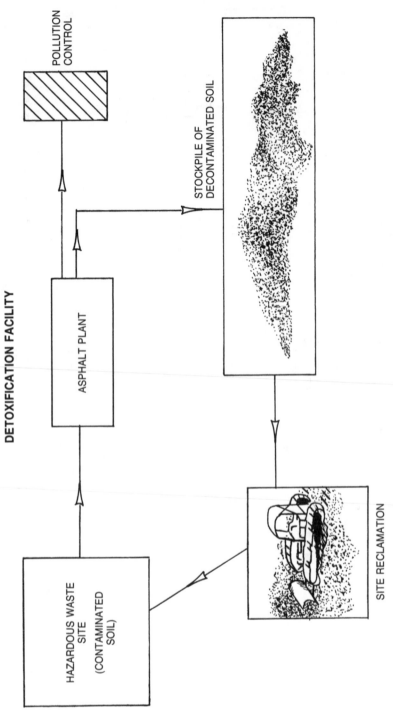

Figure 2. Proposed soil reclamation project. (*Source:* Adapted from figure by Cross/Tessitore & Associates, Orlando, FL. Used with permission.)

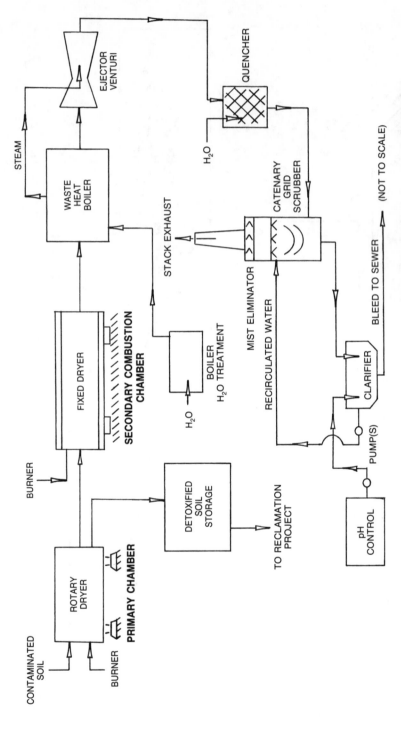

Figure 3.   Flow diagram of asphalt plant converted to hazardous waste incinerator. (*Source:* Adapted from figure by Cross/Tessitore & Associates, Orlando, FL. Used with permission.)

but it is mentioned here because it utilizes some of the better features of asphalt technology.

RESEARCH CONSIDERATIONS

Research Requirements

In order to quantitatively estimate human exposure to petroleum constituents during the asphalt batching process, the following information is required.

Excavation, Transportation, and Storage of Soils

Exposure during the excavation, transportation, and storage steps of this remedial technology cannot be quantified unless hydrocarbon volatilization rates are known. Ideally, these fluxes should be measured under a variety of field conditions with portable equipment. As an alternative, mathematical modeling could predict an average vapor flux. Accurate modeling of these processes would be difficult, due to the wide variety of conditions seen in the field.

Exhaust Emissions

The quantities and composition of exhaust gases from both a standard hot mix plant and a plant which utilizes oily soils must be determined. It is possible that this information is already in DEQE files as a result of pilot studies performed during the Henley-Lundgren approval process. If so, project coordinators should identify the appropriate contact persons in Air Quality so that this data can be acquired. Alternatively, this data would have to be acquired through a focused research project.

Process Effectiveness

The effectiveness of the rotary dryer for removal of organic compounds from soils must be examined. Presently, the process is only assumed to be effective. Again, data regarding this aspect of plant operations may already exist in the Henley-Lundgren files. Otherwise, these data can only be acquired by sampling the aggregate stream for hydrocarbon content before and after it is processed by the rotary dryer.

Asphalt Integrity

There are serious questions regarding the integrity of an asphalt product which contains miscellaneous or ungraded soils. These questions cannot be answered unless the final asphalt product is tested either at the plant or in the field. This testing program should be conducted in conjunction with an agency or organization that is thoroughly familiar with asphalt mix standards. The Asphalt Institute and the Department of Public Works are two such groups.

ACKNOWLEDGMENTS

This work was supported in part by a grant from the Office of Research and Standards, Massachusetts Department of Environmental Quality Engineering, Boston, MA, and The Environmental Institute, University of Massachusetts, Amherst, MA.

REFERENCES

1. *Principles of Construction of Hot-Mix Asphalt Pavements,* Manual Series No. 22, Asphalt Institute, 1983.
2. Massachusetts Department of Environmental Quality Engineering, 1984.
3. Joubert, Bob, N.E. Regional Engineer, Asphalt Institute. Personal communication, August 15, 1985.
4. Pagan, Chuck, Director of Research, National Asphalt Paving Association. Personal communication, August 20, 1985.
5. Kostecki, P. T. , E. J. Calabrese, and E. Garnick. "Regulatory Policies for Petroleum Contaminated Soils: How States Have Traditionally Dealt with the Problem," in *Proceedings of the Conference on the Environmental and Public Health Effects of Petroleum Contaminated Soils* (Amherst, MA: University of Massachusetts, October 30-31, 1985), John Wiley & Sons, New York, 1988.
6. Bemben, Stanley, Professor, Civil Engineering Department, University of Massachusetts, Amherst, MA. Personal communication, September 12, 1985.
7. Preslo, L. M., J. B. Robertson, D. Dworkin, E. J. Fleischer, P. T. Kostecki, and E. J. Calabrese. *Remedial Technologies for Leaking Underground Storage Tanks* (Chelsea, MI: Lewis Publishers, Inc., 1988).
8. Cross, F. L., President, Cross/Tessitore & Associates, P.A. Orlando, FL, September 15, 1986.

APPENDIX

Henley-Lundgren Approval Letter
The Commonwealth of Massachusetts
Executive Office of Environmental Affairs
Department of Environmental Quality Engineering
Central Region
75 Grove Street, Worcester, Massachusetts 01605

March 26, 1984

Henley-Lundgren Company
193 Hartford Pike
Shrewsbury, MA 01545

Re: 310 CMR 30.355 Shrewsbury
Re-Use Application
Henley Lundgren Co./Zecco. Inc.

Attn: Richard Olson

Dear Mr. Olson:

The Department of Environmental Quality Engineering, Division of Hazardous Waste, has reviewed the application for the proposed re-use of sand and gravel containing petroleum products (gasoline, #2, #4, and #6 fuel oils) at the Henley-Lundgren Company Asphalt Plant in Shrewsbury, MA.

The facility is an asphalt batching plant rated at approximately 300 tons per hour. The aggregate, including any contaminated soil, will be heated in the rotary kiln drier. The particulate matter is collected by a fabric filter collector meeting Best Available Control Technology (BACT) criteria, prior to discharge from the stack.

A pilot test including air quality stack testing was performed during 1983. The results demonstrated the capability of the system to operate in conformance with air quality requirements.

The proposal consists of the generator (Zecco, Inc.) supplying the re-user (Henley-Lundgren Co.) with approximately 800 tons (500 cu. yards) per month of contaminated soil between April 15 and November 15 of each year. The contaminated soils will be used at the Henley-Lundgren Company's asphalt batching plant to manufacture bituminous concrete products. The re-used aggregate will always be mixed with virgin aggregate, at a rate not exceeding 5% of the total mix at any time. No further processing of the soil will be required for such use.

A sloped concrete pad 75' × 75' will be constructed as described in the application, capable of storing 1000 cubic yards of oil saturated soil at any one time. The concrete pad will be diked with a bituminous concrete berm approximately 9″ in height. The storage pad will be lined with a coal tar emulsion sealer to prevent leaching of the oils and gasoline to the soils. The storage pad and soil will be completely covered by a water retardant cover when the facility is not in operation. A waste oil/gasoline collection system consisting of a drainage grate, a pre-cast concrete catch basin, and a 2000 gallon underground holding tank will be installed to collect any surface water run-off from the stock pile pad that may contain oils and gasoline. All liquid collected in the underground tank will be pumped periodically, manifested as a hazardous waste, and removed from the site by a licensed hazardous waste transporter to a location approved by the Department.

The oil contaminated soil will be transported from the waste pad by a front-end loader to a special feed hopper which will include a monitor on the mix rate to the belt feeder for the aggregate.

Under the re-use provisions of the Department Regulations, the re-used soils must not require additional treatment by the Henley-Lundgren Co. before it is re-used.

The Department believes the application complies with the provisions of 310 CMR 30.355 and hereby approves of the re-use submittal dated 12/13/83 between Zecco, Inc. (Generator) and the Henley-Lundgren Co. (Re-user) with the following provisos:

1. That during the period of plant seasonal shutdown, no soils containing oils or gasoline will be stored at the Henley-Lundgren Co., Shrewsbury, Massachusetts facility;

2. That good housekeeping always be maintained of the aggregate storage area and the portions of the plant directly involved in the transfer or storage of reusable aggregate. The area between the pad and the storage hopper will be paved to minimize contamination from spillage;

3. That Henley-Lundgren Co. and Zecco. Inc. will comply with all the requirements not herein specifically specified, but included in the submittal documents;

4a. That soil delivered under re-use will be shipped directly from a site to Henley-Lundgren Co. All shipments will be made under the direct responsibility of Zecco, Inc., acting as Generator;

4b. Limited quantities of drummed sand, soil, or speedy-dry stored off-site at Zecco, Inc. may be used if meeting all the criteria herein stated;

5. No material will be shipped until verification has been made by a representative of DEQE that such contaminated soil is the result of a spill or leak of virgin fuel oil or gasoline. Such verification must be noted on the manifest form;

6. The manifest form will be used as the recordkeeping form for this re-use application. A separate form is to be used for each site and delivery and is to include all pertinent information on nature of spill, location, type of contaminant, quantity of material, testing performed, date of spill (if known), date of receipt by re-user, name of transporter (if other than Zecco-owned vehicles;

7. All wastes will be tested to assure flash point is higher than 140 degrees Fahrenheit;

8. Only waste originating from releases in Massachusetts will be accepted;

9. Any wastes received on site that are deemed unacceptable must immediately be removed by a licensed transporter to a facility acceptable to the Department. Notification shall be made immediately to the Department of any such occurrences;

10. Records will be maintained at the Henley-Lundgren facility for at least three years;

11. Henley-Lundgren will submit to the DEQE Regional Office copies of all such deliveries on a monthly basis;

12. The storage pad catch basin and grate will be inspected daily for clogging that would prevent the free flow of run off from the aggregate stock pile area to the storage tank;

13. A quality assurance program will be established to periodically check the shipments of soil for constituents other than those approved for re-use in this application;

14. A contingency plan will be prepared by Henley-Lundgren Co. and submitted to DEQE for approval that will demonstrate what actions will be taken to prevent or contain any environmental release of contaminants from the storage pad during an emergency;

15. The Department may add further provisos or restrictions if operations so warrant.

The waste soils containing gasoline and oils to be re-used will not be considered as a hazardous waste, and are not subject to the DEQE Hazardous Waste Regulations as long as they are used in conformance with this re-use approval. However, the re-use will remain subject to other applicable federal, state, and local requirements governing the transport, storage, and management of such materials.

No changes in the amount of waste soils to be re-used or the parameters of the soil outlined in this letter, or any other criteria applicable to this written approval is permitted unless prior approval by the Department is received.

Yours truly,

Edmond G. Benoit
Deputy Regional Environmental Engineer
Air Quality and Hazardous Waste

EGB/sr
cc: William Cass
Dree Chalpin, Donovan, Joyce
Robert Jaques—Shrewsbury Board of Health
Ed Holland—Shrewsbury Town Engineer
Pat Zecco—Zecco. Inc.

CHAPTER 16

Incorporation of Contaminated Soils into Bituminous Concrete

Karl Eklund

INTRODUCTION

The by-product of many Comprehensive Environmental Response, Compensation, and Liability Act (CERCLA) and Massachusetts General Laws Chapter 21E projects is soil that is contaminated with something-or-other that shouldn't be there. That something-or-other might be completely synthetic and "unnatural" from that standpoint, or it might be unnatural because it is displaced, i.e., it has been removed from the earth in one place and moved to another place where its presence in the earth is considered inappropriate. That is the case with the many small sites that are contaminated with petroleum products.

The typical case in Massachusetts is contamination associated with underground storage of petroleum products, either from leaks in tanks, or piping, or from surface spills caused by poor management of the product during transfers. These cases very often come to notice because of the effect of Massachusetts General Laws Chapter 21E on property transfers, which makes buyers wary of buying into remediation responsibilities and lenders wary of having the security for their loans affected by the "superlien" provision. It is very common for a case of petroleum contamination to be discovered in a situation where the responsible parties are looking for a quick resolution that doesn't involve expenditures without limit in time or extent.

In many of the cases that we have come across, the environment would be best served in the most cost-effective way by some form of treatment on site such as soil-venting, perhaps in conjunction with bioremediation. However, this technology is not sufficiently well-developed to be able to be done on fixed-price contracts. This means that the responsible parties must be prepared to expend funds over an extended period; and most buyers and sellers of small properties are not comfortable with the financial mechanisms that can be set up to provide for that expenditure.

Thus, in many cases, independent of the benefit to the environment, the contaminated soil is excavated to some arbitrary limit of contamination as determined by field measurements, and the resulting hole is filled in with clean fill. This leaves a pile of contaminated soil on the site and a question as to what to do with it.

The question is a serious one both for the responsible parties and for the makers of environmental policy. In Massachusetts, as of September 1, 1987, there was no lawfully operating site that was willing to accept soil contaminated with petroleum products. Material that was disposed of had to go to one of two landfills in Maine or, if secondary contamination made that impossible, it might have to go as far as Michigan. Even then, the responsible party at the site, being the generator of the waste that was landfilled, would be responsible for any subsequent problems at the landfill and, as the Environmental Protection Agency is fond of saying, "all landfills eventually leak".

The Massachusetts Department of Environmental Quality Engineering (DEQE) has been quite aware of these problems. It has articulated a policy that the preferred channel for disposal of oily contaminated soil is in reuse as aggregate in bituminous concrete. As a contract employee of DEQE, I was a member of the committee that drafted that policy, and the committee was well aware not only of the problems inherent in landfilling, but of the political dangers (and embarrassment) of having to go out-of-state to find a lawful solution that might well not have been lawful in the Commonwealth.

After returning to the private sector I participated in founding the American Reclamation Corporation (AmRec) with the intent of recycling certain waste materials that would be solid (rather than hazardous) wastes if discarded. We have concentrated on "recycled asphalt products" or "RAP," and have a facility that is producing bituminous concrete for commercial application using RAP. As we did some of the engineering research and development for this process, it occurred to us that our techniques held some definite advantages for the recycling of oily contaminated soil over the ordinary asphalt batching plant. We have, therefore, filed an application for a permit to recycle oily contaminated soil at a facility in Massachusetts. We have conferred with DEQE on the permit and we have been given to understand that the permit is in the final stages.

BITUMINOUS CONCRETE PROCESSES

The essence of a "concrete" is viscosity or thixotropicity. At the time of application it has to be quasiliquid, like a slurry, or at least a material with a pasty plasticity that can be deformed by mechanical operations into the shape in which it is to be used. It has to be able to adhere to a surface if it is a coating, and it has to set up into a degree of hardness that is suitable for the kind and class of operation.

Unlike those concretes which are used as structural elements, bituminous concrete is generally applied as a layer on a nearly horizontal surface or as a low structure such as a curb or "speed bump." As a component in the governmental infrastructure, bituminous concrete is subject to very stringent specifications down to the particle size distribution of the aggregate, but for private and commercial use quite satisfactory performance can be achieved with a considerable variation in formulations. Bituminous concretes containing additives like rubber can be made for special uses like tennis courts and running tracks.

The use of bitumen as an adhesive has a very long history. Some of the most primitive artifacts, like spears and arrows, used naturally occurring bitumen and natural fibers to attach the stone point to the wooden shaft. It has also been used as a preservative by means of encapsulation—bitumen is found in the formulary of Egyptian mortuaries of the classic period.

Today the form of bitumen that is commonly used is asphalt, and it is produced as a coproduct with gasoline and oil in the refining of crude oils. As shown in Figure 1, asphalt is the heavy end of the selective distillation of petroleum. The essential point to remember from this diagram is that asphalt and the other petroleum products were separated from the same matrix, the mixture called crude oil. In principle, crude oil could be reconstituted by mixing asphalt, oil, kerosene, gasoline, and naphtha in the proper proportions. In practice, asphalt and the other petroleum products are miscible, if not quite in any proportion, at least over very wide ranges of proportions. If you mix asphalt and kerosene you may get a dirty-brown kerosene, or a gummy asphalt, or an intermediate oily substance; but within wide variations the mixture does not divide into phases. A small proportion of petroleum product mixed with asphalt merely produces asphalt of a slightly different specification or characterization. This is the scientific basis for our method of encapsulating oily contaminated soil in asphalt, and using that as a component in bituminous paving.

Turning asphalt into concrete involves producing a material that is plastic when it is applied and hard when it sets up, and there are two conventional ways of doing that. They are, for appropriate reasons, called the "hot mix" and "cold mix" processes.

In the hot mix process, the asphalt is liquified by heating. As in most things, the viscosity of asphalt decreases with increasing temperature and it is a liquid at a temperature sufficiently below its flash point to be safe to use. The melted asphalt is mixed with aggregate and kept hot during the mixing stage. It is then transported, still hot, to the workplace, or it is stored in heated silos until it can be transported.

Obviously this hot liquid asphalt cannot be mixed with cold, wet aggregate.

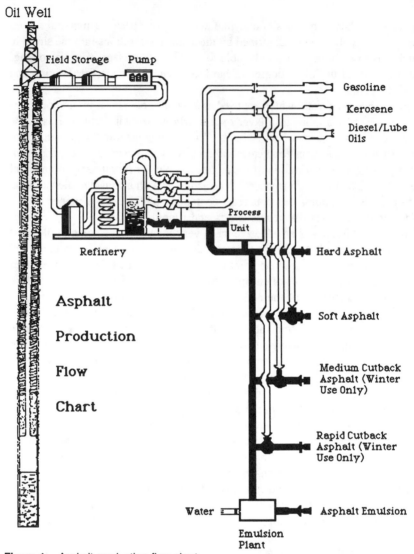

Figure 1. Asphalt production flow chart.

If the aggregate is cold enough it will immediately cause the asphalt to cool, producing a lump of bituminous concrete where a plastic mass is desired. If the aggregate is wet there will be explosions of steam when it mixes with the hot asphalt. Hot mix plants, therefore, incorporate a rotary kiln to dry the aggregate and heat it to a few hundred degrees F. It then mixes with the asphalt without difficulty.

When the aggregate is soil contaminated with petroleum products the kiln drying process introduces a number of problems, or at least constraints. The conventional asphalt plant drying kiln is heated with an open flame directed into the rotating barrel from the outlet end. Obviously the first constraint must be that the contamination level is sufficiently low so the open flame does not ignite the petroleum or cause the vapors to explode. In Massachusetts this has been done by limiting the hot plants to an admixture of 5–10% of contaminated soil in their aggregate.

As the aggregate progresses down the rotary kiln it gradually heats up to the 500–800°F that it reaches at the outlet of the kiln. This produces a gradual distillation of the contamination, so that the light ends are driven off near the entry to the kiln, where there may be no open flame. They are then typically drawn off by exhaust fans, and pass through the air pollution control system (usually a cloth filter or "baghouse"), and are exhausted into the atmosphere. Since these unburned hydrocarbon fumes are not collected by the baghouse, they can contribute significantly to the air pollution produced by the hot mix plant.

As the contaminated soil proceeds down the kiln, the heavier components are distilled off. If these are not ignited by the open flame heater, they will be drawn off by the exhaust fans and cooled in the plenum to the baghouse. There they can combine with the airborne particulates ("fines") and produce a kind of bituminous concrete that adheres to the baghouse filters, making them difficult to keep clean.

Those heavy components that remain with the aggregate are no problem because they combine with the liquid asphalt in the mixing part of the operation.

The operation of the hot mix system is shown schematically in Figure 2, where the wavy arrows indicate the parts of the system that produce hydrocarbon vapors as an air pollutant.

In contrast, the cold asphalt process is shown in Figure 3.

ASPHALT EMULSION

The advantages of the cold process are so obvious that it would appear that the deck is stacked. The disadvantages become more evident when one asks how you can mix aggregate with an asphalt that is sufficiently hard that it will set up into a usable paving material. The secret is the use of water as a lubricant.

When the concrete is mixed, the asphalt is in the form of an emulsion in which the particles of asphalt are kept suspended in the liquid and separated from each other and the aggregate by a film of water. Under pressure, the film of water is expelled and the asphalt comes into contact with itself and the aggregate. In the process it cements the aggregate into a hard concrete that is essentially identical to the hot mix bituminous concrete.

This is shown schematically in Figure 4.

The "secret" to a workable cold process bituminous concrete is in the emulsion. At AmRec (and, before AmRec was founded, at Environmental Restoration Engineering and Ashland Industrial Fuels, Inc.) we have been experimenting with various formulations of asphalt emulsion for recycling asphalt products for

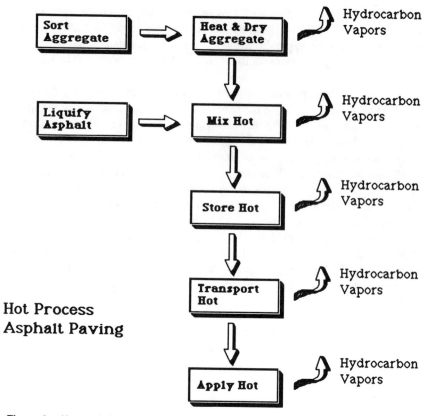

Figure 2. Hot asphalt process.

several years. Some of these emulsions incorporated oils as well as the usual asphalt, water, and emulsifying agents because the oil served to soften the recycled asphalt and make for better adhesion to other additives and fillers. We are able to use this backlog of workable formulations to adjust a particular batch of cold process concrete to almost any degree of soil contamination found in practice. If necessary, we can combine aggregates from different sites to bring the composite level of contamination into the range where we can incorporate it into one of our proprietary mixtures.

This process works for soils contaminated with oils (i.e., the range that includes No. 2 to No. 6 fuel oils and most lubricating oils). Because the admixture of lighter petroleum products (such as kerosene and gasoline) with asphalt generates hydrocarbon vapors when applied hot, these asphalt mixtures (called "cutback" asphalt by the trade) are only allowed for use as winter-service coldpatch. Until we can get an exception to the Massachusetts air quality regulations allowing the use of cutback asphalts in cold process paving, we will restrict the processing of soils contaminated with gasoline to paving that will be applied as winter coldpatch.

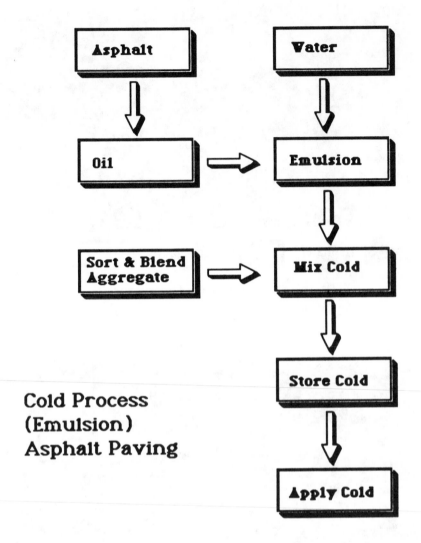

Figure 3. Cold asphalt process.

ENVIRONMENTAL EFFECTS

We are still in the process of conducting tests on the environmental effects of bituminous paving, but we have some preliminary results that are interesting. We leached our paving, commercial coldpatch, commercial hotpatch, and old bituminous concrete in distilled water under controlled conditions. After a week the amount of oil and grease in the leachate was near the detection limit for the method, and the largest amount detected was for the commercial hotpatch, at about 2 ppm.

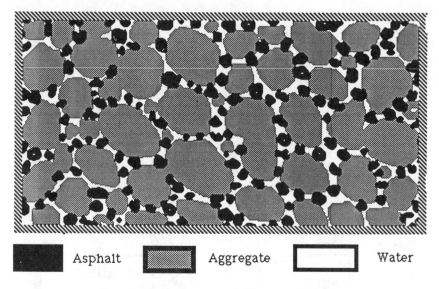

	Asphalt		Aggregate		Water

Asphalt Emulsion Concrete

Above: as mixed

Below: after compression

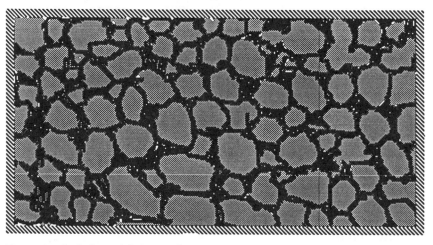

Figure 4. Asphalt emulsion concrete.

We also formulated some experimental paving using up to 6% waste oil in the emulsion. Although waste oil has an appreciable lead content, when leached in an acid solution simulating "acid rain" for a week, the leachate contained less than 3 ppm of lead.

All of these tests confirm that the petroleum contamination in the soil is combined with the asphalt in the emulsion to produce a mixture that will not separate and release the oil back into the environment.

We are presently trying to determine what happens to the water that is released during the compression of the mixture into concrete, but the amount of water in the concrete mix is so small that we have not been able to capture it under working conditions. It can be expected, however, that if any escapes in liquid form it will be simply water and emulsifying agent. This is no more dangerous to the environment than the residue from dishwashing, which is also water and emulsifying agents.

CONCLUSION

The conclusion is that the incorporation of soils contaminated with petroleum products as aggregate in a cold mix emulsion-system bituminous paving is an environmentally benign method of recycling the contaminated soil. It is greatly superior to landfilling untreated contaminated soil, and has significant advantages over incorporation into hot mix bituminous concrete. While it has disadvantages compared to in situ treatment because of the added transport and management required, it can be recommended where in situ treatment is impractical.

ACKNOWLEDGMENTS

This work could not have taken place without the guidance of Nathan Wiseblood, P.E., who is the walking encyclopedia of asphalt, and without the enthusiastic support of John Glynn, Jr. Some of the research was done by Fred Hooper, staff chemist of Ashland Industrial Fuel, Inc. My own involvement with the asphalt encapsulation technique began at DEQE.

BIBLIOGRAPHY

1. Asphalt Technology and Construction Practices, Manual ES-1, 2nd Ed., The Asphalt Institute, College Park, MD, 1983.
2. A Basic Asphalt Emulsion Manual, Manual 19, 2nd Ed., The Asphalt Institute, College Park, MD, 1986.

CHAPTER 17

Low Temperature Stripping
of Volatile Compounds

Luis A. Velazquez and John W. Noland

Contamination of soils from past operations involving volatile organic compounds (VOCs), such as from leaking underground fuel storage tanks, has become a major environmental concern. This contamination, if allowed to remain
in the soil, can migrate to contaminate underlying groundwater. Roy F. Weston,
Inc. (WESTON) has been performing studies on the removal of these contaminants. Low temperature thermal stripping of the VOCs from the soil was
found to be an efficient method for removal of these contaminants. Further work
in the field of low temperature thermal treatment (LT³) has resulted in WESTON's
design of an innovative low temperature process, LT³.

TECHNOLOGY DESCRIPTION

WESTON's LT³ process is a demonstrated technology that provides evaporation of the VOCs, but does not require heating the soil matrix to combustion temperatures. The heart of the technology is the thermal processor, an indirect heat
exchanger (hot screws), that is used to dry and heat the contaminated soils up
to 450°F, consequently stripping the VOCs from the soil. The indirect heating
of the soil minimizes gas flows and reduces the size of the offgas handling equipment. Once the organic contaminants are vaporized, they can either be destroyed
through high temperature incineration in an afterburner, or recovered through

condensation or adsorption on activated carbon. WESTON has a U.S. patent pending for this technology.

Some of the advantages of the LT³ process are:

- lower capital and operating costs than incineration
- process is considerably smaller and more mobile than an incinerator with the same throughput
- process can be used for product recovery, which is not possible for incineration
- process results in a processed soil that is suitable for onsite backfill. This eliminates the long-term liability of landfill disposal or onsite containment or storage.
- low temperature minimizes the chance of hazardous heavy metal emissions
- process can be designed to be compact and fully mobile (i.e., fully mounted on three flatbed trailers.)

LT³ PROCESS DEVELOPMENT

The LT³ process was developed by WESTON under its current contract with the U.S. Army Toxic and Hazardous Materials Agency (USATHAMA). Bench scale studies of the low temperature thermal stripping concept identified the method as a feasible treatment process. A pilot scale field demonstration of the process using VOC-contaminated soils became the next phase in the development of the technology.[1] In May 1985, USATHAMA contracted WESTON to perform this phase of the work. A site with soil having various VOC contamination (see Table 1) was selected for the study.

Table 1.   Pilot Study, Feed Soil VOCs Concentrations.

VOC	Average (ppm)	Maximum (ppm)
Dichloroethylene	83	470
Trichloroethylene	1,673	19,000
Tetrachloroethylene	429	2,500
Xylene[a]	64	380
Other VOCs	14	88
Total VOCs	2,263	22,438

[a]Xylene is not classified as a VOC since its boiling point is approximately 140°C. However, it was included in the study to evaluate the effectiveness of this technology on higher boiling point semivolatile compounds.

WESTON's responsibilities for the project included:

- design of the pilot system

- preparation of test and safety plans
- environmental permitting
- equipment selection
- equipment installation and start-up
- performance of the test program
- demobilization and site closure
- preparation of a technical report

Results of this project indicated the following:

- A comparison of the VOCs measured in the feed soil to the VOCs measured in the processed soil and stack gas yielded the following destruction and removal efficiencies:
  —greater than 99.99% removal of VOCs was evidenced in the soils
  —no VOCs were detected in the stack gas, indicating a destruction and removal efficiency (DRE) of 100% for the overall system
- Stack emissions were in compliance with all federal and state regulations (including VOCs, HCI, CO, and particulates).

The demonstrated process represents a significant breakthrough in the treatment of soils contaminated with VOCs. The process represents a unique mix of proven techniques combined in an innovative way to provide an efficient and cost-effective method of treatment. Based on the unqualified success of this demonstration and on the increasing demand for mobile thermal treatment, WESTON, through its affiliate, Weston Services, Inc. (WSI), has begun to provide mobile LT³ systems. The first of these systems was scheduled to be commissioned in October 1987.

MOBILE LT³ SYSTEM

The mobile LT³ system is designed to handle 15,000 lb per hr of contaminated soil based on 20% soil moisture and 1% (i.e., 10,000 ppm) VOCs. The system is comprised of equipment assembled on three flatbed trailers. The feed system trailer is 35 ft long. Each of the other two trailers is 42 ft long. All three trailers are 8 ft wide and 55 in. from the ground to the top of the bed. The total height of the trailers, with the equipment assembled, is under 13.5 ft. Weight and weight distribution is designed so as not to require special road permits. Interconnecting ducts, pipes, conveyors, power and control wires, and equipment that would extend beyond the allowable limits for trailers, such as the afterburner stack, are disconnected and shipped on the trailers during transportation. The process schematic of the WSI mobile LT³ system is shown in Figure 1. System layout, along with the equipment supplied on each of the trailers, is shown in Figure 2.

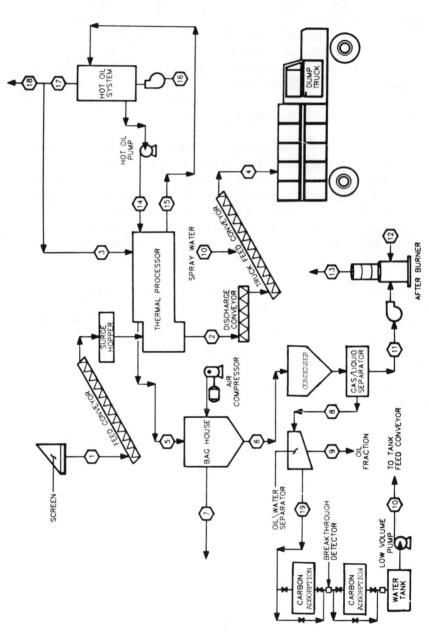

Figure 1. Process schematic mobile low temperature thermal treatment.

Equipment

Trailer No. 1

1.A   Simplicity Screen w/Hopper & Feed Screw
1.B   Drag Conveyor on Truck for Mobilization
1.C   Oil Storage Tank
1.D   Tool Box
1.E   Discharge Screw on Truck for Mobilization
1.F   Truck Feed Conveyor on Truck for Mobilization
ALSO   Interconnecting Ducts, Hoses, Wires, and Pipes Secured on This Trailer for Mobilization

Trailer No. 2

2.A   Surge Hopper
2.B   Holo-Flytes
2.C   Baghouse
2.D   Air Compressor
2.E   MCC
2.F   Control Panel

Trailer No. 3

3.A   Hot Oil System
3.B   Condenser
3.C   Gas/Liquid Separator
3.D   Oil/Water Separator
3.E   Afterburner
3.F   I.D. Fan
3.G   Carbon Adsorption Units
3.H   Breakthrough Detectors
3.I   Recycle Water Tank & Pump

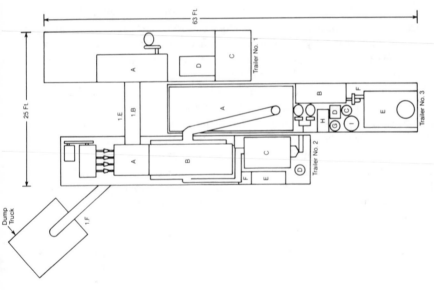

Figure 2.   Mobile low temperature thermal treatment.

MOBILE LT³ PROCESS DESCRIPTION

Excavated soil is fed onto the 4 ft×10 ft vibrating screen permanently assembled on Trailer No. 1. The soil is screened to a maximum soil topsize of 1 in. to protect the mechanical downstream equipment. This screened soil is conveyed from the hopper, directly under the screen, to a drag flight conveyor. The drag flight conveys the soil up to a surge hopper that completely covers the feed inlet of the top thermal heating unit. This creates a "live bottom" hopper effect to prevent bridging of the soil in the surge hopper. The hopper has a capacity of 100 ft³ (approximately 45 min of operating capacity). It is designed with the sides hinged at the bottom so they can be folded onto each other for system shipment.

As previously stated, the heart of the LT³ system is the thermal processor. The thermal processor consists of two indirect heating thermal units (hot screws). These units are arranged in a "piggyback" or series configuration. Each unit is equipped with four screws that are 20 ft long and 18 in. in diameter. Hot oil is circulated through the shafts, flights, and trough of the screw conveyors to indirectly heat the soil from the surge hopper to a maximum of 450°F measured at the outlet of the second unit. Water is sprayed on the treated soil at the truck feed conveyor for cooling and dust control. This moistened treated soil is fed into a dump truck for transport and backfill. The thermal processor is operated under negative pressure to prevent fugitive emissions.

A 6 million BTU/hr hot oil heater is used to provide the required heat for the process. The oil is heated to 640°F and then circulated through the thermal processor. The oil is heated by the combustion offgases from the burner of the heater. A portion of these offgases are directed through the enclosures above the screws to provide an inert atmosphere in the units (i.e., to avoid reaching the lower explosive limit [LEL] of contaminants in air) and to maintain a discharge offgas temperature of 280°F to avoid the formation of condensate in the units.

Processor offgases consisting of water vapors, VOCs, and the used portion of the offgases from the oil heater are passed through a fabric filter (or baghouse) to reduce particulate emissions to less than 0.08 grains per standard dry ft³ corrected to 12% CO_2. Particulate removed in the baghouse is returned to the surge hopper for treatment. Gases from the baghouse enter a two-stage condenser which knocks out, then cools, the water to a minimum of 140°F.

The gas fraction from the condenser is drawn by the Induced Draft (I.D.) fan and directed into an afterburner. The I.D. fan provides the vacuum required to maintain a negative pressure at the processor. A gas/liquid separator is provided between the gas outlet of the condenser and the I.D. fan in the event there is too much free moisture carryout. The afterburner thermally treats the gases by providing a minimum residence time of 2 sec at 1800°F. The heat of combustion of the VOCs provides some of the heat required for this thermal treatment. Auxiliary fuel (propane) is also burned in the afterburner to maintain a minimum of 1800°F in the afterburner during operation. This is a necessary safety consideration, since the VOC concentration in the feed soil is expected to vary.

The cooled liquid fraction from the condenser is pumped to an oil/water separator to remove any oil present. The oil fraction is collected for offsite treatment and disposal. The water fraction is directed through a two-stage carbon adsorption system which removes any other contaminants in the water. A monitor (i.e., total hydrocarbon analyzer) continuously samples the water from the first carbon adsorption unit to detect breakthrough. In the event of a breakthrough, the units are reversed and the spent unit is replaced. The treated water is collected and sprayed onto the treated soils at the truck feed conveyor for cooling and dust control.

The net result is moist, VOC-decontaminated soil suitable for onsite backfill. This eliminates the long-term liability of landfill disposal or onsite containment or storage.

OPERATIONAL REQUIREMENTS

The following site pad and utilities are required:

- The three trailers require a 26 ft×63 ft area.
- electrical: 460 V/3 phase/60 Hz/300 amps
- propane: 430 lb/hr
- water: not required
- One senior onsite manager and one technician/operator are required per shift.

LT³ PROCESS APPLICATIONS IN THE UNDERGROUND STORAGE TANK (UST) MARKETPLACE

There are over two million underground storage tanks in the United States now used to store gasoline (see Figure 3). Literally billions of gallons of engine fuels are stored in these tanks by farms, retail gasoline stations, fleet users, and the military. Refineries and airports generally use above-ground tanks for fuel storage. Some experts estimate that between 75,000 and 100,000 underground storage tanks are now leaking.

On a national basis, a statistical analysis of the underground storage tank population indicated that roughly 5% of the tanks are failing. On a regional basis, the number can be considerably higher. The Maine Department of Environmental Protection estimates that 25% of the state's 10,000 retail gasoline storage tanks have failed. The state of New York estimates that 19% of its 83,000 underground gasoline tanks are leaking.

The environmental disasters associated with leaking underground storage tanks are not limited to contamination resulting from the accidental spillage of extremely toxic materials. A substantial portion of the problem results from gradual leakage during routine storage of motor fuels. For instance:

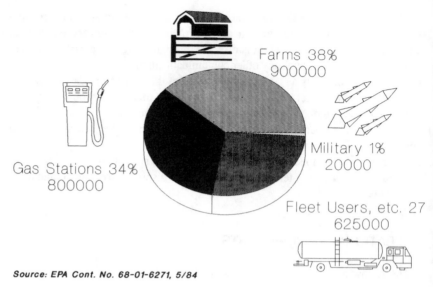

Source: EPA Cont. No. 68-01-6271, 5/84

Figure 3. Underground fuel tanks in the U.S.

- Underground storage tanks owned by one petroleum company were found in 1978 to be leaking gasoline at a Long Island service station in East Meadow, New York. An estimated 30,000 gal of gasoline leaked into groundwater supplies, causing odor and safety problems for 27 families living in 25 homes. This company subsequently bought 23 of the homes at 150% of their market value, and settled with the two remaining home owners. With some suits still pending, total costs to date are estimated to be between $5,000,000 and $10,000,000.
- In 1980, a gasoline leak was traced to a service station in the Northglenn suburb of Denver, Colorado. A federal court convicted the petroleum company in June 1981, forcing it to purchase 41 of the homes in the affluent neighborhood at 2.2 times their appraised value. Estimated losses to the petroleum company are approximately $10,000,000.

These are but two examples of environmental disasters where the mobile LT^3 system can be used to remediate contaminated soil problem.

PROCESS DEVELOPMENT AREAS

Research is continuing to determine variables that affect the low temperature stripping of the organics from various soils. Preliminary studies indicate low temperature thermal treatment may be applicable for the removal of relatively high boiling point organics (i.e., up to 600°F). DRE and economic evaluations are being performed using stripping agents (e.g., water, solvent) used not only to enhance the low temperature volatilization of the organic low temperature boilers but also of the high temperature boilers, to determine other applications for the LT³, and to establish the most efficient, cost-effective methods of operation.

Another research area where the application of the LT³ is being used is in the recovery of the contaminant. The vaporized contaminants, stripped from the soil from the LT³ process, are being processed and condensed for recovery.

SUMMARY

Low temperature thermal stripping has proved to be an effective method of treatment for the removal of VOCs from soils. The LT³ is an innovative technology based on the low temperature thermal stripping principle which is efficient, practical, cost-effective, and versatile.

REFERENCE

1. "Task 11. Pilot Investigation of Low Temperature Thermal Stripping of Volatile Organic Compounds (VOCs) from Soil," Report No. AMXTH-TE-CR-86074, Roy F. Weston, Inc., June 1983.

CHAPTER **18**

Recovery Techniques and Treatment Technologies for Petroleum and Petroleum Products in Soil and Groundwater

James Dragun

INTRODUCTION

Many industrial and commercial operations and activities depend on the use of petroleum and its products. Indeed, petroleum and its products are the backbone of American industry. They are the main chemical component which industry and commerce utilizes to produce the many household appliances and consumer items that the typical American enjoys, yet takes for granted.

Petroleum and its products enter into soil and groundwater primarily through a few operations and through accidents. These include railcar and trucking accidents, pipeline ruptures, failure of underground storage tanks, and road oiling for dust suppression.

The presence of petroleum and its products in soil may or may not pose a risk, depending on the extent of exposure to humans and biota. In cases where unacceptable levels of exposure to soil containing petroleum or its products exist, remediation of the problem is necessary.

At the present time, there are several physical, chemical, and biological methods that can remediate the problem of petroleum and its products in soil and groundwater. These can be divided into two general groups: recovery techniques and treatment technologies. Recovery techniques retrieve petroleum or its products from the soil without significantly disturbing or removing the soil. Treatment technologies degrade or immobilize petroleum or its products, and usually require the removal and processing of soil.

Some of these methods are well-established; others are in the developmental stage. Because many field conditions, as well as the volume and properties of petroleum and its products, affect the migration and recovery of bulk petroleum and its products in soil, no single technique or approach works under all situations. Also, one technique can be feasible and cost-effective in one soil and be very difficult to implement in another.

The first section of this chapter provides a general overview of the forms of petroleum and its products that can be found in soil. The second section deals with the recovery techniques and treatment technologies that can partially, and in some cases totally, remove petroleum and its products from soil. The feasibility, design, costs, application, and limitations of these methods are discussed elsewhere in this book in much greater detail.

FORMS OF PETROLEUM AND ITS PRODUCTS IN SOIL AND GROUNDWATER

Similar to water, petroleum and petroleum products which are released in bulk quantities at the soil surface can penetrate through the soil surface. If the released quantity is large, downward migration occurs with all soil pores being saturated with petroleum or petroleum product.

Petroleum possesses a specific gravity or bulk density that is less than that of water; as a result, it will behave in a different manner than petroleum products with specific gravities greater than that of water. Downward migration of petroleum will eventually cease because: (a) the mobile petroleum is transformed into residual saturation; (b) it will encounter an impermeable bed; or (c) it will reach, but not penetrate through, the water table.[1]

As a mass of petroleum migrates beyond a unit mass of soil, a small amount of the total petroleum mass will remain attached to these soil particles via capillary forces. This small mass of petroleum that is retained by soil particles is immobile and is known as *residual petroleum* or *residual saturation*.[2] The maximum amount of petroleum that can be retained by a soil is known as the *residual saturation capacity*. If the migrating mass of petroleum is small relative to the soil surface area, the migrating mass of petroleum will be eventually exhausted as it is converted into residual saturation. When conversion is complete, downward migration ceases.[1]

If the migrating petroleum mass reaches the water table, it will begin spreading laterally over the water table. This layer of petroleum assumes the shape of a "pancake" and is commonly known as the *pancake* layer or as *free-floating* petroleum.

Migrating masses of bulk quantities of a petroleum product with specific gravities greater than that of water are sometimes known as *dense hydrocarbons* or *dense product.* They will eventually cease downward migration for two reasons. First, the migrating mass of product may be transformed into residual saturation. Second, it will encounter an impermeable bed. It is most important to note that these petroleum products react differently when they reach the water table; they will continue to migrate downward after they encounter the water table. This phenomenon is sometimes referred to as "density flow."[1]

Petroleum and petroleum products are generally not miscible with water. However, they are comprised of numerous individual chemicals that are soluble to some extent in water. When percolating soil water or groundwater contact residual saturation, a pancake layer, or a migrating mass of a bulk quantity of petroleum or its products, some of the chemicals are released from the petroleum or product and are now dissolved in soil water or groundwater.[1] These are referred to in this text as *dissolved petroleum* and *dissolved product.*

Petroleum and petroleum products are comprised of numerous individual chemicals that possess relatively high vapor pressures. These chemicals prefer to exist in the vapor state and not in the solid or liquid state. Because all soils contain pores, and because soils in the unsaturated zone are not always filled with water, these chemicals are released from residual saturation or from the pancake layer into the air-filled pores. These chemicals are referred to in this chapter as *vapor phase* petroleum and products.

It is most important to recognize that the presence in soil of one or more of the organic chemicals which comprise petroleum and its products is not necessarily a result of a spill or leak. Many chemicals which comprise petroleum and its products are naturally-occurring and can be found in soil at high parts per billion (ppb) and low parts per million (ppm) concentrations.[1] These include alkanes, alkanoic acids, anthracene, benzene, benzofluorene, benzoic acid, carbazole, ethylbenzene, methanethiol, methanol, naphthalene, perylene, phenanthrene, toluene, and xylene.[1]

RECOVERY TECHNIQUES FOR RESIDUAL SATURATION IN SOIL

There are two in situ techniques and one technique using excavated soil that can be utilized to partially or totally recover residual petroleum and its products from soil. *Soil venting* is an in situ technique that enhances the migration of the volatile constituents of petroleum and its products through soil air by artificially inducing differential pressures.[3] The volatile constituents migrate to an extraction well, where they are removed from the airstream by cold traps and/or carbon adsorption.

In situ flushing is a relatively new technique in which water percolating through unsaturated-zone soil carries petroleum and its products downward to the groundwater. The groundwater is collected and treated. Surfactants can be added to the percolating water in order to enhance its ability to dissolve petroleum and its products.[4]

Low temperature stripping utilizes a thermal processor to dry and heat soils up to 450°C; as a result, the volatile constituents of petroleum and its products are stripped from the soil.[5] Then, they can either be destroyed through high temperature incineration in an afterburner, or recovered through condensation or adsorption on activated carbon. In addition to recovering residual saturation, this method can be utilized to recover free-floating and dense petroleum and its products from excavated soil.

TREATMENT TECHNOLOGIES FOR SOIL WITH RESIDUAL SATURATION, FREE-FLOATING PETROLEUM AND ITS PRODUCTS, AND DENSE PETROLEUM PRODUCTS

There are several treatment technologies that can be utilized for soil with residual saturation, free-floating petroleum and its products, and dense petroleum products. All but two require the excavation, handling, and processing of soil containing petroleum or its products.

Natural degradation of the organic chemicals comprising petroleum and its products occurs in soil. It occurs via chemical reactions which include hydrolysis, oxidation, and reduction; also, it occurs via microbiological reactions.[1] These reactions occur at a rate dependent upon the physical, chemical, and biological properties of the soil.

The primary objective of *land treatment* is the microbiological degradation of petroleum or its products in soil as it resides at the land surface.[1,6] It is one of the most commonly utilized treatment technologies for soil containing petroleum or its products.[7] The petroleum-soil mixture or the product-soil mixture is either (a) spread on the soil surface, or (b) injected or incorporated into the upper few inches of soil by mechanical manipulation, such as plowing and tilling. Nutrients may be added to stimulate microorganism metabolism. Also, the addition of mutant microorganisms appears to be a promising approach to enhance the degradation rate of recalcitrant organic chemicals.

Composting also utilizes microorganisms to degrade petroleum and its products which reside in a waste pile.[1] In open composting systems, the waste is either (a) stacked in a low, narrow, elongated heap known as a windrow, or (b) stacked in a low-lying, rectangular-based pile known as a static pile. In a closed system, the waste resides in an enclosed digester. Oxygen usage, temperature, and nutrients can be manipulated to control the degradation rate.

A number of treatment technologies generally bind the petroleum or petroleum products in a relatively insoluble matrix. These include *asphalt cement* and *bituminous cement*.[8,9] These technologies require the mixing of soil with a binding agent and processing of the mixture to create an insoluble matrix. The presence of chemicals other than petroleum or its products can significantly affect the ability of these technologies to bind petroleum or petroleum products.[10]

Vitrification is one in situ thermal treatment technology which converts soil into a chemically inert and stable glass and crystalline matrix; it destroys organic constituents in the soil by pyrolysis.[11]

In situ biodegradation is the second in situ treatment technology that results in the degradation of petroleum or its products in soil. It transforms the volume of contaminated soil and groundwater into an in situ treatment system by enhancing soil microorganisms' ability to degrade petroleum or its products.[1,12] Dissolved nutrients and oxygen are introduced through water injection wells, circulated through the contaminated zone, and extracted through a recovery well.

IN SITU RECOVERY TECHNIQUES FOR FREE-FLOATING PETROLEUM AND PETROLEUM PRODUCTS

Two techniques can be utilized to recover free-floating bulk petroleum and its products residing in the pancake layer.

Interceptor trenches are best utilized whenever free-floating bulk petroleum and its products exist in a soil with a water table near the land surface. This technique utilizes a trench or ditch, constructed across the leading edge of the pancake layer, to intercept the petroleum or its products.[2] The petroleum or petroleum product migrates to the trench and floats at the water surface within the trench. The petroleum is either skimmed or pumped out of the trench.

Recovery wells are utilized extensively to recover free-floating bulk petroleum and its products, especially whenever the petroleum or petroleum product lies a substantial distance below the land surface. This technique utilizes a groundwater extraction well to extract water out of this well and to lower the water table in the vicinity of the well.[2] This results in petroleum or petroleum products, which are in the pancake layer, to migrate into the groundwater extraction well. Once in the well, the petroleum or petroleum product can be pumped out.

IN SITU RECOVERY TECHNIQUES FOR DENSE PETROLEUM PRODUCTS

The depth of vertical migration of dense petroleum products is not limited by the water table. As a result, they can migrate substantial distances below the land surface.[1]

Recovery wells are the primary means of recovering dense products that cannot be excavated cost-effectively. These wells are placed in the center of the mass of the product in order to intercept and extract as much of the product as possible. Lowering the water table in the vicinity of these wells cannot enhance the recovery of the dense product because dense products do not float on the water table.

RECOVERY TECHNIQUES AND TREATMENT TECHNOLOGIES FOR DISSOLVED PETROLEUM AND ITS PRODUCTS

The most extensively applied recovery techniques involve the extraction of groundwater by pumping or by gravity flow into drains and trenches.[7,13] The dissolved petroleum or petroleum product is treated at the land surface, using more traditional unit processes such as *air-stripping* and *carbon adsorption,* or *biodegradation*-based processes such as activated sludge, aerobic lagoons, anaerobic lagoons, facultative lagoons, rotating biological filters, and trickling filters.[13]

Natural degradation of the dissolved petroleum and its products occurs in soil. It occurs via chemical reactions which include hydrolysis, oxidation, and reduction; also, it occurs via microbiological reactions.[1] These reactions occur at a rate dependent upon the physical, chemical, and biological properties of the soil.

SUMMARY

Petroleum and its products can exist in soil in several forms. These include residual saturation, free-floating petroleum, dissolved petroleum, and vapor phase petroleum.

There are two in situ techniques—soil venting and in situ flushing—and one technique using excavated soil—low temperature stripping—that can be utilized to partially or totally recover residual petroleum and its products from soil.

There are several treatment technologies that can be utilized for soil with residual saturation, free-floating petroleum and its products, and dense petroleum products. These include natural degradation, land treatment, composting, incorporation into binding agents and fixatives, vitrification, and in situ biodegradation.

In situ recovery technologies for free-floating petroleum and petroleum products include interceptor trenches and recovery wells. The primary in situ recovery technology for dense petroleum product is the recovery well.

The most extensively applied recovery techniques for dissolved petroleum and its products involve the extraction of groundwater by pumping or gravity flow into drains and trenches. Treatment usually involves traditional unit processes.

REFERENCES

1. Dragun, J. *The Soil Chemistry of Hazardous Materials* (Silver Spring, MD: Hazardous Materials Control Research Institute Press, 1988).
2. *Underground Spill Cleanup Manual.* API Publication 1628 (Washington, D.C.: American Petroleum Institute, 1980).
3. Hoag, G. "The Use of Soil Venting to Remediate Contaminated Soils," paper presented at the Second Conference on Environmental and Public Health Effects of Soils Contaminated with Petroleum Products, University of Massachusetts, Amherst, September 28–30, 1987.

4. Nash, J. H., and R. P. Traver. "Field Studies of in Situ Soil Washing," *see* Chapter 13, this volume.

5. Velazquez, L. A., and J. W. Noland, "Low Temperature Stripping of Volatile Compounds," *see* Chapter 17, this volume.

6. Lynch, J., and B. R. Genes. "Land Treatment of Hydrocarbon Contaminated Soils," *see* Chapter 14, this volume.

7. Preslo, L., M. Miller, W. Suyama, M. McLearn, P. Kostecki, and E. Fleischer. "Available Remedial Technologies for Petroleum Contaminated Soils," *see* Chapter 10, this volume.

8. Kostecki, P. T., E. J. Calabrese, and E. J. Fleischer. "Asphalt Batching of Petroleum Contaminated Soils as a Viable Remedial Option," *see* Chapter 15, this volume.

9. Eklund, K. "Incorporation of Contaminated Soils into Bituminous Concrete," *see* Chapter 16, this volume.

10. Cullinane, M. J., Jr., and R. M. Bricka. "An Evaluation of Organic Materials That Interfere with Stabilization/Solidification Processes," *see* Chapter 11, this volume.

11. Timmerman, C. L., J. L. Buelt, and V. F. FitzPatrick. "In Situ Vitrification Processing of Soils Contaminated with Hazardous Wastes," *see* Chapter 12, this volume.

12. Raymond R. L. "Biodegradation of Petroleum Products in Contaminated Soils," paper presented at the Second Conference on Environmental and Public Health Effects of Soils Contaminated with Petroleum Products, University of Massachusetts, Amherst, September 28–30, 1987.

13. Schneiter R. W., J. Dragun, and T. G. Erler. "Remedial Action," *Chem. Eng.* 91:73–78 (1984).

PART IV

Risk Assessment/Risk Management

CHAPTER 19

Risk Assessment for Soils Contaminated with Petroleum Products: An Overview

Barbara D. Beck

The Second Conference on Environmental and Public Health Effects of Soils Contaminated with Petroleum Products, held at the University of Massachusetts in September 1987, represents the third major conference on soil contamination in New England in the past two years, and the second meeting focused on petroleum contamination in particular. It is clear from the growing number of attendees at each of these meetings that soil contamination represents an environmental exposure of increasing public health concern. The conference identified some of the critical issues in assessing and managing health risks from soil contaminants today.

One of the major complexities in evaluating health risks from soil is the identification and quantification of the important exposure routes. Such an evaluation includes an analysis of two very broad categories of information: site usage, and fate and transport of the chemicals in the environment.

Site usage is important in defining the exposed population. For contaminated soils in residential areas, children would generally represent the primary population of concern. This is particularly true with nonvolatile contaminants such as polycyclic aromatic hydrocarbons or certain metals which are relatively immobile. Concern for children as the exposed population is a reflection of their relatively high rate of soil ingestion, particularly in the 2- to 6-year-old range.[1]

In contrast, for industrial sites, workers may represent the most exposed population and the relevant exposure routes would be via dermal contact and inhalation of volatilized contaminants and of windblown dust. Ingestion is assumed to be fairly low for the worker population, although there is much uncertainty regarding adult soil ingestion rates. Inhalation of windblown dust is not likely to be a very important route of exposure under typical conditions. This is a consequence of the small contribution of soil particles to respirable (<3 μm mass median aerodynamic diameter) particulates.[2]

Evaluation of exposure to chemical contaminants in soil by ingestion requires estimation of soil ingestion rates by children and adults. Recent studies by Binder and coworkers[3] and by Clausing and coworkers,[4] using fecal analysis of soil-specific tracers, indicate average soil ingestion rates on the order of 100 mg/day. These studies are based on several assumptions, such as the uniqueness of these tracers for soil and nonabsorption of the tracers in the gut. Uncertainties about these parameters could bias the results either upward or downward. The study presented by Dr. Calabrese in Chapter 25 represents a refinement of the above studies by providing an analysis of nonsoil sources of the tracers and absorption rates for the tracers in adults.

The studies by Binder et al. and Clausing et al. do not provide information on the role of pica among children in determining soil ingestion rates. Accurate estimates of the percent of children with pica, the persistence of pica in individual children over time, the role of socioeconomic factors on determining pica, and the quantities of soil ingested by pica children are limited. This information would be important in estimating the full range of soil ingestion rates among children.

It should also be noted that estimates of adult soil ingestion rates are also limited and usually based upon comparison to rates in children, using the assumption that adult rates must be less than those of children. There are no data to actually verify this assumption. The paper of Clausing et al.[4] provides an estimate of average tracer levels of aluminum and titanium in adult feces. Using their methodology and assuming 25 g dry weight of feces per day, adult ingestion rates of 220 and 300 mg soil per day are calculated. These values are quite implausible. It is likely that nonsoil sources of aluminum and titanium contributed to the fecal levels. Still, the calculation is useful in that it serves to demonstrate difficulties with the tracer method as well as the lack of quantitative information on adult ingestion rates.

Information on soil ingestion rates often are not all that is needed to estimate the body burden of a chemical from soil over time. Estimates of bioavailability of contaminants are frequently required. For example, the effect of soil composition on determining bioavailability has been evaluated for tetrachlorodibenzodioxin (TCDD). Bioavailability of TCDD can vary from approximately 0.5% to nearly 50%,[5] depending upon the soil matrix. In Chapter 24, Dr. Abdel-Rahman shows that similar soil matrix effects exist for benzene as well. He presents data to demonstrate that soil characteristics can influence not only the total dose of benzene, but also plasma-time curves and excretion patterns. Matrix effects were observed for both dermal and oral routes. Development of short-term bioassays or chemical tests, such as leaching studies, would be very helpful in defining the body burden over time from chemical contaminants in soil.

While the above discussion has focused primarily on direct contact with contaminated soils, offsite migration and exposure via contaminated groundwater or from ingestion of fish contaminated by bioconcentration of chemicals in sediments represents another route of exposure. In the home, exposure could occur in several ways, including ingestion of contaminated water and inhalation of volatiles released from water during bathing, showering, and other domestic water use. Inhalation can, depending upon the physical characteristics of the individual chemical, be a more important route than direct ingestion.[6] The physical and chemical properties of the contaminant, soil, chemistry, and geological characteristics of a site are also important determinants of the relevant routes of exposure. In Chapter 22, Dr. Paustenbach evaluates the relative importance of the different exposure routes, using an approach originally developed for TCDD contaminated soil.

Given the importance of site characteristics in determining exposed population, routes of exposure, bioavailability of contaminants, and other important variables, it is clear that there is a need for site-specific approaches in designing cleanup procedures aimed at reducing health risks. In Chapter 21, Dr. Kostecki describes seven approaches that have been developed by state and federal agencies for assessing risks of soil contaminants. The effect of different variables, such as environmental fate or pharmacokinetic factors, are evaluated. It appears unlikely that a single health-based level for an acceptable concentration of a contaminant could be developed that would be applicable to all sites. Mr. Ibbotson discusses procedures to derive cleanup guidelines for trace organic compounds at two petroleum contaminated sites based on acceptable daily intakes or on cancer risk assessment in Chapter 26. While similar methods could be applied to other sites, Mr. Ibbotson shows that it is inappropriate to directly apply the cleanup levels to other sites.

REFERENCES

1. Calabrese, E. J., P. T. Kostecki, and C. E. Gilbert. "How Much Soil Do Children Eat? An Emerging Consideration for Environmental Health Risk Assessment," *Comments on Toxicol.* 1:229–241 (1987).
2. Dzubay, T. G. "Chemical Element Balance Method Applied to Dichotomous Sampler Data," *Ann. N. Y. Acad. Sci.* 338:126–144 (1980).
3. Binder, S., D. Sokal, and D. Maughan. "Estimating Soil Ingestion—The Use of Tracer Elements in Estimating the Amount of Soil Ingested by Young Children," *Arch. Environ. Health* 41:341–345 (1986).
4. Clausing, P., B. Brunekreef, and J. H. van Wijnen. "A Method for Estimating Soil Ingestion by Children," *Int. Arch. Occup. Environ. Health* 58:73–82 (1987).

5. Beck, B. D. "Overview: Assessing Health Risks From Contaminated Soils," *Comments on Toxicol.* 1:171–175 (1987).
6. Murphy, B. L. "Total Exposure From Contaminated Tap Water," Paper 87-98-2, 80th Annual Meeting of the Air Pollution Control Association, New York, NY, June 21–26, 1987.

CHAPTER 20

A Methodology for Evaluating the Environmental and Public Health Risks of Contaminated Soil

Dennis J. Paustenbach

INTRODUCTION

Millions of yards of soil throughout the United States have been contaminated with numerous types of liquid and solid wastes containing metals, radioactive materials, gasoline, solvents, used oils, and sludges. The potential environmental hazards due to contamination of runoff water and pollution of groundwater, as well as the possible adverse effects on biota, fish, and wildlife, have been studied for a number of years.[1,2] The potential hazards posed by contaminated soil have received less study.[3-12] In light of the thousands of hazardous waste sites which have been identified in the United States alone and the presence of contaminated soil in residential areas, a scientifically valid and generalized approach to evaluating the human and environmental health risk is needed.[10,13]

Although the health hazards of living close to a hazardous waste site have been a concern during much of the past 10 years, in the main, epidemiology studies have not shown that these sites produce adverse health effects in humans who live near them.[5,12,151] Certainly, under some circumstances, direct contact with soil contaminated with sufficiently high levels of particular toxicants could present a hazard to humans and wildlife, but such situations appear to be uncommon. The primary hazard or nuisance is the leaching of toxicants from the soil into ground-water and the loss of odorous, gaseous, fugitive emissions.

Exposure to contaminated soil can occur through skin contact, inhalation, or ingestion. Although an important route of exposure, inhalation of contaminated dust, by itself, will rarely represent a significant health hazard.[10] Humans can be exposed as a result of their occupation, recreational activities, or through exposure in and around the home.[14,15] Occupational exposure will normally be limited to those who are involved in the remediation of hazardous waste sites.[16-19] Exposure of persons within the community can occur through the inhalation of dust derived from soil at a contaminated site, through accidental uptake via the ingestion of dust from hand-to-mouth contact due to poor hygienic practices, and through dermal absorption of contaminated soil which has fallen out of the air onto the skin, i.e., deposition.[3,9,10] The direct ingestion of contaminated soil by adults will generally not be a concern, since most adults do not intentionally eat dirt. The ingestion of soil by children ages 2–6 who have mouthing tendencies needs to be considered in any risk assessment, but generally toddlers will not have access to hazardous waste sites. When soil within a residential area is contaminated, its ingestion by toddlers will almost certainly be the primary hazard.[3,6,7,13]

Although some have claimed that it is difficult or impossible to estimate—either prospectively or retrospectively—the dermal, inhalation, and oral uptake of chemicals to which humans have been exposed in the environment or at work, there appears to be little basis for these claims. It is true that there will be varying degrees of uncertainty in the estimates of likely exposure. However, there is no reason why a reasonably accurate approximation cannot be made using currently available data. In an attempt to evaluate the magnitude of the health hazard, estimates of exposure have been made for: [a] pesticide applicators;[20-26] [b] children exposed to lead via dust and dirt;[27-54] [c] workers exposed to vapors and dusts;[55-57] [d] vegetation, wildlife, and humans exposed to nuclear fallout;[58-60] [e] persons involved in accidental chemical releases;[61,62] and [f] those exposed in Vietnam.[63-66] Each of these studies contains information which is valuable for developing exposure estimates for nearly any situation.

The work conducted to date should contain a sufficient amount of information to develop exposure estimates which are within a factor of 3–10 fold (too high or too low) of the actual exposure.[65] These quantitative estimates should be much more valuable in understanding the potential health risks of an activity than such routinely used descriptions as small, moderate, or large. In contrast, such estimates which contain, for example, more than one order of magnitude of uncertainty, should be valuable since they considerably narrow the universe of possible values for exposure which could span as much as 14 orders of magnitude (0.0000000000001 g/kg/day to 5 g/kg/day). In light of the apparent acceptance of the uncertainties surrounding risk estimation models used to interpret cancer bioassays, which can vary by 3–4 orders of magnitude (depending on the model used and the bounding technique),[63-64] the degree of uncertainty inherent in a high quality exposure estimate should be acceptable for most situations.

Post hoc estimates of exposure can also be valuable in interpreting and conducting epidemiological studies. In this process, observed adverse health effects in a population can be compared against the exposure levels, and a dose/response relationship can be inferred. Certainly, the procedures which have been used in attempts to understand the risk to Vietnam veterans exposed to 2,3,7,8-tetra-chlorodibenzo-p-dioxin (TCDD),[63-66] agricultural workers exposed to pesticides,[20-26] the citizens of Seveso exposed to TCDD,[62] the residents of communities whose soil or buildings were contaminated by TCDD,[3,4,6,12] the residents of Hiroshima and Nagasaki following their exposure to radiation,[67] the miners exposed to radon,[61] and the environmental uptake of cadmium[68] should be adequate to distinguish insignificant vs significant exposure.

Contaminated soil can pose not only a direct hazard to humans but also indirect hazards. For example, we know that some chemicals in the soil can leach into groundwater, as well as into streams via runoff.[10,15,69-72] The contamination of streams can pose a threat to wildlife, e.g., birds, fish, deer, rabbits, and other species.[70] This can be a genuine problem in areas where pesticides are incorrectly applied to farm land.[71] Finally, the ingestion of contaminated soil by grazing animals which are a source of food can be a significant, if not the most important, source of exposure to the soil contaminant.[116-128,150]

Thus far, no United States federal regulatory agency has promulgated a standard for contaminated soil, but recent proposed regulations acknowledge the potential risks.[11,73-74] The Environmental Protection Agency (EPA) has suggested ways to assess the hazard posed by direct contact with contaminated soil.[75-76] The potential hazards posed by soil that has been sprayed or amended with sludges have recently received attention, and approaches for evaluating it have been offered.[8,9,77] In addition, contaminated soil often dictates the degree of necessary cleanup at Superfund sites. Usually, the existing land use or the use of legal restrictions to prevent certain future uses (e.g., pasture land, shopping center, industrial site, or residential area) will dictate the level of necessary cleanup.

This chapter discusses specific physical, physiological, and chemical parameters needed to estimate the uptake of a given chemical present on contaminated soil. Special emphasis is placed on the risk assessments of TCDD-contaminated soil which were conducted by the Centers for Disease Control (CDC)[3] and, to a lesser degree, by the EPA,[75] since they have become the benchmarks against which other assessments of contaminated soil are compared.[4,6,7,10] Alternative assumptions to those used in these assessments are discussed. Although there are many important factors to consider when evaluating the human health hazards posed by contaminated soil, this chapter will address some of the most critical ones (Figure 1).

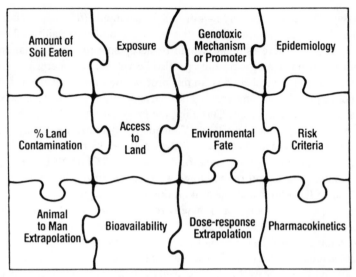

Figure 1.   Environmental risk assessments of contaminated soil must consider a number of critical factors. To understand fully the risks, each of the factors, like the pieces of a puzzle, must be considered. (From: Paustenbach et al.[6]).

INGESTION OF SOIL BY CHILDREN

The exposure estimation procedure is perhaps the most important aspect of risk assessments of sites having contaminated soil. In CDC's assessment of dioxin in soil,[3] their estimates indicated that the primary route of exposure for humans was soil ingestion (Table 1). CDC predicted that, for a residential site containing soil contaminated with TCDD, about 95% of the average lifetime uptake of TCDD would occur as a result of soil ingestion; about 3% of the lifetime dose would be absorbed through the skin as a result of contact with contaminated soil (associated with gardening and poor hygiene); and no more than 2% of the total dose would be due to inhalation of TCDD-contaminated dust. Other environmental risk assessments[4,6] have also indicated that soil ingestion by toddlers is the primary hazard in residential settings. Paustenbach[10] showed that for those living 70 years in an area having TCDD-contaminated soil, inhalation was the predominant route of entry, although the level of contamination will usually present

Table 1.   CDC Estimates of TCDD Uptake by Route of Exposure (1 ppb in Soil).

Route	Average Daily Dose (fg/kg/day)	% Lifetime Uptake
Ingestion	606	95
Dermal uptake of soil	20	3
Uptake of soil inhaled	10	2

From Paustenbach et al.[6] with permission from *Regulatory Toxicology and Pharmacology.*

a negligible hazard. In its calculations, CDC assumed that 10,000 mg/day of soil would be ingested by children aged 1.5 to 3.5 years, and that during other periods ingestion would be much less, depending on age (Table 2). Persons between 5 and 10 years were assumed to ingest 1,000 mg of soil/day, and those aged 17–70 were expected to ingest 100 mg/day through incidental exposure.[3] If these assumptions are used, the resulting estimates indicate that about 80% of the entire lifetime dose of nonvolatile and nonwater-soluble chemicals which are present in contaminated soil occurs during the first 5 years of life (i.e., chemicals whose environmental half-life is at least 10 years).

Table 2. Ingestion of Dirt[a] (CDC's 1984 Assumptions).

Age Group	Soil Ingested (mg/day)	% Lifetime Uptake
0–9 months	0	0
9–18 months	1,000	2.6
1.5–3.5 years	10,000	70.0
3.5–5 years	1,000	5.2
5–70 years	100	22.6

[a]Adjusted for seasonal variations.
From Paustenbach et al.[6] with permission from *Regulatory Toxicology and Pharmacology*.

In light of the critical role which soil ingestion can play when estimating human exposure to contaminated soil, a survey of the literature was undertaken to identify the typical amount consumed by children and adults.[10] The research efforts to evaluate lead uptake by children due to ingestion of contaminated soil, paint chips, dust, and plaster provided the best source of information.[27-54] Walter and co-workers[27] estimated that a normal child typically ingests very small quantities of dust or dirt between the ages of 0–2, the largest quantities between ages 2–7, and nearly insignificant amounts thereafter. In the classic text by Cooper,[28] it was noted that the desire of children to eat dirt or place inedible objects in their mouths "becomes established in the second year of life and has disappeared more or less spontaneously by the age of four to five years." A study by Charney et al.[29] also indicated that mouthing tends to begin at about 18 months and continue through 72 months, depending on several factors such as nutritional and economic status as well as race. Work by Sayre et al.[30] indicated that ages 2–6 years are the important years, but that "intensive mouthing diminishes after 2–3 years of age."

An important distinction that is often blurred is the difference between the ingestion of very small quantities of dirt due to mouthing tendencies and the disease known as pica. Children who intentionally eat large quantities of dirt, plaster, or paint chips (1 to 10 g/day) and, as a consequence, are at greater risk of developing health problems, can be said to suffer from the disease known as pica.[28] If the craving is for dirt alone, this disease is known as geophasia. It is this disease, rather than pica, which is of primary concern in areas of contaminated soil. The incidence of pica has often been misquoted because some of the best studies were conducted in children who already suffered from lead poisoning due, in part, to pica. For example, in such populations, the incidence of pica has been reported at about 20%.[28] In contrast, in the general population, actual pica occurs in about 1–10% of the population, while geophasia is much more rare.[30,31] In light of the potential size of this group, for those soils where access to children is not restricted, risk estimates for those with geophasia should be identified.

Duggan and Williams[32] have summarized the literature on the amount of lead uptake due to ingestion of dust and dirt. Based on their review, a quantity of 50 μg of lead was the best estimate for daily ingestion of dust by children. The field studies by Clausing et al.[50] and Calabrese et al.[51] support their conclusions. Assuming, on the high side, an average lead concentration of 1000 ppm, this would indicate an ingestion of 50 mg/day of soil and dust. Lepow and co-workers[33-34] estimated a rate of ingestion equal to 100–250 mg/day (specifically, 10 mg ingested 10–25 times a day). Barltrop[35] also estimated that the potential uptake of soil and dust by a toddler is about 100 mg/day. In a Dutch study, the amount of lead on hands ranged from 4–12 ng.[36] Assuming maximum lead concentrations of 500 ng/g (the levels were typically lower) and assuming the child ingested the entire contents adsorbed to his hand on 10 separate occasions, the amount of ingested dirt would equal 240 mg. Thus, to eat 10,000 mg of soil per day, as suggested by CDC, the child would have to place his hand in his mouth 410 times daily, a rate which seems improbable.[78]

A report by the National Research Council which addressed the hazards of lead[37] suggested a figure of 40 mg/day. Day et al.[38] suggested a figure of 100 mg/day (based on eating soiled candy), and Bryce-Smith[46] estimated 33 mg/day. In its document addressing lead in air, EPA[47] assumed that a child ate 50 mg/day of household dust, 40 mg/day of street dust, and 10 mg/day of dust derived from their parents' clothing (i.e., a total of 100 mg/day). More recently, Hawley[4] reviewed the available literature on soil ingestion and concluded that uptake by toddlers was approximately 100 mg/day, but he developed a more complex lifetime exposure schedule than had been proposed by others (Table 3). Bellinger et al.[48] suggested a figure of 20 mg/day. Based on all these data, a value of 100 mg/day was identified by LaGoy[52] for purposes of risk assessment.

Rather than estimate the amount of soil eaten by children through visual observations, the better approach is to measure the presence of a nonmetabolized

Table 3.   Other Estimates for Ingestion of Soil by Humans.

Age Group (years)	Lepow et al.[34a] (mg/day)	Duggan and Williams[32] (mg/day)	Baltrop[35] (mg/day)	Hawley[4] (mg/day)	Clausing et al.[50] (mg/day)	Calabrese et al.[51] (mg/day)
0–2	—[b]	—	—	Negligible	—	—
2–6	100	50	100	90	56	<50
6–18	—	—	—	21	—	—
18–70	—	—	—	57	—	—

[a]Used in EPA TCDD risk assessment.
[b]—Indicates the researchers did not discuss these age groups.
From Paustenbach et al.[6] with permission from *Regulatory Toxicology and Pharmacology*.

"tracer" chemicals present in the soil in the children's feces and/or urine. Binder et al.[49] conducted a preliminary or pilot study which employed a rigorous experimental approach involving the analysis of trace elements in children's stool samples. Their data indicated that children 1–3 years of age ingest about 180 mg/day of soil (geometric mean), based on the quantity of silicon, aluminum, and titanium found in the feces. The limitations of this study and the difficulties encountered in the interpretation of the titanium data have been reviewed by Clement Associates.[7]

In a more recent study, Clausing et al.[50] studied the amount of soil eaten by 24 hospitalized and nursery school children. They analyzed the amount of aluminum, titanium, and acid-soluble residue in the feces of children aged 2–4. The data were normally distributed. They found an average of 105 mg/day of soil in the feces of nursery school children, and 49 mg/day in those hospitalized. Even with the limited number of children in the study, the difference between the two groups was significant ($p < 0.01$). If the value for the hospitalized children is assumed to be the background due to intake of these substances from nonsoil sources (e.g., diet and toothpastes), the estimated average amount of soil ingested by this group of nursery school children would be 56 mg/day. This value is in the lower range of the estimates in the literature, and supports the use of 50 mg/day as a reasonable daily average uptake of soil by toddlers (ages 2–4 or 1.5 to 3.5).

The most thorough and rigorous study, to date, was recently completed by Calabrese et al.[51] They quantitatively evaluated six different tracer elements in the stools of 65 school children, ages 2–4. They attempted to evaluate children from diverse socioeconomic backgrounds. This study was more definitive than prior investigations because they analyzed the diet of the children, assayed for the presence of tracers in the diapers, assayed house dust and surrounding soil, and corrected for the pharmacokinetics of the tracer materials. This study was especially impressive in light of the number of tracer substances and high level of compliance in the volunteers. Their results showed that most children ingest less than 50 mg/day of soil and dust.

When this published information on soil ingestion is considered, the data indicate that the best estimate of soil ingestion by children (ages 1.5 to 3.5 or ages 2–4) is about 50 mg/day. A 100 mg/day figure was used by the EPA in its risk assessment[75] and in the EPA *Superfund Health Assessment Manual.*[74] Depending on the situation and the chemical, as will be shown later in this chapter, the use of 50 mg/day for ages 1.5 to 3.5, rather than 10,000 mg/day, can significantly change the risk estimates for contaminated areas.

Soil Ingestion by Adults

For the majority of persons beyond the age of 5–6, the daily uptake of dirt due to intentional ingestion will be quite low.

With the exception of some lower income persons who eat clay, especially black women, adults will not intentionally ingest dirt or soil.[53] However, one potential route by which adults may ingest dirt is fruits and vegetables. It has been shown that nearly all soil ingested from crops will be due to leafy vegetables.[47,60] Interestingly, investigations at nuclear weapons trials have shown that particles which exceed 45 μm are seldom retained on leaves.[58] This is consistent with the results observed with granular pesticides.[59] In addition, the superficial contamination by the smaller particles is readily lost from leaves—usually by mechanical processes or rain, and certainly by washing.[59-60] As a result, unless the soil contaminant is absorbed into the plant, surface contamination of plants by dirt will rarely present a health hazard.

It has been estimated that the deposition rate of dust from the ambient air in rural environments is about 0.012 μg/cm^2/day,[47] assuming that rural dust contains about 300 μg/g of lead (the substance for which these data were obtained). EPA[47] also estimated that even at relatively high air concentrations (0.45 mg/m^3 total dust), it is unlikely that surface deposition alone can account for more than 0.6 to 1.5 μg lettuce/g dust (2–5 μg/g lead) on the surface of lettuce during a 21-day growing period. These data suggest that the daily ingestion of dirt and dust by adults is unlikely to exceed about 0–5 mg/day, even if all of the 137 g of leafy and root vegetables, sweet corn, and potatoes consumed by adult males each day were replaced by family garden products.[47]

In EPA's document on lead,[47] they estimated that persons could take up 100 μg/day due to surface contamination using worst-case assumptions. The actual uptake by adults from vegetables should actually be much less, and probably negligible, since this estimate assumes that all of the suspended dust is contaminated with lead, that persons don't wash the vegetables, that garden vegetables are eaten throughout the year rather than only during the growing season, and that persons eat vegetables only from their own garden.

A second potential way to ingest dirt is through poor personal hygiene. It has been suggested that the primary route of uptake will be through the accidental ingestion of dirt on the hands,[4] and that may be of special concern to smokers who tend to have more frequent hand-to-mouth contact. It is true that before the importance of this route of entry was recognized, persons who worked in lead factories between 1890 and 1920 probably received a large portion of their body burden of this chemical due to poor hygiene; however, such conditions are now rare.[79] The exposure experience of agricultural workers who apply or work with pesticide dusts should be a more useful resource for estimating the potential human uptake of soil from poor personal hygiene.[21-25,55-57] Due to the frequency and degree of exposure to these chemicals during pesticide manufacture or application, these surrogate data can be expected to overpredict the likely uptake of soil from the hands of persons who live on or near sites having contaminated soil. Most of the published studies on pesticides have involved liquids like the organophosphates rather than "soil-like" particles. However, studies of the exposure of persons who apply granular pesticides might be more useful in defining upper boundary estimates of dermal exposure for those who are exposed to contaminated soil[20,22] than estimates based on dusty workplaces.

In general, the incidental ingestion of contaminated soil due to poor personal hygiene indicates that this should not constitute a significant hazard. However, this route of entry remains important in industrial settings. For example, Knarr and co-workers[22] showed that the maximum likely uptake of granular and liquid pesticides by applicators via all routes of exposure is in the range of 2–20 mg/day, and that this was consistent with the results of the other investigators. These estimates would appear to be overestimates of actual conditions at hazardous waste sites, since persons involved in remediation will generally wear personal protective equipment when working with the contaminated dirt. For those persons who live offsite, the contribution of contaminated dirt from the site to the overall airborne dust level and the resulting deposition of soil or house dust onto skin should not be markedly less. By analogy to the study of dusty highways, the deposition of dust decreases dramatically with distance from the road especially for very large particles.[80,81] Even having considered the contribution of poor hygiene and soil-contaminated food,[82] the 100 mg/day figure used by CDC to estimate soil uptake by adolescents and adults seems unlikely, and a figure of 10 mg/day seems more reasonable and justifiable, based on consideration of all the available, relevant data.

EXPOSURE FROM DERMAL CONTACT

Quantitative estimates of the dermal uptake of chemicals from dusts or soil contain more uncertainty than estimates for other routes of entry. In CDC's assessment of TCDD contaminated soil, they assumed that dermal exposures would follow "an age-dependent pattern of deposition similar to soil ingestion" as shown in Table 4.[3] For TCDD, CDC assumed that dirt would remain on the hand for a period long enough to bring about 1% absorption—the percent absorption determined in rats exposed for 24 hr[75,83,84] A 24-hr duration is almost certainly longer than what is likely for humans under normal conditions; therefore, this would represent a value beyond "worst-case." A more likely scenario is that persons would be exposed for 4–8 hr/day, with some degree of washing at the end of this period. Generally, the pure liquid chemical will be absorbed to a larger degree than when present on a carrier or medium such as soil.[75,83-91] For chemicals less strongly bound to the soil, such as those which are water-soluble, the rate of release, availability, and absorption through the skin might be much greater

Table 4. Amount of Soil Deposited on Skin (CDC Assumptions).

Age Group	Soil on Skin (day)
0–9 months	0 g
9–18 months	1 g
1.5–3.5 years	10 g
3.5–5 years	1 g
5–70 years	100 mg

From Paustenbach et al.[6] with permission from *Regulatory Toxicology and Pharmacology*.

than 1%.[92,93] For many risk assessments, such as those involving soil contaminated with petroleum products, dermal bioavailability of soil-bound chemicals of 2-10% may be reasonable.

In both assessments of dioxin on soil, CDC and EPA correctly assumed that the opportunity for dermal exposure would be affected by weather conditions, and consequently they made adjustments.[3,75] In the CDC calculations, it appears that for the midwestern states, CDC assumed that persons would come into contact with soil for about 180 days/year for 64 years (70 minus 6). This estimate is almost certain to overestimate the actual average exposure, since: (1) not everyone in the community gardens; (2) many persons wear gloves when working intimately with dirt; (3) gardeners work directly with the soil primarily during planting and weeding only; (4) most people do not garden each day; and (5) the number of days of precipitation during the gardening season further diminishes the frequency of exposure.

EPA, in its risk assessment,[75] used an alternative set of assumptions for dermal exposure which, because they were based on actual field investigations, seem more realistic than CDC's assumptions. EPA cited the work of Roels et al.,[45] who showed that about 0.5 mg of soil per cm^2 of skin adheres to a child's hand after playing in and around the home. This is similar to that reported by Day et al., who observed that 5-50 mg of dirt transferred from a child's hand to a sticky sweet. Sayre et al.[30] analyzed the amount of lead on children's hands due to house dust. Assuming a value of 500 $\mu g/g$ and 2000 $\mu g/kg$ for the lead concentration in rural and urban house dust, their data indicate dust uptake due to mouthing tendencies at about 100 mg/day if all dust on both sides of the hand were ingested or absorbed through the skin. Uptake can also be estimated by multiplying the appropriate absorption rate for a given chemical and the amount of dirt on the skin using table values for skin surface area,[94] as will be illustrated later in the discussions of Cases I and II.

EXPOSURE FROM INHALATION

Although some persons have been concerned about inhaling dangerously high levels of chemicals from waste sites, either as a vapor or as a contaminant of ambient dust, the EPA, CDC, and numerous independent scientists have shown that inhalation usually will not constitute a route of entry which will adversely affect human health.[2,3,6,9,81] It is worthwhile to note that even though a health hazard may not exist at many of these sites, the presence of odorous chemicals in the soil or the water solubility of the chemical contaminant (posing a threat to groundwater) will frequently dictate the necessary level of cleanup. The degree of inhalation hazard will generally be dictated by the volatility of the chemical, the proximity of the population to the waste site, and the amount of dust generated at the site (for nonvolatile contaminants). In the CDC assessment of dioxin, they assumed that the average air concentration of total suspended particulates

(TSP) was 0.14 mg/m³, and that 100% of this amount (by weight) was respirable. From this, they calculated that exposure to dust containing 1 ppb of TCDD would result in an average lifetime daily uptake of 10 femtograms(fg)/kg/day— roughly 2% of the amount taken up by ingestion, according to the CDC calculations (Table 1).

When actual field data are considered, inhalation is actually a less important route of entry than has often been assumed. For example, the EPA has collected data which indicate that the average concentration of total suspended particulates (TSP) in Missouri is about one-half the level assumed by CDC, or about 0.070 mg/m³.[95] For other geographical locations, a concentration of 0.15 mg/m³ for urban environments and 0.10 mg/m³ for rural environments are conservative estimates of TSP for use in risk assessment; however, site-specific information should be used whenever possible. Due to the many surveys conducted by EPA during the past 15 years, site-specific data are often available. In those rare situations where vehicular traffic on bare contaminated soil or gravel is possible, the resulting increased human exposure to higher levels of dust should be considered.[81]

To assess the health hazard posed by airborne particles, three size categories should be considered: total suspended particulates (TSP), inhalable, and respirable particles. Generally, TSP are those having an aerodynamic mean diameter of under 30 μm. The respirable fraction are those less than 10 mm.

For most sites, only a fraction (about 30%) of the inhalable particles from the soil are respirable (less than 10 μm aerodynamic diameter).[95] Schaum[75] estimated that no more than 50% of the TSP should be respirable and Cowherd et al.[96] confirmed that only about 30% of TSP are respirable. It is important to remember that not all of the airborne particles will be derived from contaminated soil. In one study, about 83% of the nonrespirable particles obtained during sampling were from crustal material (e.g., soil), but only 47% of the respirable particles were from soil.[95] More exhaustive studies conducted by the U.S. EPA have suggested that the portion of inhalable dust derived from soil is even less than these figures.[97]

BIOAVAILABILITY

Dermal Absorption

Bioavailability is an important parameter in any risk assessment of contaminated soil. Throughout this discussion, bioavailability will be used to describe the percentage of a chemical in soil which is absorbed by humans (as suggested in animal studies) following exposure via inhalation, ingestion, or dermal contact.[98] For dioxin contaminated soil, there are a number of parameters which are likely to influence the degree of dermal and oral bioavailability, including aging (time following contamination); soil type (e.g., silt, clay, and sand); co-contaminants (e.g., oil and other organics); and the concentration of contaminant on the soil.[98] The bioavailability of a chemical on soil will be very much affected by its chemical and physical properties.[93] Depending on the chemical of concern and the above-mentioned variables, the dermal bioavailability of a given soil contaminant can vary between 0.05% to 50%.[6,91,98] Large molecular weight chemicals will often bind to the soil and be less water-soluble, while smaller molecules will frequently be water-soluble, volatile, and highly bioavailable.[92,93]

Most dermal bioavailability data for contaminated soil have been obtained in animals or in in-vitro test systems. This introduces a significant source of uncertainty. In the interest of conservatism, safety factors have often been applied to dermal bioavailability data obtained in animals when estimating human uptake. This uncertainty factor is probably unnecessary in most cases, since human skin has generally been shown for a diverse class of chemicals to be about tenfold less permeable to xenobiotics than the skin of rabbits and rats.[99,100] More recently, Shu et at.[98] have reviewed the oral bioavailability studies on TCDD and the uncertainties in the research.

Ingestion

The difficulties of assessing the oral bioavailability of a chemical contaminant on soil are clearly demonstrated by the research surrounding TCDD. The data of Poiger and Schlatter[83] suggested that, as the time of contact between the soil and TCDD increased (known as aging), the oral bioavailability decreased. McConnell et al.[89] studied Missouri soil contaminated with TCDD, and they concluded that TCDD absorption from soil by test animals is highly efficient, but they had difficulty in arriving at an exact percentage for bioavailability. The data of Lucier et al.[90] suggested that bioavailability was dose-dependent: 24% at 1 μg/kg and 50% at 5 μg/kg TCDD. In a 1985 abstract, Umbreit et al. reported the oral bioavailability of TCDD as less than 0.05% for a New Jersey manufacturing site. This work was subsequently published,[91] wherein they also reported oral bioavailability of 21% for a salvage yard in Newark. More recently, Shu et al.[98] have reviewed the oral bioavailability studies on TCDD, and many of the uncertainties were discussed. Shu and co-workers found that the oral bioavailability of TCDD in soil was about 42% in the rat. The bioavailability of other chemicals in soil have not been studied to a great degree, but inferences can be reached by evaluating soil-adsorption data,[71,72,101] volatility information,[2,69] and soil binding studies.[92,102]

Inhalation

The laboratory assessment of the bioavailability of inhaled particles has received very little attention, since it is generally assumed that respirable particles will be 100% absorbed. In CDC's analysis of TCDD (Kimbrough et al.[3]), it was assumed that 100% of the TCDD present on all the inhaled particles would be retained and absorbed in the respiratory tract. In contrast, the EPA assessment[75] assumed that only 25% of the inhaled particles would be absorbed in the lower airways, since at least 50% of the particles would be nonrespirable (especially by weight). For most risk assessments it is probably acceptable to assume that 100% of those particles which remain in the alveoli will be absorbed. However, it must be noted that of the total suspended particulates, usually no more than 50% are respirable (i.e., particles less than 10 μm). Of these, about 50% of the respirable particles are deposited in the upper airways and ultimately swallowed, while the rest reach the alveoli or are expired.

The bioavailability of the large particles (between 10 and 50 μm) is dictated by the oral route, since they are taken up the mucocilliary escalator and then swallowed.[6,75] In any assessment of soil contaminated by a chemical, the difference in bioavailability between those particles which are caught in the upper airways and swallowed versus those that reach the lower airways should be considered.

BIOLOGIC HALF-LIFE

When deciding what level of soil contamination is acceptable, it is often necessary to consider the half-life of the substance in humans and wildlife.[2] For many of the persistent chemicals, the concentration present in wildlife can often exceed that detected in soil or water. The concentration of DDT, chlordane, and the dioxins in some fish can be sevenfold greater than the water solubility.[103] Also, nondetectable levels in the environment can often bioaccumulate via the food chain to detectable levels in human tissue.[104] The steady-state body burden of persistent chemicals in humans can be estimated using basic pharmacokinetic principles and this has been illustrated in a recent publication.[19] The potential hazard posed by chemicals which have biologic half-lives in excess of 12 hrs is that repeated exposure to innocuous levels may produce body burdens which are ultimately problematic.

EFFECTS OF DISTRIBUTION AT THE SITE

Kimbrough et al.[3] observed that of all the assumptions used in their risk analysis, the "most prominent of these is the assumption of uniform levels of contamination throughout the living space (environment)." CDC was especially sensitive to the issue of uniformity of contamination and accessibility to the contaminated soil, and generated an interesting plot which illustrates the relationship between the magnitude of the risk and the percent of the surface area of the land that has been contaminated (Figure 2). The concept of averaging the surface soil levels when conducting the risk assessment is important in all assessments of contaminated soil, regardless of the chemical. Of course, when averaging is used, criteria should be established for defining so-called "hot spots." These are localized areas where the level of soil contamination is much higher than the

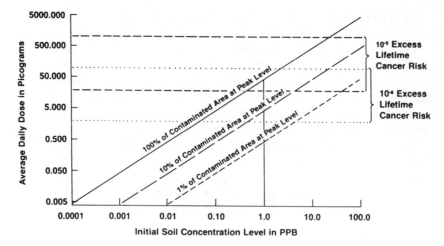

Figure 2. Effect of the degree prevalence of dioxin soil contamination at a residential site versus the estimated risk (from Kimbrough et al.[3]). This plot is based on an acceptable daily lifetime dose of 633 fg/kg/day (CDC's best estimate of a dioxin uptake which might increase the individual cancer risk by 1 in 1,000,000).

surrounding ones. These localized areas of "higher than normal" contamination should be removed before an average is calculated or the soil in the hot spot should be mixed with uncontaminated soil until the concentration is below the level of concern.

SOIL SAMPLING AND DATA ANALYSIS

In most assessments of areas with potentially contaminated soil, an emphasis has been placed on locating the areas having the greatest concentration, rather than obtaining a representative sampling of the site. Regrettably, the results of such quantitative sampling are often used to develop risk assessments and remedial action plans. Since such nonrandom sampling is not representative of the average level of contamination at the site, it is virtually impossible to develop an accurate estimate of average exposure. The more appropriate approach to gathering data useful for conducting a health hazard assessment is to divide the contaminated site into equal quadrants,[105] and to collect a statistically valid number of samples on a random basis. Samples should be collected at depths of 0–6", 6–12", 12–36", and at similar intervals until hitting bedrock. This will allow for a gross assessment of the depth of contamination. If water-soluble chemicals are involved, a 24–48" sample is usually worth collecting during the initial survey. Having collected and assayed these samples, the data should be interpreted using lognormal statistics if the data are lognormally distributed (as is usually the case). This approach does not allow unusually high values to dominate the overall results and is appropriate for statistical analysis of environmental samples.[106]

RESIDENTIAL VERSUS INDUSTRIAL SITES

Often, the differences in the conditions of human exposure between residential, industrial, agricultural, and recreational sites are overlooked by those who conduct risk assessment or those agencies responsible for setting cleanup guidelines. As noted in CDC's publication, "in all of these scenarios (factories, farms, residential sites), decisions must be made on a site-specific basis."[3] The most prominent difference between a residential and an industrial setting, with respect to the hazard to human health, is the lack of access to industrial sites by the public, and, in particular, by children. In the absence of such access, at least 95% of the potential exposure to relatively nonvolatile substances, such as dioxin or the heavy metals, is eliminated (Table 1).

ENVIRONMENTAL FATE OF THE TOXICANT

It is important to recognize that the degree of human exposure (ingestion, inhalation, and dermal) is almost always a function of the concentration of the contaminant at the soil surface (top 0–1 cm) rather than of the concentration at lower depths. This is significant since, when exposed to sunlight, many chemicals degrade, are rapidly volatilized, or are metabolized by soil microbes.[73] The end use of the site will dictate whether it is necessary to decrease the contaminant concentration at depths of 3, 12, 24, or 48 inches. In the case of TCDD on soil, research suggests that within 18 months, nearly all of the TCDD in the top 1/2 cm of soil is no longer present due to photodegradation (if in the presence of a solvent) and volatilization.[107] However, dioxin which is not exposed to sunlight appears to have a half-life near to or in excess of 10 years.[3,13] In contrast, volatile chemicals may persist for only days or weeks at the soil surface.

A chemical's environmental half-life is important when estimating risk. For example, the effect of environmental half-life on the anticipated average lifetime daily uptake of TCDD via soil is illustrated in Table 5. Using the CDC's assumptions for potential oral, dermal, and inhalation exposure to soil, if the half-life at the soil surface were 1 year, the soil concentration was 100 ppb, and the bioavailability was 30%, the maximum anticipated average lifetime daily dose would be 12 pg/kg/day for 70 yr of exposure. In contrast, if the half-life were 12 yr, the average dose would increase to 64 pg/kg/day using CDC's assumptions. More importantly, using the alternative assumptions which will be discussed in Case I, the same conditions of exposure, and assuming a 1- and 12-year environmental half-life for TCDD, the projected uptakes are 0.14 and 1.6 pg/kg/day, respectively. The importance of a chemical's half-life on the risk estimates is particularly important when children are not exposed to the soil (e.g., an industrial site). In these settings, such as occupational exposure, the average lifetime daily uptake by adults (ages 20–65) drops dramatically with decreasing environmental half-life (Table 5). For example, using CDC's assumptions for adult exposure

Table 5. Estimated Lifetime Average Daily Dose (pg/kg/day) of TCDD as a Function of Initial Soil Concentration and Environmental Half-Life.

Exposure Period (years)	Half-Life of TCDD in Soil (years)	100 ppb							
		CDC				Alternative			
		30%		10%		30%		10%	
		CDC[a]	ALT[b]	CDC	ALT	CDC	ALT	CDC	ALT
0→70	1	12	11	4.1	3.7	0.56	0.14	0.48	0.06
0→70	6	53	51	19	17	3.2	1.0	2.6	0.51
0→70	12	64	62	24	21	43	1.6	3.6	0.93
20→65	1	55×10^{-8}	55×10^{-8}	23×10^{-8}	23×10^{-8}	7.5×10^{-8}	72×10^{-8}	7.5×10^{-8}	7.2×10^{-8}
20→65	6	31×10^{-2}	0.31	0.13	0.13	0.045	0.042	0.045	0.042
20→66	12	1.8	1.8	0.79	0.77	0.27	0.25	0.27	0.25

[a] CDC (Center for Disease Control) estimates are based on the assumptions used by Kimbrough et al.[3] for oral, dermal, and inhalation exposure and their assumptions for bioavailability. The various assumptions for the degree of oral intake and oral bioavailability used in the calculations are shown in table heading. The initial average soil concentration was 100 ppb TCDD.

[b] ALT (alternative) estimates are based on the assumptions described in this chapter for oral, dermal, and inhalation exposure and the different assumptions for bioavailability for each route of entry. The various assumptions for the degree of oral intake and oral bioavailability used in the calculations are shown in the table heading. The initial average soil concentration was 100 ppb TCDD.

(ages 20–65) to contaminated soil, the estimated uptake of TCDD is 55×10^{-8} pg/kg/day if the half-life is only 1 year, but the risk is 9 orders of magnitude higher when the half-life is 12 years.

AVIAN HAZARD

Some sites, especially where contaminated soil is present over a large geographical area, may present a hazard to various bird species.[108] When evaluating the potential avian hazard, a number of factors must be considered. Among the most important parameters is the environmental half-life of the chemical and the avian toxicity. To fully understand the hazard one must also consider the opportunity for acute and chronic exposure, the size of the contaminated area, the number of exposed species, the migratory or nonmigratory nature of the birds, the biologic half-life of the substance, the concentration of the contaminants within the top 1″ of soil, its water solubility and lipid solubility, and the presence of soil-dwelling insects.

Although an evaluation of the avian hazard posed by contaminated soil is a complex one, relevant information can be drawn from the avian experience with granular pesticides. Numerous poisoning incidents have been studied wherein seeds, insects, plants, and other media have become contaminated with both short- and long-lived pesticides. Approaches used to evaluate sites contaminated with pesticides are invaluable sources of information for developing a strategy to assess sites having contaminated soil.[108, 148-150] The historical data on chlordane, heptachlor, diazanon, and carbofuran constitute a valuable data base from which one can develop avian assessments.

APPLYING THESE CONCEPTS

As noted in both the CDC and EPA assessments for dioxin, the process of setting standards for soil cleanup should be done on a site-specific basis. This task, however, need not be a process so complex as to be impractical for regulatory agencies. Certainly, an algorithm which incorporates all of the important variables could be developed, as discussed by Hawley,[4] Schaum,[78] and Paustenbach[10] and these calculations can easily be handled by a desktop computer.

The following case studies illustrate how a generalized methodology for evaluating contaminated soil could be applied, on a site-specific basis, incorporating data on exposure, bioavailability, % land contamination, and environmental fate. These examples illustrate the effect of altering only a few of the critical assumptions in an environmental risk assessment and the importance of site-by-site analyses. In developing a comprehensive risk assessment, it is also important to acknowledge the uncertainties, as recommended by the National Academy of Science.[112] Kimbrough et al.[3] recognized the fragility and uncertainty of many of the assumptions which they used, and specifically noted that "it must be stressed that the exposure assessments used, in estimating risks for carcinogenicity and reproductive health effects, contain critical assumptions that are not likely to be actually encountered." Other critical issues, including the use of classic cancer risk models as opposed to the safety factor approach to estimate carcinogenic potency in humans at low doses (which depends on genotoxicity) and the incorporation of the available epidemiology data, should be examined in any comprehensive analysis of soils contaminated by suspected carcinogens.[6,11,12]

CASE I (A RESIDENTIAL SITE)

In this example, a hypothetical residential site will be evaluated. For the sake of this discussion, 10% of the soil surface in this community is assumed to be contaminated with 10 ppb TCDD. For several years, many state and federal agencies have adopted a 1.0 ppb cleanup level for TCDD in soil (found in any location) based on CDC's original health assessment for residential sites.[3] This case study will illustrate how indiscriminant adoption of this or any other guideline, without a thorough site-specific analysis, is likely to improperly characterize the health hazards.

Using the methods described by the CDC and the EPA, a preliminary evaluation of the need for cleanup at a given site can be readily conducted. In this example, the CDC assessment procedure is used but a few of their values are updated. Specifically, in these calculations the following parameters are used: (1) a soil uptake of 100 mg/day (ages 2–6), and negligible soil ingestion for the rest of the lifetime; (2) 10% oral and 1% dermal bioavailability; (3) quantitative assumptions on dermal exposure, (4) actual field data on the concentration of respirable dust in air, and (5) corrections for the differences in bioavailability between particles that are swallowed, as opposed to those retained in the lung (Table 6).

As shown in Table 7, using the alternate assumed values, the oral dose would not be expected to be greater than 2.8 fg/kg/day. If a 40% oral bioavailability were assumed, the average daily dose increases to 17.0 fg/kg/day. CDC's assumed values would suggest an oral uptake of about 606.5 fg/kg/day. The alternate method estimated dermal uptake of 2.7 fg/kg/day, in contrast to 20 fg/kg/day using CDC's assumed values. Finally, CDC assumptions would suggest that 10 fg/kg/day might be inhaled, versus 5.6 fg/kg/day.

In summary, use of the alternative assumptions predicts that the most likely uptake of TCDD will, in this example, be 11.1 fg/kg/day. If a 40% oral bioavailability were assumed, the average daily dose increases to 17.0 fg/kg/day. In contrast, the assumptions in the CDC assessment led to an estimate for total TCDD uptake of 636.5 fg/kg/day (which CDC considered acceptable when recommending the 1 ppb guideline). This analysis illustrates the critical nature of certain assumptions, and it shows that remediation would not be indicated at this site. Of course, in any isolated instances where children or adults might be exposed to areas of

Table 6. Exposure Assumptions Used in Case I Calculations (Lifetime Average).

Route	CDC	Alternate
Oral	0.21 g/day ingested (lifetime average) 30% absorption	0.0028 g/day ingested (lifetime average) 10% absorption
Dermal	0.21 g/day of soil on the skin Dose is weighted by age over lifetime 1% absorption	0.5 mg soil/cm² skin 1400 cm² exposed surface 8 hr/day, 90 days/year 1% absorption
Inhalation	Total suspended particulates is 0.14 mg/m³ 15 m³/day 10% in lungs 100% bioavailable	Total suspended particulates is 0.075 mg/m³ 15 m³/day Of particles inhaled, assume 50% ingested and 50% in lungs Bioavailability weighted for differences in absorption for respirable and nonrespirable particles Only 50% of the particles are from the contaminated site

From Paustenbach et al.[6] with permission from *Regulatory Toxicology and Pharmacology*.

Table 7. Residential Site[a] (Case I).

Route	TCDD Uptake (fg/kg/day)	
	CDC	Alternate
Oral	606.5	3.1
Dermal	20.0	2.7
Inhalation	10.0	5.6
Total	636.5	11.4 (17.0)[b]

[a]Ten percent of land contaminated (10 ppb TCDD).
[b]Using a 40% oral bioavailability (based on Shu et al., 1988b).
From Paustenbach et al.[6] with permission from *Regulatory Toxicology and Pharmacology*.

limited size which had TCDD soil concentrations in excess of 500–1000 ppb (e.g., hot spots), cleanup of these would need to be considered.[6]

CASE II (INDUSTRIAL SITE)

The methodology used by CDC,[3] Paustenbach et al.,[6] and Eschenroeder et al.,[9] can also be used to assess the potential hazards at an industrial site. In this example, assume that 20% of the surface area of site has soil which has been contaminated with 100 ppb TCDD. Because it is an industrial site, children have virtually no access. Is remediation necessary? The assumptions and factors used here are shown in Table 8.

Because children are not exposed, the oral route of exposure is a much smaller factor. As shown in Table 9, calculations based on the CDC-assumed values in

Table 8. Exposure Assumptions Used in Case II Calculations (Lifetime Average).

Route	CDC	Alternate
Oral	Not relevant	Not relevant
Dermal	Exposure is almost same as oral exposure 15 m³/day 100% in lungs	0.5 mg/soil 1400 cm² exposed skin 8 hr/day, 40 years exposure
Inhalation	0.14 mg/m³ (dust) 15 m³/day 100% in lungs	Total suspended particulates are 0.075 mg/m³ (Springfield, Mo.) Of particles inhaled, assume 50% ingested and 50% in lungs
	100% bioavailable	50% of respirable particles from contaminated soil Adjust bioavailability for ingestion and inhalation

From Paustenbach et al.[6] with permission from *Regulatory Toxicology and Pharmacology*.

Table 9. Industrial Site[a] (Case II).

Route	TCDD Uptake (fg/kg/day)	
	CDC	Alternate
Oral	0	0
Dermal	390	78
Inhalation	200	57
Total	590	135

[a]20% of land contaminated (100 ppb TCDD).
From Paustenbach et al.[6] with permission from *Regulatory Toxicology and Pharmacology*.

dicate that an average lifetime dose of 590 fg/kg/day TCDD could potentially be absorbed by those who work at this site (lifetime average). This value is still less than 636 fg/kg/day, the daily dose which CDC considered acceptable (Table 1). In addition, use of the alternative assumptions indicates that the more likely daily uptake is 135 fg/kg/day. Since the daily intake is less than 636 fg/kg/day, these calculations suggest that 20% contamination of soil by 100 ppb TCDD is safe even using the criteria proposed by CDC. Also, at many industrial sites it may be acceptable to permit soils contaminated with TCDD at levels greater than100 ppb to remain in place, especially where soil erosion is not a problem and where the contamination is at the soil surface (where it is available for photodegradation).

UPTAKE BY PLANTS

The extent to which crops will take up (absorb) appreciable quantities of a chemical from contaminated soil must be evaluated on a chemical-by-chemical basis. Field studies where crops have been grown in contaminated soil have been conducted on numerous chemicals. Inferences for other chemicals can be drawn based on the physical and chemical properties of the contaminant and the soil/contaminant matrix; however, actual field data are much preferred. Interestingly, the amount of soil contaminant taken up by a plant will vary with plant species. For example, oats and soybean plants grown to maturity in soil contaminated with 60 ppb TCDD showed less than 1 ppb of TCDD in the seeds.[113] Wipf et al.[114] failed to detect any measurable 2,3,7,8-TCDD in the flesh of fruits and vegetables collected from the contaminated area in Seveso during 1977–1979, although the TCDD concentration in the soil was 10 ppb. For many chemicals, only the tubers (potatoes, carrots, onions, etc.) seem to absorb or bioconcentrate them. For other plants, the primary routes of contamination are the volatilization of the chemical from the soil with subsequent absorption by the leaves closest to the soil, or by retention of dust which has been deposited on the leaves.

If an estimate of the concentration of a chemical in a plant is known, the quantitative uptake by humans can be calculated through use of the USDA's food tables. Usually, the concentration of xenobiotics in various plants and food is not available unless the xenobiotic is a pesticide.

Generally, common industrial solvents would not be expected to be present at significant levels in most crops due to their volatility, absorption to soil particles, metabolism, or degradation in soil. However, those chemicals which are stable or highly lipid-soluble may be present and available for ingestion by those animals who eat soil while grazing.[8] Many of the stable compounds found on soil (except perhaps the metals) will be taken up in the fat or the liver of the animal, simply due to lipid solubility or binding. Other chemicals, if present, will usually be metabolized and eliminated. From recent estimates of the amount of fat contained in the various classes of meat in the U.S., one can often calculate dose.[115]

UPTAKE BY GRAZING ANIMALS

As noted by Fries,[8] when the primary route of human exposure to a contaminant is through food, the persons most likely to receive the highest exposures are farmers who eat their own products, as well as their direct customers. Exposure of the general population will generally be several orders of magnitude lower because agricultural products are distributed widely, and food containing low levels of these chemicals will be diluted with uncontaminated food.

If a chemical grown in contaminated soil is absorbed by plants and not digested, it is a straightforward exercise to calculate uptake (lb. eaten per day×conc. in plant÷body weight). The pharmacokinetics of the chemical in the animal must also be considered in the calculations, since biologic half-life can markedly affect the steady-state body burdens. For dioxin and any other chemicals which are not taken up by plants, animal feed is not an important route of animal exposure. Field experience with polybrominated biphenyls (PBB) indicates that dust on the surface of plants gathered during harvest of forage crops makes a negligible contribution to residues in feed as harvested. The concentration was less than 1% of that present in soil.[116]

Interestingly, the intentional ingestion of soil by grazing animals may often represent the primary route of uptake. This phenomenon has been studied in cattle and sheep under a variety of conditions.[116-128] Generally, soil ingestion is inversely related to the availability of forage. The amount is as low as 1–2% of dry matter intake during periods of lush plant growth, and it rises as high as 18% in periods when forage is sparse.[8] Under New Zealand conditions, where animals can graze 365 days a year, average soil intake was about 6% of the estimated dry matter intake for cattle and 4.5% of the dry matter intake for sheep.[6] In the United States, a reasonable estimate of soil uptake by grazing dairy cows is 5% of dry matter for sparse lands and less than 1% for those which are occasionally on pasture. For nonlactating beef cattle, a reasonable estimate is 2% of dry matter and for a lactating cow, while 3% is a conservative estimate. For a nonlactating cow of 350 kg, this represents soil ingestion of up to 0.6 lb/day in a diet of 20 lb/day feed. The amount of soil ingested is reduced more than 50% when animals are offered harvested feed as a supplement to pasture.[8,125]

Farm animals also ingest soil when they are confined to unpaved holding areas.[123,124] Under typical U.S. farm conditions, lactating cows may consume as much as 2–3% of their dry matter intake as soil, whereas nonlactating cattle, who have greater exposure and are less intensely fed, may consume up to 2% of their dry matter intake from this source. Pigs, due to their habit of rooting, will tend to have higher values, which can be as great as 8% of dry matter intake.[124] Since pigs never subsist on pasture alone, this value may be considered an upper limit for the amount of soil ingested by pigs.

When estimating the amount of a chemical (especially persistent ones) which could be present in meat or milk, several basic factors must be considered: (a) amount of soil ingested by the animal (yearly average); (b) the fat:diet ratio, e.g. 5:1 for TCDD in milk fat and body fat; (c) the amount of soil ingested should be averaged over a year when animal access is less than a full year; (d) the half-life of the chemical in growing cattle; (e) the environmental degradation of the soil contaminant in the top 2–4 cm; (f) rate of metabolism by the animal; and (g) soil bioavailability in the animal.[123-127]

Fortunately, the practical significance of soil ingestion as a potential route of animal exposure is greatly mitigated under the U.S. agricultural conditions.[8] Lactating dairy cows are rarely pastured, and some form of supplemental feeding is almost always employed. Consequently, it is unlikely that soil ingestion would ever exceed 1% of dry matter intake in actual farm situations. Fries[8] has noted that, even though cattle raised for beef might often be on pastures with little other feed, it is the general practice to fatten these animals in feed lots before slaughter. This period of time may be as long as 150 days, and animals can gain as much as 60% to 70% in body weight.[117] During these periods, the contaminant will also be reduced by dilution in the expanding body fat pool.

Most hogs destined for slaughter are confined and would never be exposed to contaminated soil. Thus, as noted by Fries,[8] only cull-breeding cattle and pigs might be expected to go directly from soil to slaughter, and these are not the animals ordinarily used for home consumption by farmers who raise them. In short, for most domestic animals in the United States, few will ever be exposed to contaminated soil without having been provided with ample feed. Due to the many factors that need to be considered, the setting of an acceptable soil level should be done on a site-specific basis.

Environmental Persistence

For those chemicals in soil which are not rapidly degraded, the potential for accumulation in the food chain is possible.[15] Specifically, soils contaminated with chemicals whose bioconcentration factors (BCF) exceed 1000 need to be assessed for the potential of the soil to enter streams or lakes via runoff.[129] Even for those substances whose BCF is between 100 and 1000, it may be necessary to consider both the concentration and the total amount of contaminant in soil if it could be eroded into water.

There are numerous examples of situations where even relatively small amounts of soil from sites contaminated with the PCBs, dioxins, furans, DDT, kepone, and heptachlor eventually contaminated streams to a level that was easily detected in nearby fish. Existing environmental fate models should be used to set standards for chemical contamination of streams to protect aquatic life and wildlife.

Human Biologic Half-Life

Often, chemicals which are long-lived in the environment also have long biologic half-lives; that is, the time needed for 50% of the absorbed substance to be eliminated from a human or other biologic species. Those chemicals with biologic half-lives longer than 2–3 months may require some degree of review, since it will require about 1 year for steady-state levels to be reached in living organisms. However, chemicals with human half-lives of elimination in excess of 3–5 years clearly require special attention, since it is possible that hazardous levels could be reached following repeated exposure to fairly small quantities.

It is important to recognize that it is relatively easy to estimate the acceptable daily intake for humans and other animals, even for such long-lived or stable chemicals using pharmacokinetic principles.[19] An important consideration when assessing soil is to understand that if a contaminant has a long environmental half-life and a long biologic half-life, that the chemical needs to be kept out of the food chain. For soils, this means that limited quantities of contaminated soil should be allowed to enter streams or lakes via run-off or to enter the food chain via grazing animals.[8]

Leaching into Water

The possibility that water-soluble chemicals in soil can enter water via runoff or leaching has been recognized for a number of years.[2,128,130] In general, chemicals which are water-soluble are those of primary concern. Chemicals which do not bind very strongly to plants or soil represent the next most important class of chemicals. Finally, those chemicals which have a long half-life in the environment are often a concern. The erosion of contaminated soil in runoff can present a special hazard for the persistent chemicals.

There are numerous factors to consider when water is contaminated by a liquid- or soil-bound contaminant. Some chemicals are bound to sediment, and are subsequently taken up by bottom-feeding fish and mollusks.[70] Others can pass through clay or rock-like layers of subsurface material which often separate subsurface waters from deep aquifers (so-called "drinking water aquifers"). Another consideration is that fish or other aquatic organisms can take up the contaminant from the water and concentrate it in their adipose tissue or skin.[2,73] Finally, sufficiently high levels of chemicals in water can affect the chemistry and biology of lakes and streams, and this can cause adverse effects on algae and other organisms.

Uptake by Fish

For many of the water-soluble chemicals, the hazard to fish will be the major one.[70] In recent years a number of models have been developed which allow us to predict the potential hazard to fish posed by chemicals released into the environment,[2,70] and these should be useful in assessing the potential hazard posed by contaminated soils. Fish and mollusks, due to their ability to bioconcentrate water-soluble chemicals, will usually be the species most at risk if a chemical is a potent toxicant in aquatic species. Often, the bioconcentration factor (BCF) is useful for predicting the peak steady-state level of a toxicant in fish.[131,132] Unfortunately, for those chemicals which are very poorly water-soluble or very lipophilic, the simple BCF formula may "break down"; e.g., it can dramatically overestimate the actual BCF observed in fish.

Hazard to Wildlife

The potential hazard of contaminated soil to deer, earthworms, birds, squirrels, and other wildlife has become a greater concern in recent years.[133,134] In general, only a limited number of studies have assessed the various factors which can adversely affect the health of wildlife in the field, especially as a result of contaminated soil. It would appear that those chemicals which are long-lived in the environment (low water solubility, poorly metabolized, and resistant to degradation) are those of primary concern. The studies of DDT and PCB in the environment probably represent the best data base for predicting the effect of rather long-lived chemicals on various wildlife.

One of the most thorough evaluations of the potential environmental hazard posed by contaminated soil was that conducted at Eglin Air Force Base in Florida.[135-138] In their studies, over 300 biological species (plants and animals) were analyzed for TCDD. Histopathology was also conducted in the majority of the animals captured. These researchers found that the beachmouse was the best species for evaluating the maximum level of biomagnification, since they lived underground, reproduced frequently, and soil ingestion (due to preening) was a major route of uptake.[138] They showed that the higher species (deer, opossum, and rabbit) usually do not eat sufficient quantities of soil to reach significant tissue levels of TCDD (soil levels less than 1 ppb). In this study, no dioxin was found in their tissues, even though they spent months on soils containing about 80 ppt (geometric mean) with a range of 10–1500 ppt.

In contrast, cotton rats and beachmouse were shown to have liver concentrations of 10–210 ppt and 300–2900 ppt (30–50% of total body burden of TCDD is in liver). Numerous insects were also assayed, and these showed levels of 40–238 ppt, while birds (insect predators) had levels in the range of 50–440 ppt in the liver. Young et al.[138] concluded that for insects or animals the closeness of the relationship to the soil seemed to dictate the degree of uptake. He noted that the beachmouse and cotton rat burrow in the soil, but that the deer, opossum, and rabbit do not. He was unable to identify whether the predators of insects actually received their body burdens due to ingestion of the insects or through preening of the soil with which they came into contact.

EXPOSURE/RISK CRITERIA

Although a considerable amount of effort is needed to properly identify the target species and the degree of exposure to a toxicant due to the presence of contaminated soil, the evaluation of the hazard is not complete until the absorbed dose is compared with some standard or guideline. These are usually derived from studies involving animals in accidentally exposed populations. Guidelines set for chemical carcinogens are sometimes called risk criteria.

For much of the past 50 years, the safety factor approach has been applied to toxicity data collected in rodents to identify the dose at which humans would not be expected to be adversely affected.[111] Until recently, the approach was also used for carcinogens, developmental toxicants, and all systemic agents. Although much less complex than the mathematical models that were later developed for estimating the risk of exposure to carcinogenic agents, limits based on safety factors have generally been effective at preventing disease among exposed persons.[110-111]

The safety or uncertainty factor approach is a simple one. Traditional toxicological procedures define a safe level of exposure for humans as some arbitrary fraction of the dose level at which no effects were observed in a group of test animals, i.e., no observed effect level (NOEL). When evaluating food additives and pesticides, where toxic effects other than cancer are a concern, an acceptable daily intake has often been established by dividing the NOEL by a safety factor of 100. The purpose of the safety factor is to account for the possibility that humans may be up to 10 times more sensitive than the animal species tested, and that a 10-fold variation in sensitivity within the human population may exist.[139-141] The magnitude of this factor may be modified depending on the chemical and pharmacokinetic properties of the test compound, the severity of the adverse effect, and the quality of the available toxicological data.

Because genotoxic carcinogens may pose some degree of risk even at very low doses, an approach to establishing risk exposure limits or criteria other than those set through the use of safety factors was developed. These models are used to estimate the cancer risk at doses to which humans might be exposed due to water, food, air, or soil contamination. The purpose of so-called cancer models is to identify a virtually safe dose (VSD) based on the extrapolation data obtained in animal tests. Because of the statistical and biological problems inherent in the identification of a true no-effect level, most mathematical models for carcinogens have eliminated the concept of a threshold i.e., a dose where no response would be expected.

The setting of standards, exposure criteria, or risk criteria, for contaminated soil is a difficult task since site-specific factors have a significant influence on what degree of contamination would pose a hazard. For this and other reasons, only a handful of standards or regulations for contaminated soil have been established by state or federal agencies. The approximately six to ten soil standards or guidelines typically address trichloroethylene, methylene chloride, benzene, petroleum fuels, lead, PCB, and 2,3,7,8 TCDD. To set such limits, the toxicological approaches discussed above should be used with consideration given to the ten or more other site-specific factors discussed throughout this chapter.

DISCUSSION

Selection of the most appropriate assumptions is critical in any environmental risk assessment since often as few as one, two, or three assumptions or factors can dramatically influence the results. For example, in the health assessment of dioxin in soil the most critical parameters are the quantity of soil ingested by children and adults, the mathematical modeling of the bioassay data, consideration of dioxin's lack of genotoxicity, and dioxin's bioavailability in a soil matrix. Other assumptions used in the assessment, while important, cannot alter the results by 2–3 orders of magnitude.

When evaluating the hazard posed by soil contaminated with other chemicals, numerous additional factors must be considered. For example, does the chemical bind tightly to soil; does it become more difficult to desorb over time; is its water solubility quite low; is it photolabile; and, most importantly, is there enough contaminant present to affect the exposed animals or humans?

Regulatory criteria regarding the acceptability of risk for nonubiquitous soil contaminants will probably be different than for ubiquitous soil contaminants like lead or arsenic. For example, because fewer than 200 persons are known to live in residential areas where soil is contaminated with significant levels of dioxin, while millions live in neighborhoods having relatively high lead or arsenic soil levels, higher levels of risk may be more acceptable for the former than for the latter.[142] For example, any regulatory position regarding the risk criteria to be used for food additives (to which everyone is exposed), such as 1 in 1,000,000, would probably be inappropriate for dioxin- or PCB-contaminated soil, in light of the relatively small number of persons exposed to the latter toxins and their tendency not to migrate from the soil (assuming erosion is controlled).[143,144]

Most environmental and occupational regulations that have been promulgated thus far accept cancer risks in the region of 1 in 1,000 to 1 in 1,000,000, even where thousands of persons may be exposed since: (a) it is believed that cancer risk models tend to overestimate the true risks (based on human data); (b) because the cost/benefit criteria indicate that lower risks are impractical; or (c) because the aggregate lifetime risk for the exposed population is close to, or less than, one additional case of cancer.[143,144] For example, EPA has recently found the maximum individual risks and total population risks posed by a number of radionuclide and benzene sources to be too low to be considered "significant." Specifically, benzene emissions from maleic anhydride process vents created maximum individual risks of 7.6 in 100,000, and an aggregate yearly cancer incidence of twenty-nine thousandths of one case.[145] Radionuclides from facilities of the Department of Energy would expose a person who accrued lifetime exposure to a plant's most concentrated emissions to a risk of 1 to 8 in 10,000, so that in the aggregate, only eight-hundredths of an additional cancer case would be predicted to occur yearly.[146] EPA found these risks to be insignificant, and eventually withdrew the proposed regulations.

The issue of the potential numbers of people affected should influence any risk management decision. Travis et al.[144] conducted a retrospective examination of the level of risk which triggered regulatory action in 132 cases. They considered three variables: (1) individual risk (an upper-limit estimate of the probability that the most highly exposed individual in a population will develop cancer as a result of a lifetime exposure); (2) population risk (an upper-limit estimate of the number of additional incidences of cancer in the exposed population); and (3) population size. The findings of Travis et al.[136] can be summarized as follows:

1. Every situation presenting with an individual lifetime risk above 4×10^{-3} received regulatory action. Those with values below 1×10^{-6} remained unregulated.

2. For small populations, regulatory action never resulted for individual risks below 1×10^{-4}.
3. For effects resulting from exposures to the entire U.S. population, a risk level below 1×10^{-6} never triggered action; a risk level above 3×10^{-4} always did.

Consequently, regulatory agencies have taken different action, depending on the magnitude of the risk and the size of the population.

Travis et al. summarized their conclusion as follows:[144]

> Perhaps the most surprising aspect of our study is the consistency found among federal agencies' methods in the use of cancer risk estimates for regulatory decisions. With the possible exception of FDA decisions concerning de minimus risks, the history of federal decision making indicates that all agencies are fairly consistent in their implicit definitions of de manifestis and de minimus levels of risk. If the above three guidelines were adopted explicitly, consistency with past decisions would be maintained and the process of regulatory decision making would be simplified considerably.

Perhaps the most serious problem which has plagued risk assessments of the past 10 years, and one that is likely to affect assessments of contaminated soil, is the unreasonableness of some risk estimates due to the cumulative effect of several worst case exposure assumptions. Overly zealous risk assessors have often repetitively utilized parameters for persons in 95% of the population. Compounding 95% confidence limits does not result in a 95% confidence limit on the exposure estimate, but rather this produces a much higher confidence limit. For example, if an upper bound on the intake of a chemical in drinking water were calculated assuming that the water is contaminated at the 95% percentile, and if a person drinks large quantities of water (the 95th percentile), and if he were physically small (the 5th percentile), then the upper bound is not a 95% upper bound, but rather a much higher confidence limit. The maximum level of confidence here is almost 100% (specifically, $[1-(0.05)^3]=0.99875$).

Of course, the specific confidence level depends on the distribution of each of the various parameters. By way of illustration, if the concentration was normally distributed with means of 5.5 mg/L and variance 7.5, if drinking water consumption was normally distributed with a mean of 1 L and variance of 0.55, if body weight was normally distributed with a mean of 78 kg and variance of 160, and if concentration, consumption, and weight were independent, then the confidence level of the corresponding upper bound would be 99.82% and not 95%. In this example, instead of 5% of the population exceeding the upper bound on intake, only 0.18% would. In other words, instead of 1 in 20 persons exceeding the upper bound on intake (the presumed goal of worst-case assumptions), only 1 person in 555 would. Applying this concept to procedures which are supposed to ensure that an individual's chance is no greater than 5% of exceeding a cancer risk of 1 in 1,000,000, the individual's actual chance is no greater than 5% of exceeding a cancer risk of 1 in 2,000,000!

In summary, soil contaminated by chemicals, including heavy metals, can pose a health hazard to exposed persons and to wildlife if the concentration and exposure are sufficiently great. However, our experience suggests that the hazard posed by direct exposure to contaminated soil, with the exception of certain well-acknowledged hazards, will frequently be quite low. When evaluating such risks, numerous factors must be considered, but special emphasis must be placed on the assumptions and factors used in estimating exposure. Due to the diversity of situations where soil is contaminated by any number of different chemicals, no generic guideline can be established for a given chemical or class of chemicals, but the procedures suggested here should be most helpful in developing a standardized process for assessing the environmental and human health hazards.

REFERENCES

1. Chanlett, E. *Environmental Protection* (New York: McGraw-Hill, 1979), pp. 66–148.
2. Bergmann, H. L., R. A. Kimerle, and A. W. Maki. *Environmental Hazard Assessment of Effluents* (New York: Pergamon Press, 1986).
3. Kimbrough, R., H. Falk, P. Stehr, and G. Fries. "Health Implications of 2,3,7,8-tetrachlorodibenzo-p-dioxin (TCDD) Contamination of Residential Soil," *J. Toxicol. Environ. Health* 14:47–93 (1984).
4. Hawley, J. "Assessment of Health Risk from Exposure to Contaminated Soil," *Risk Anal.* 5:289–302 (1985).
5. Grisham, J. W., Ed. *Health Aspects of the Disposal of Waste Chemicals* (New York: Pergamon Press, 1986).
6. Paustenbach, D. J., H. P. Shu, and F. J. Murray. "A Critical Analysis of Risk Assessments of TCDD Contaminated Soil," *Regul. Toxicol. Pharmacol.* 6:284–307 (1986).
7. Nesbit, I. C. T., P. LaGoy, and C. Schultz. "The Endangerment Assessment for the Smuggler Mountain Site, Pitkin County, Colorado: A Case Study," in *The Risk Assessment of Environmental and Human Health Hazards*, D. J. Paustenbach, Ed. (New York: John Wiley & Sons, 1988).
8. Fries, G. F. "Assessment of Potential Residues in Foods Derived from Animals Exposed to TCDD Contaminated Soil," *Chemosphere* 16 (8–9): 2123–2128.
9. Eschenroeder, A., R. J. Jaeger, J. J. Ospital, and C. P. Doyle. "Health Risk Assessment of Human Exposures to Soil Amended with Sewage Sludge Contaminated with Polychlorinated Dibenzodioxins and Dibenzofurans." *Vet. Hum. Toxicol.* 28:435–442 (1986).
10. Paustenbach, D. J. "Assessing the Potential Environmental and Human Health Risks of Contaminated Soil," *Comments Toxicol.* 1:185–220 (1987).
11. Kostecki, P. T., and E. J. Calabrese. *Petroleum Contaminated Soils: Public and Environment Health Effects.* (New York: John Wiley & Sons, 1988).
12. Lowrance, W. *Assessment of Health Effects at Chemical Disposal Sites* (Los Altos, CA: Wm. Kaufmann, Inc., 1981).

13. "An Approach to Estimating Exposure to 2,3,7,8-TCDD," EPA Exposure Assessment Group, Washington, DC (1988).
14. Patterson, D. G., R. E. Hoffmann, L. L. Needham, D. W. Roberts, J. L. Bagby, J. R. Pirkle, H. Falk, E. J. Sampson, and V. N. Houk. "2,3,7,8-Tetrachlorodibenzo-p-dioxin Levels in Adipose Tissue of Exposed and Control Persons in Missouri. An Interim Report," *JAMA* 25b:2683–2686 (1986).
15. Menzer, R. E., and J. O. Nelson. "Water and Soil Pollutants, in *Casarett and Doull's Toxicology,* 3rd ed. Klaassan, Amdur, and Doull, Eds. (New York: Macmillan Publishing Co., 1986), pp. 825–853.
16. Sawyer, C. "Contingency Plans," in *Protecting Personnel at Hazardous Waste Sites,* S. Levine, Ed. (Ann Arbor MI: Ann Arbor Science, 1985), pp. 327–349.
17. *Occupational Safety and Health Guidance Manual for Hazardous Waste Site Activities.* NIOSH Pub. 85–115, Government Printing Office, Washington, DC (1985).
18. "Regulations for Workers at Hazardous Waste Sites." *Federal Register,* Occupational Safety and Health Administration (1987).
19. Leung, H. W., and D. J. Paustenbach. "Establishing Acceptable Levels of Occupational Exposure: A Study of 2,3,7,8 TCDD," in *The Risk Assessment of Environmental and Human Health Hazards: A Textbook of Case Studies,* D. J. Paustenbach, Ed. (New York: John Wiley & Sons, 1988).
20. Devine, J. M., G. B. Konishita, R. P. Peterson, and G. L. Picard. *Arch. Environ. Contam. Toxicol.* 15:113–119 (1986).
21. Knaak, J., Y. Iwata, and K. T. Maddy. "Evaluating the Human Health Risks Posed by Exposure to Pesticide Treated Fields," in *The Risk Assessment of Environmental and Human Health Hazards: A Textbook of Case Studies,* D.J. Paustenbach, Ed. (New York: John Wiley & Sons, 1988).
22. Knarr, R. D., G. L. Cooper, E. A. Brian, M. G. Kleinschmidt, and D. G. Graham. "Worker Exposure During Aerial Application of a Liquid and a Granular Formulation of Ordram Selective Herbicide to Rice," *Arch. Environ. Contamin. Toxicol.* 14:523–527 (1985).
23. Durham, W. F., H. R. Wolfe, and J. F. Armstrong. "Exposure of Workers to Pesticides," *Arch. Environ. Health* 14:622–633 (1967).
24. Maddy, K. T., R. G. Wang, J. B. Knaak, C. L. Liao, S. C. Edmiston, and C. K. Winter. "Risk Assessment of Excess Pesticide Exposure to Workers in California," in *Dermal Exposure Related to Pesticide Use.* (Washington, DC: American Chemical Society, 1985), pp.445–465.
25. Popendorf, W. J., and R. C. Spear. "Preliminary Survey of Factors Affecting the Exposure of Harvesters to Pesticide Residues," *Am. Ind. Hyg. Assoc. J.* 35:374–380 (1974).
26. Wojeck, G. A., H. N. Nigg, R. S. Braman, J. H. Stamper, and R. L. Rouseff. "Worker Exposure to Paraquat and Diquat," *Arch. Environ. Contam. Toxicol.* 11:661–669 (1982).
27. Walter, S. D., A. J. Yankel, and I. H. von Lindern. "Age Specific Risk Factors for Lead Absorption in Children," *Arch. Environ. Health* 35:53–58 (1980).
28. Cooper, M. *Pica* (Springfield IL: Charles C. Thomas, Publishers, 1957), pp. 60–74.
29. Charney, E., J. Sayre, and M. Coulter. "Increased Lead Absorption in Inner City Children: Where Does the Lead Come From?" *Pediatrics* 65:226–231 (1980).

30. Sayre, J. W., E. Charney, J. Vostal, and B. Pless. "House and Hand Dust as a Potential Source of Childhood Lead Exposure," *Am. J. Dis. Child* 127:167–170 (1974).

31. Millican, F. K., E. M. Layman, R. S. Lourie, L. Y. Takahashi, and C. C. Dublin. "The Prevalence of Ingestion and Mouthing of Nonedible Substances by Children," *Clinical Proc. Child. Hosp.* 28: 207–214 (1962).

32. Duggan, M. J., and S. Williams. "Lead-in-Dust in City Streets," *Sci. Total Env.* 7:91–97 (1977).

33. Lepow, M. L., L. Bruckman, R. A. Robino, S. Markowitz, M. Gillette, and J. Kapish. "Role of Airborne Lead in Increased Body Burden of Lead in Hartford Children," *Environ. Health Pers.* 6:99–101 (1974).

34. Lepow, M. L., L. Bruckman, M. Gillette, A. Markowitz, R.Robino, and J. Kapish. "Investigations into Sources of Lead in the Environment of Urban Children," *Environ. Res.* 10:415–426 (1975).

35. Barltrop, D. "Sources and Significance of Environmental Lead for Children," in *Proc. Inter. Symp. Env. Health Aspects of Lead.* Commission of European Communities, Center for Information and Documentation, Luxembourg, 1973, pp. 675–681.

36. Brunekreef, B., D. Noy, K. Biersteker, and J. Boleij. "Blood Lead Levels of Dutch City Children and Their Relationship to Lead in the Environment," *J. Air Poll. Control Assoc.* 33:872–876 (1973).

37. *Lead in the Human Environment.* Washington DC: National Research Council, Washington, 1980.

38. Day, J. P., M. Hart, and M. S. Robinson, "Lead in Urban Street Dust," *Nature* 253:343–345 (1975).

39. Page, A. L., and T. J. Ganje. "Accumulation of Lead in Soils for Regions for High and Low Vehicle Traffic Density," *Environ. Sci. Technol.* 4:140–142 (1970).

40. Vostal, J. J., E. J. Taves, W. Sayre, and E. Charney. "Lead Analysis of House Dust: A Method for the Detection of Another Source of Lead Exposure in Inner City Children," *Environ. Health Pers.* 7:91–97 (1974).

41. Barltrop, D. "The Prevalence of Pica," *Am. J. Dis. Child* 112:116–123 (1966).

42. Walter, S. D., A. J. Yankel, and I. H. Von Lindern, "Age Specific Risk Factors for Lead Absorption in Children. *Arch. Environ. Health* 35: 53–58 (1980).

43. Calabrese, E. "How Much Soil Do Children Eat? An Emerging Consideration for Environmental Health Risk Assessment," *Comments Toxicol.* 1:229–241 (1987).

44. Lin-Fu, J. S. "Vulnerability of Children to Lead Exposure and Toxicity," *New England J. Med.* 289:1289–1296 (1973).

45. Roels, H., J. P. Buchet, and R. R. Lauwerys. "Exposure to Lead by the Oral and Pulmonary Routes of Children Living in the Vicinity of a Primary Lead Smelter," *Environment.* 22:81–94 (1980).

46. Bryce-Smith, D. "Lead Absorption in Children," *Phys. Bull.* 25:178–181 (1974).

47. *Air Quality Criteria for Lead.* Vol. II. EPA-600/8-83-028A, U.S. Environmental Protection Agency, Washington, DC (1984b).

48. Bellinger, D., A. Leviton, M. Rabinowitz, M., H. Needleman, and C. Waternaux. "Correlates of Low-Level Lead Exposure in Urban Children at Two Years of Age," *Pediatrics.* Submitted for publication, 1988.

49. Binder, S., D. Sokal, and D. Maughan. "Estimating Soil Ingestion: The Use of

Tracer Elements in Estimating the Amount of Soil Ingested by Young Children,'' *Arch. Environ. Health.* 41:341–345 (1987).

50. Clausing, O., A. B. Brunekreef, and J. H. van Wijnen. "A Method for Estimating Soil Ingestion by Children," *Inter. Arch. Occup. Environ. Health* 59:73–82 (1987).

51. Calabrese, E. J., C. E. Gilbert, P. T. Kostecki, R. Barnes, E. Stanek, P. Veneman, H. Pastides, and C. Edwards, "Epidemiological Study to Estimate How Much Soil Children Eat." A report by the Division of Public Health, University of Massachusetts, Amherst, MA, 1988.

52. LaGoy, P. "Estimated Soil Ingestion Rates for Use in Risk Assessment," *Risk Anal.* 7:355–359 (1987).

53. Vermeer D. L., and D. A. Frate, *Amer. J. Clin. Nutrition* 32:2129–2135 (1979).

54. Gallagher, J. E. J., P. C. Elwood, K. M. Phillips, B. E.Davies, and D. T. Jones. "Relation Between Pica and Blood Lead in Areas of Differing Lead Exposure," *Arch. Dis. Child.* 59:40–44 (1984).

55. Wolfe, H. R., and J. F. Armstrong. "Exposure of Formulating Plant Workers to DDT," *Arch. Environ. Health* 23:170–176 (1971).

56. Wolfe, H. R., D. C. Armstrong, and W. F. Durham. "Exposure of Mosquito Workers to Fenthion," *Mosquito News.* 34:263–267 (1974).

57. Wolfe, H. R., D. C. Staiff, J. F. Armstrong, and J. E. Davis. "Exposure of Fertilizer Mixing Plant Workers to Disulfoton," *Bull. Environ. Contam. Toxicol.* 20:79–86 (1978).

58. Romney, E. M., N. G. Lindberg, H. A. Hawthorne, B. B. Bystrom, and K. H. Larson. "Contamination of Plant Foliage with Radioactive Nuclides," *Ann. Rev. Plant Phys.* 14:271–279 (1963).

59. Martin, W.E. "Loss of Sr-90, Sr-89 and I-131 from Fallout of Contaminated Plants," *Radiation Bot.* 4:275–281 (1964).

60. Russell, R. S. "Entry of Radioactive Materials into Plants," in *Radioactivity and Human Diet,* R. S. Russell, Ed. (New York: Pergamon Press Inc., 1966).

61. Bresson, G., P., J. Lombard, and F. Fagnani."Occupational Radiological Risks in Uranium Mining," in *Inter Conference on Radiation Hazards in Mining* (New York: Society of Mining Engineers, 1981), pp. 111–114.

62. di Domencio, A., V. Silano, G. Viviano, and G. Zapponi. "Accidental Release of 2,3,7,8-Tetrachlorodibenzo-p-dioxin (TCDD) at Seveso, Italy. Part V. Environmental Persistence of TCDD in Soil," *Ecotox. Env. Safety* 4:339–345 (1980).

63. Stevens, K. M. "Agent Orange Toxicity: A Quantitative Perspective," *Human Toxicol.* 1:31–39 (1981).

64. Gough, M. *Dioxin, Agent Orange: The Facts* (New York: Plenum Publishing Corporation, 1986).

65. Paustenbach, D. J. "A Validated Approach to Estimating the Uptake of Non-Volatile Chemicals by Humans Due to Exposure to Environmental Contamination: Case Studies Involving Dioxin," *J. Toxicol. Environ. Health.* Submitted for publication, 1988.

66. Young, A. "Determination and Measurement of Human Exposure to the Dibenzo-p-dioxins," *Bull. Environ. Contam. Toxicol.* 33:702–709 (1984).

67. Land, C. E., in *Perceptions of Risk.* (Washington DC: National Council on Radiation Protection and Measurements, 1980) pp. 169–185.

68. Rupp, E .M., D. C. Parzyck, P. J. Walsh, R. S. Booth, R. J. Raridon, and B. L. Whitfield. "Composite Hazard Index for Assessing Limiting Exposures to Environ-

mental Pollutants: Application Through a Case Study," *Environ. Sci. Technol.* 12:802–807 (1978).

69. Haque, R., Ed. *Dynamics, Exposure and Hazard Assessment of Toxic Chemicals* (Ann Arbor MI: Ann Arbor Science, 1980).

70. Rand, G. M., and S. R. Petrocelli. *Fundamentals of Aquatic Toxicology* (New York: McGraw-Hill, 1985).

71. Bailey, G. W., and J. L. White. "Factors Influencing the Adsorption, Desorption, and Movement of Pesticides in Soil," *Residue Rev.* 32:29–92(1970).

72. Thibodeaux, L. J. *Chemodynamics, Environmental Movement of Chemicals in Air, Water and Soil* (New York: John Wiley & Sons, 1979).

73. Conway, R. *Environmental Risk Assessment* (NY: Van Nostrand, 1982).

74. *Superfund Health Assessment Manual.* Office of Emergency and Remedial Response, U.S. Environmental Protection Agency. Produced by ICF. Clement Inc., under EPA contract 68–01–6872, 1986.

75. Schaum, J. *Risk Analysis of TCDD Contaminated Soil.* Office of Health and Environmental Assessment, U.S. Environmental Protection Agency, Washington D.C., 1983.

76. *Endangerment Assessment Handbook.* Office of Emergency and Remedial Response, U.S. Environmental Protection Agency, Washington, D.C., 1985.

77. Keenan, R. "Assessing the Health Hazard of TCDD Contaminated Sludge as a Soil Amendment," in *Risk Assessments of Human Environmental Health Hazards: A Textbook of Case Studies,* D.J. Paustenbach, Ed. (New York: John Wiley & Sons, 1988).

78. Lipsky, D. *Assessment of Potential Public Health Impacts of Predicted Emissions of PCDD and PCDF from Brooklyn Naval Yard Resource Recovery Facility* in *The Risk Assessment of Environmental and Human Health Hazards,* D. J. Paustenbach, Ed. (New York: John Wiley & Sons, 1988).

79. Sicherman, B. *Alice Hamilton: A Life in Letters* (Cambridge MA: Harvard University Press, 1984), pp. 158–160.

80. Cannon, H. L., and J. M. Bowles. *Science* 137:765–766 (1962).

81. Motto, H. L., R. H. Daines, D. M. Chilko, and C. K. Motto. "Lead in Soils and Plants: Its Relationship to Traffic Volumes and Proximity to Highways. *Environ. Sci. Technol.* 4:231–237 (1970).

82. Markin, G. P. "Translocation and Fate of the Insecticide Mirex Within a Bahia Grass Pasture Ecosystem," *Environ. Pollut.,* 26:227–241 (1981).

83. Poiger, H., and C. Schlatter. "Influence of Solvents and Absorbents on Dermal and Intestinal Absorption of TCDD," *Food Cosmet. Toxicol.* 18:477–481 (1980).

84. Shu, H., P. Teitelbaum, A. S. Webb, L. Marple, B. Brunck, D. Dei Rossi, J. Murray, and D. J. Paustenbach. "Bioavailability of Soil-Bound TCDD: Dermal Bioavailability in the Rat," *Fund. Appl. Toxicol.* 10:335–343 (1988a).

85. Philippi, M., V. Krasnobagew, J. Zeyer, and R. Huetter. "Microbial Metabolism of TCDD Under Laboratory Conditions," in *Microbial Cultures and Soil Under Laboratory Conditions. Fems Symp* 12:2210–2233 (1981).

86. Huetter, R., and M. Philippi. "Studies on Microbial Metabolism of TCDD under Laboratory Conditions," in *Chlorinated Dioxins and Related Compounds, Impact on the Environment.* O. Hutzinger, R. W. Frei, E. Merian, and F. Pocchiari, Eds. (New York: Pergamon Press, 1982), pp. 87–93.

87. Silkworth, J., D. McMartin, A. DeCaprio, R. Rej, P. O'Keefe, and L. Kaminsky. "Acute Toxicity in Guinea Pigs and Rabbits of Soot from a Polychlorinated Biphenyl-

Containing Transformer Fire," *Toxicol. Appl. Pharmacol.* 65:425-429 (1982).

88. Van den Berg, M., K. Olie, and O. Hutzinger. "Uptake and Selective Retention in Rats of Orally Administered Chlorinated Dioxins and Dibenzofurans from Fly-Ash and Fly-Ash Extract," *Chemosphere* 12:537-544 (1984).

89. McConnell, E., G. Lucier, R. Rumbaugh, P. Albro, D. Harvan, J. Hass, and M. Harris, "Dioxin in Soil: Bioavailability after Ingestion by Rats and Guinea Pigs," *Science* 223:1077-1079 (1984).

90. Lucier, G. W., R. C. Rumbaugh, Z. McCoy, J. Hass, R. D. Harvan, and P. Albro. "Ingestion of Soil Contaminated with 2,3,7,8 Tetrachlorodibenzo-p-dioxin (TCDD) Alters Hepatic Enzyme Activities in Rats," *Fund. Appl. Toxicol.* 6:364-371 (1986).

91. Umbreit, T. H., E. J. Hesse, and M. A. Gallo, "Acute Toxicity of TCDD Contaminated Soil from an Industrial Site," *Science* 232:497-499 (1986).

92. Kenaga, E. "Correlation of Bioconcentration Factors of Chemicals in Aquatic and Terrestrial Organisms with Their Physical and Chemical Properties. *Ecotoxicol. Env. Safety* 4:26-38 (1980).

93. Lambert, S. M. "Functional Relationship Between Sorption in Soil and Chemical Structure," *J. Agric. Food Chem.* 15:572-576 (1967).

94. Snyder, W. S. *Report of the Task Group on Reference Man.* International Commission of Radiological Protection, No. 23 (New York: Pergamon Press, 1975).

95. Trijonis, J., J. Eldon, J. Gins, and G. Berglund. *Analysis of the St. Louis RAMS Ambient Particulate Data.* Produced by Technology Service Corporation under EPA contract 68-02-2931 for the Office of Air, Noise, and Radiation of the U.S. Environmental Protection Agency. EPA report 450/4-80-006a, 1980.

96. Cowherd, C., G. Muleski, P. Englehart, and G. Gillette. *Rapid Assessment of Exposure to Particulate Emissions from Surface Contamination Sites,* EPA Control No. 68-01-06861, 1968.

97. *Air Quality Criteria for Particulate Matter and Sulfur Oxides.* Vol. 2. EPA-600-8-82-029bF (Research Triangle Park, NC: U.S. Environmental Protection Agency, Office of Environmental Criteria and Assessment, 1984), pp. 5-106 to 5-112.

98. Shu, H., D. J. Paustenbach, J. Murray, L. Marple, B. Brunck, D. Dei Rossei, A. S. Webb, and T. Teitelbaum. "Bioavailability of Soil-Bound TCDD: Oral Bioavailability in the Rat," *Fund. Appl. Toxicol.* 10: 648-654 (1988b).

99. Bartek, M. J., and J. A. LaBudde. "Percutaneous Absorption, In Vitro in *Animal Models in Dermatology,* H. Maibach, Ed. (New York: Churchill Livingston, 1975), pp. 103-120.

100. Wester, R. C., and P. K. Noonan. "Relevance of Animal Models for Percutaneous Absorption," *Int. J. Pharm.* 7:99-110 (1980).

101. Kaminsky, L. S., A. P. DeCaprio, J. F. Gierthy, J. B. Silkworth, and C. Tumasonis. "The Role of Environmental Matrices and Experimental Vehicles in Chlorinated Dibenzodioxin and Dibenzofuran Toxicity," *Chemosphere* 14:685-695 (1985).

102. Hamaker, J. W., and J. M. Thompson. "Adsorption," in *Organic Chemicals in the Soil Environment,* Vol. I. C.A.I. Goring and J. M. Haymaker, Eds. (New York: Marcel Dekker, 1972), pp. 51-122.

103. Edwards, C. A. "Insecticide Residues in Soils," *Residue Rev.* 13:83-132 (1966).

104. Williams, D. T., G. L. LeBel, and E. Junkins. "A Comparison of Organochlorine Residues in Human Adipose Tissue Autopsy Samples from Two Ontario Municipalities," *J. Toxicol. Environ. Health* 13:19-29 (1984).

105. Castanho, M. "Methods of Soil Sampling," *Comments Toxicol.* 1:221–227 (1987).
106. Rappaport, S. M., and J. Selvin. "A Method for Evaluating the Mean Exposure from a Log-Normal Distribution," *Am. Ind. Hyg. Assoc. J.* 48:374–379 (1987).
107. Puri, R. K., T. E. Clevenger, S. Kapila, and A.F. Yanders. "Studies of the Physico-Chemical Parameters Affecting Translocation of Polychlorinated Dibenzo-p-dioxins in Soil," Abstract TF-09. Seventh International Symposium on Chlorinated Dioxins and Related Compounds," Las Vegas, 1987.
108. DeWeese, L. R., L. C. McEwen, G. L. Hensler, and B. E. Petersen. "Organo-chlorine Contaminants in Passeriformes and Other Avian Prey of the Peregrine Falcon in the Western United States," *Environ. Toxicol. Chem.* 5:675–693.
109. Baes, C. F. III, R. D. Sharp, A. Sjoreen, and R. Shor. "A Review and Analysis of Parameters for Assessing Transport of Environmentally Released Radionuclides Through Agriculture," ORNL-5786. U. S. Department of Energy (Oak Ridge, TN: Oak Ridge National Laboratory, 1984).
110. Stokinger, H. E. "Threshold Limit Values. Part I," in *Dangerous Properties of Industrial Materials Report* (New York: Van Nostrand-Reinhold, 1961), pp. 8–13.
111. Dourson, M. L., and J. F. Stara. "Regulatory History and Experimental Support of Uncertainty (Safety) Factors," *Regul. Toxicol. Pharmacol.* 3:224–238(1983).
112. *Risk Assessment in the Federal Government: Managing the Process,* National Academy of Sciences, National Academy Press, Washington, DC, 1983.
113. Isensee, A. R., and J. R. Jones. "Absorption and Translocation of Root and Foliage Applied 2,4-Dichlorophenol, 2,7-Dichlorodibenzo-*p*-dioxin, and 2,3,7,8-Tetra-chlorodibenzo-*p*-dioxin." *J. Agric. Food Chem.* 19:1210–1214 (1971).
114. Wipf, H. K., E. Homberger, N. Neimer, U. B. Ranalder, W. Vetter, and J. P. Vuilleumeir. "TCDD Levels in Soil and Plant Samples from the Seveso Area," in *Chlorinated Dioxins and Related Compounds: Impact on the Environment,* O. Hutzinger et al., Eds. (New York: Pergamon Press, 1982), pp. 115–126.
115. Breidenstein, B. C., *Contribution of Red Meat to the U.S. Diet* (Chicago IL: National Livestock and Meat Board, 1984).
116. Fries, G. F., and L. W. Jacobs. "Evaluation of Residual Polybrominated Biphenyl Contamination Present on Michigan Farms in 1978," *Mich. State Univ. Agric. Exp. Sta. Res.* Report 477, 1986.
117. Healy, W. B. "Ingestion of Soil by Dairy Cows," *N. Z. J. Agric. Res.* 11:487–499 (1968).
118. Healy, W. B., T. W. Cutress, and C. Michie. "Wear of Sheep's Teeth, IV. Reduction of Soil Ingestion and Tooth Wear by Supplementing Feeding," *N. Z. J. Agric. Res.* 10:201–209 (1967).
119. Healy W. B., and K. R. Drew. "Ingestion of Soil by Hoggets Grazing Swedes," *N. Z. J. Agric. Res.* 13:940–944 (1970).
120. Healy, W. B., and T. G. Ludwig. "Wear of Sheep's Teeth. 1. The Role of Ingested Soil," *N. Z. J. Agric. Res.* 8:737–752 (1965).
121. Mayland, H. F., A. R. Florence, R. C. Rosenau, V. A. Lazar, and H. A. Turner. "Ingestion by Cattle on Semiarid Range as Reflected by Titanium Analysis of Feces," *J. Range Manage.* 28: 48–452 (1975).
122. Thornton, I., and P. Abrahams. "Soil Ingestion as a Pathway of Metal Intake into Grazing Livestock," in *Proceedings of the International Conference on Heavy Metals in the Environment* (Edinburgh: CEP Consultants, 1981), pp. 167–172.
123. Fries, G. F., G. S. Marrow, and P. A. Snow. "Soil Ingestion by Dairy Cattle,"

J. Dairy Sci. 65:611–618 (1982a).

124. Fries, G. F., G. S. Marrow, and P. A. Snow. "Soil Ingestion by Swine as a Route of Contaminant Exposure," *Environ. Toxicol. Chem.* 1:201–204 (1982b).

125. Fries, G. F. "Potential Polychlorinated Biphenyl Residues in Animal Products from Application of Contaminated Sewage Sludge to Land," *J. Environ. Qual.* 11:14–20 (1982).

126. Jensen, D. J., and R. A. Hummel. "Secretion of TCDD in Milk and Cream Following the Feeding of TCDD to Lactating Cows," *Bull. Environ. Contam. Toxicol.* 29:440–446 (1982).

127. Fries, G. F. "Bioavailability of Soil-Borne Polybrominated Biphenyls Ingested by Farm Animals," *J. Toxicol. Environ. Health* 16: 565–579 (1985).

128. Jones, D., S. Safe, E. Morcom, C. Coppock, and W. Ivie. "Bioavailability of Tritiated 2,3,7,8 TCDD Administered to Holstein Dairy Cows," *Chemosphere* 16 (8–9): 1743–1748.

129. Halter, M. T., and H. E. Johnson, in *Aquatic Toxicology and Hazard Evaluation* (Philadelphia, PA: American Society for Testing and Materials, 1977), pp. 178–196.

130. Munro, I. C., and D. R. Krewski. "Risk Assessment and Regulatory Decision Making," *Food Cosmet. Toxicol.* 19:549–560 (1981).

131. Macek, K. J., S. R. Petrocelli, and B. H. Sleight. "Considerations in Assessing the Potential for, and Significance of, Biomagnification of Chemical Residues in Aquatic Food Chains," *Aquat. Toxicol.*, L. L. Marking and R. A. Kimerle, Eds., American Society for Testing and Materials, ASTM STP 667, 1979, pp. 251–268.

132. Spacie, A., and J. Hamelink. "Bioaccumulation" in *Fundamentals of Aquatic Toxicology,* G. M. Rand and S. R. Petrocelli, Eds. (New York: McGraw-Hill, 1985).

133. Reinecke, A. J., and R. G. Nash. "Toxicology of TCDD and Bioavailability by Earthworms," *Soil Biol. Biochem.* 16:45–49 (1984).

134. Oliver, B. G. "Uptake of Chlorinated Organics from Anthropogenically Contaminated Sediments by Obigochaete Worms," *Can. J. Fish Aquat. Sci.* 41:878–883 (1984).

135. Young, A. L., C. E. Thalken, and W. E. Ward. "Studies of the Ecological Impact of Repetitive Aerial Application of Herbicides on the Ecosystem of Test Area C-52A, Eglin Air Force Base, Fl.," *Air Force Technical Report AFATL-TR-75-142, Air Force Assessment Laboratory Eglin AFB, Florida,* 127 p. (Springfield, VA: Natl. Tech. Info. Services, 1975).

136. Young, A. L., and L. G. Cockerham. "Fate of TCDD in Field Ecosystems: Assessment and Significance for Human Exposures," in *Dioxins in the Environment,* M. A. Kamrin and P. W. Rodges, Eds. (New York: Hemisphere Publishing Corp., 1985), pp. 153–171.

137. Thalken, C. E., and A. L. Young. "Long-Term Field Studies of a Rodent Population Continuously Exposed to TCDD," in *Human and Environmental Risks of Chlorinated Dioxins and Related Compounds,* R. E. Tucker, A. L. Young, and R. Gray, Eds. (New York: Plenum Publishing Corp., 1983).

138. Young, A. L., L. G. Cockerham, and C. E. Thalken. "A Long Term Study of Ecosystem Contamination with TCDD," *Chemosphere.* 16: (8–9) 1791–1816 (1987).

139. Leung, H. W., and D. J. Paustenbach. "Establishing Occupational Exposure Limits: Case Studies of Different Classes of Chemicals," in *The Risk Assessment of Environmental and Human Health Hazards: A Textbook of Case Studies* (New York: John Wiley & Sons, 1988).

140. Lehmann, A. J., and O. G. Fitzhugh. "100-Fold Margin of Safety," *Q. Bull.- Assoc. Food Drug Off. U. S.* 18:33–35(1954).
141. Lehmann, A. J., F. A. Vorhes, et al. *Appraisal of the Safety of Chemicals in Foods, Drugs, and Cosmetics.* Published by the Association of Food and Drug Officials of the United States, Washington DC, 1959.
142. "Quantitative Risk Assessment" in *Food Safety Assessment* (Washington DC: Nutrition Foundation, Food Safety Council, 1980).
143. Rodricks, J. V., S. N. Brett, and G. C. Wrenn. "Risk Decisions in Federal Regulatory Agencies," *Regul. Toxicol. Pharmacol.* 7:307–320(1987).
144. Travis, C. C., S. A. Richter, C. E. Crouch, R. Wilson, and E. D. Klema. "Cancer Risk Management," *Environ. Sci. Technol.* 21:415–420 (1987).
145. "National Emission Standards for Hazardous Air Pollutants; Benzene Emissions from Maleic Anhydride Plants, Ethylbenzene/Styrene Plants, and Benzene Storage Vessels; Proposed Withdrawal of Proposed Standards," 49 *Federal Register* 8386, 8388 (March 6, 1984) ["Benzene NESHAPS Withdrawal Notice"]. Environmental Protection Agency (EPA) (1984).
146. "National Emission Standards for Hazardous Air Pollutants; Regulation of Radionuclides; Withdrawal of Proposed Standards," 49 *Federal Register* 43906, 43910 (Oct. 31, 1984) ["Radionuclides NESHAPS Withdrawal Notice"]. Environmental Protection Agency (EPA) (1984d).
147. Leung, H. W., F. J. Murray, and D. J. Paustenbach. "A Proposed Occupational Exposure Limit for 2,3,7,8 TCDD," *Am. Ind. Hyg. Assoc. J.* In press, 1988.
148. Stickel, L. F. "Pesticide Residues in Birds and Mammals," in *Environmental Pollution by Pesticides* C. A. Edwards, Ed. (London and New York: Plenum Press, 1973).
149. Korschgen, L. J. "Soil-Food-Chain-Pesticide Wildlife Relationships in Aldrin-Treated Fields," *J. Wildlife Management* 34:186–199(1973).
150. Rand, G. M. "Avian Hazard Posed by Pesticides," in *The Risk Assessment of Environmental and Human Health Hazards,* D. J. Paustenbach, Ed. (New York: John Wiley & Sons, 1988).
151. Fries, G. F., and D. J. Paustenbach. "Assessment of Potential TCDD Risks Due to Ingestion of Foods Derived from Animals Exposed to Particulates from Municipal Incinerators," *Toxicol. Environ. Health* (submitted).
152. Ames, B. N. "Six Common Errors Relating to Environmental Pollution," *Regul. Toxicol. Pharmacol.* 7:379–383.

CHAPTER 21

Review of Present Risk Assessment Models for Petroleum Contaminated Soils

Paul T. Kostecki, Edward J. Calabrese, and Holly M. Horton

Seven risk assessment methodologies were evaluated as a preliminary means of developing a risk assessment methodology that can be applied to petroleum contaminated soils. By determining the strengths and weaknesses of these methodologies, a stronger risk assessment process can be developed.

A risk assessment methodology should include variables which clearly represent the hazard level, exposure level, and the level of risk at a particular site. Criteria used to assess the strengths and weaknesses of the hazard, exposure, and risk analyses are discussed in the following section.

CRITERIA USED TO ASSESS RISK ASSESSMENT METHODOLOGIES

Hazard Analysis

The toxicities of contaminants at a site are determined during the hazard analysis. Animal studies and human studies, when available, are used to derive toxicity values based on dose-response relationships and statistical models. The manner in which toxicity values are derived should be compared among the different methodologies. If the toxicity values are based on already existing standards, incorporation of safety factors or modifiers may be necessary.

Chemicals at a site can have diverse short-term and long-term effects. In order to account for these varying effects over time, noncarcinogenic acute and chronic toxicity should be determined as well as carcinogenic chronic effects. In determination of acute toxicity values for children, body weight factors may be used to modify toxicity values originally derived for adult toxicity.

Pharmacokinetic factors, when available, can account for differences in absorption and metabolism due to different exposure routes (i.e., water ingestion versus air inhalation of a chemical). Without inclusion of pharmacokinetic factors, the same toxicity value used for different exposure routes may not adequately represent the hazard level. For example, toxicity due to ingestion of a chemical may be grossly different from toxicity due to inhalation of the same chemical.

Since many sites are contaminated with a multitude of chemicals, a procedure for evaluation of mixtures of chemicals at a site is important. A standard procedure in a risk assessment methodology would aid in identification of the most toxic chemicals, which can then represent the overall toxicity of a mixture. In this manner, fewer chemicals need to be individually evaluated, thereby streamlining the risk assessment process.

Exposure Analysis

The exposure analysis evaluates the means by which humans encounter chemicals originating from a contaminated site. Exposure routes may include air and dust inhalation, water ingestion, soil ingestion, dermal absorption, and ingestion of crops, livestock, or fish which have been exposed to a contaminant from the site. Not all of these exposure routes will be important for a particular site; however, a useful risk assessment methodology would have the capacity to include any of these factors when relevant for a specific site.

Environmental fate analyses may include variables such as air transport, surface water or groundwater transport, and bioconcentration of chemicals in plants, fish, and livestock. This type of analysis is important because it can indicate not only present exposure routes but future exposure routes as the chemicals travel through the environment. Within environmental fate analyses, chemical interactions are also important and may include factors such as solubility, vapor pressure, half-life, and soil adsorption characteristics. Site-specific characteristics such as geology, climate, soil type, and location of aquifers, surface waters, and runoff areas have a direct influence on the environmental fate of chemicals at a site. A specific environmental fate model may be used in this step which will dictate the types of measurements and analyses necessary.

Risk Analysis

In the risk analysis, the human exposure level for a chemical and the toxicity resulting from such exposure is compared to a critical toxicity value for the chemical. The critical toxicity value generally represents an acceptable exposure level for the chemical. If the critical toxicity value is exceeded, remedial actions are recommended.

The risk analysis should include additivity of the hazard level for chemicals which have the same toxic effect (i.e., erythrocyte hemolysis, nerve damage, liver damage, etc.). When two or more chemicals exert the same effect, the level of risk should be reduced so that a safe level is determined for all of the chemicals. Similarly, the total carcinogenic risk for all carcinogens of concern at a site should be considered so that the total cancer risk is acceptable. In cases where synergistic effects are known, this information can also be incorporated into a risk analysis.

Humans may be exposed to one chemical from multiple sources due to air, dust or water transport; and fish, crop, or livestock uptake. Exposure via different media should be considered in order to evaluate the cumulative toxicity from all media.

All of the criteria used to evaluate the seven risk assessment methodologies are listed in Table 1. These criteria attempt to evaluate the most important variables involved in the hazard, exposure, and risk analyses.

Seven methodologies were reviewed and include: the California Site Mitigation Decision Tree,[1] Rosenblatt et al.,[2] EPA Superfund,[3] Ford and Gurba,[4] New Jersey Method as proposed by Stokman and Dime,[5] the State of Washington Cleanup Policy,[6] and California's Leaking Underground Fuel Tank (LUFT) Field Manual.[7] Each methodology is described and critically evaluated. A summary table follows each methodology (Tables 2–7) along with a final tabular summary (Table 9), directly comparing each of the seven methodologies.

Table 1. Criteria Used to Evaluate Risk Assessment Methodologies.

Hazard Analysis
  1. Derivation of toxicity values
  2. Noncarcinogenic acute toxicity
  3. Noncarcinogenic chronic toxicity
  4. Carcinogenic chronic toxicity
  5. Assessment of mixtures
  6. Body weight factor
  7. Pharmacokinetic factors

Exposure Analysis
  1. Air inhalation
  2. Dust inhalation
  3. Water ingestion
  4. Soil ingestion
  5. Dermal absorption
  6. Crop uptake
  7. Livestock uptake
  8. Fish uptake
  9. Environmental fate
  10. Half-life factor
  11. Site-specific factors

Risk Analysis
  1. Additivity of the hazard level for chemicals having a common toxic effect
  2. Additivity of the hazard level for multimedia exposure
  3. Synergistic effects

CALIFORNIA SITE MITIGATION DECISION TREE

Summary

The California Site Mitigation Tree[1] consists of four major steps: (1) identification of toxic substances, (2) determination of Applied Action Levels (AALs), (3) environmental fate modeling, and (4) risk analysis.

The identification of toxic substances involves review of chemical information concerning substances at the site. The available data are evaluated for quality and adequacy. The chemicals of interest are classified as carcinogens or noncarcinogens.

AALs are defined as media-specific levels of a substance which, if exceeded, present a significant risk. The AALs are derived from No Observed Adverse Effect Levels (NOAELs) for noncarcinogens and from risk factors for carcinogens. Standard body weight, standard intake, pharmacokinetic, and uncertainty factors are included in the derivation of the AALs.

Mathematical models are utilized to determine the environmental fate of a chemical of interest. Soil adsorption, bioconcentration, and emission rates are evaluated in this step. Exposure levels determined from the environmental fate analysis are used in the risk analysis.

The risk analysis involves comparison of exposure levels with AALs for specific chemicals and media. Multimedia exposures and additive toxic effects are taken into account. If exposure levels are found to exceed the AAL for any chemicals, a significant risk is deemed to exist.

Strengths and Weaknesses

The first step in the California risk assessment process involves identification and separation of the toxic substances from the nontoxic substances at the site. Pertinent toxicity information is reviewed for the chemicals of interest. Guidelines are offered for evaluating the quality and adequacy of data, and the chemicals are classified as carcinogens or noncarcinogens.

Problems with the toxicity analysis include: (1) no rating system is offered for ranking chemicals having a variety of toxic effects and physical/chemical properties, (2) it is unclear how the review of toxicity information is integrated into the risk assessment methodology, and (3) no process is offered for narrowing down chemicals from a complex mixture of substances.

AALs are determined in the second step of the risk assessment. AALs are media-specific levels of a substance which, if exceeded, presents a significant risk. AALs are specific for chemical and media, and can be applied to carcinogens or noncarcinogens. AALs for noncarcinogens are derived from NOAELs, while AALs for carcinogens are derived from risk factors for specific chemicals. The NOAEL or risk factor is used to determine a Maximum Exposure Level (MEL). MELs are modified by standard intake, pharmacokinetic, and uncertainty factors to determine the AAL:

Epidemiological studies which determine a NOAEL are considered to be the best basis for development of an AAL.

$$\text{MEL (mg/day)} = \frac{\text{NOAEL (mg/kg/day)}}{\text{Uncertainty Factor (10)}} \times \text{adult body weight (70 kg)}$$

The uncertainty factor of 10 is designed to protect the more sensitive members of the population. AALs are then developed for specific media using the MEL:

$$\text{AAL}_{\text{water}}(\text{mg/L}) = \frac{\text{MEL (mg/day)}}{\text{Average daily intake of water}} \times \text{Pharmacokinetic Factors (PF)}$$
$$\text{(2 L)}$$

$$\text{AAL}_{\text{air}}(\text{mg/m}^3) = \frac{\text{MEL (mg/day)}}{\text{Average daily intake of water}} \times \text{Pharmacokinetic Factors (PF)}$$
$$\text{(20 m}^3\text{/day)}$$

Pharmacokinetic factors (PF) may include differences in absorption, distribution, metabolism, and excretion due to different routes of exposure. AALs are not determined for exposures via soil ingestion or dermal absorption.

If epidemiological studies are unavailable for the chemical of interest, NOAELs from long-term animal bioassays can be used. An uncertainty factor of 100 is used to account for uncertainties in animal to human extrapolation:

$$\text{NOAEL}_{\text{human}}(\text{mg/kg/day}) = \frac{\text{NOAEL}_{\text{animal}}(\text{mg/kg/day})}{100} \times \text{PF}$$

AALs may be derived from Threshold Limit Values (TLVs), which are designed to protect the worker from toxic agents in occupational settings. TLVs are time-weighted average exposures based on an eight-hour workday and five-day work week. MELs derived from TLVs only utilize TLVs based on chronic effects, and the TLV is extrapolated to a 24-hour day and seven-day week. An uncertainty factor of 10 to 100 is applied:

$$\text{NOAEL (mg/day)} = \text{TLV (mg/m}^3) \times \frac{8 \text{ hours}}{24 \text{ hours}} \times \frac{5 \text{ days}}{7 \text{ days}} \times \frac{47 \text{ years}}{70 \text{ years}} \times \frac{20\text{m}}{\text{day}}$$

$$\text{MEL (mg/day)} = \frac{\text{NOAEL (mg/day)}}{\text{Uncertainty Factor (10 to 100)}}$$

AALs for air and water exposure are then derived from the TLV for air exposure. The validity of this type of extrapolation should be questioned. Extrapolations from air to water exposures and from occupational to lifetime exposures may introduce significant errors. TLVs account for intermediate exposures, while the AALs must account for lifetime, continuous exposure.

In cases where no chronic toxicity data is available, NOAELs can be estimated by extrapolation from subchronic toxicity data. In subchronic studies, animals are exposed to a toxic agent for a fraction of their lifespan (usually 90 days). An uncertainty factor of 10 is used to account for uncertainties in extrapolation from subchronic to chronic data:

$$NOAEL_{chronic}(mg/kg/day) = \frac{NOAEL_{subchronic}(mg/kg/day)}{Uncertainty\ Factor\ (10)}$$

$$NOAEL_{human}(mg/kg/day) = \frac{NOAEL_{animal}(mg/kg/day)}{Uncertainty\ Factor\ (100)} \times PF$$

NOAELs may also be developed from acute toxicity studies which establish an LD_{50} value. In this case, an uncertainty factor of 1000 is used.

$$NOAEL_{chronic}(mg/kg/day) = \frac{Oral\ LD_{50}(mg/kg)}{1000}$$

In case of little or no toxicity data, AALs can be determined using structure activity analyses. This type of analysis involves a regression analysis which compares a specific predictor variable with toxicity. Examples of predictor variables are octanol/water partition coefficients, molecular weight, solubility, density, boiling point, vapor pressure, etc.

Problems with this type of analysis include the fact that the predictor variable must be carefully chosen and be closely related to the toxic properties of the specific chemical. If the predictor variable does not adequately represent the toxicity of the agent, significant errors may result.

The development of AALs for carcinogens is based upon epidemiological studies or long-term animal studies which delineate a level of risk:

$$MEL\ (mg/day) = \frac{Risk\ Factor\ (mg/day)}{Level\ of\ risk\ (10^{-6}\ or\ less)}$$

It is unclear how the formula for determination of the MEL establishes a safe level of exposure. The term 'risk factor' is defined as the probability of additional cancers which could result from the exposure of a population to a substance of concern.

The environmental fate of a chemical is analyzed in detail using various mathematical models. Environmental fate is evaluated for the factors of soil adsorption, bioconcentration, and emission rates. The exposure levels of the specific chemicals derived from environmental fate modeling are then used in the risk analysis.

The risk analysis compares the AAL for a chemical in a specific media with concentration levels of the chemical at the site. Three types of comparisons are utilized in the risk analysis.

The first comparison compares the level of exposure in one media to the AAL for the same media:

$$\frac{\text{concentration in one medium}}{\text{AAL in one medium}}$$

The second comparison evaluates cumulative exposure to one chemical over different media:

$$\sum_{medium=1}^{m} \frac{\text{concentration}_m}{\text{AAL}_m}$$

The third comparison evaluates cumulative exposure to different chemicals which exert the same toxic effect over different media:

$$\sum_{substance=1}^{s} \sum_{medium=1}^{m} \frac{\text{concentration}_{m,s}}{\text{AAL}_{m,s}}$$

If any of the three ratios exceeds 1, risk management should be initiated.

The risk analysis is strong in its consideration of multiple exposure pathways and cumulative effects for one toxic endpoint. However, only air and water exposure pathways are assessed. In addition, if NOAELs are unavailable for certain chemicals it may be difficult to assess the risk posed by such chemicals using this risk assessment methodology. Alternatives for such a situation include use of TLVs or structure-activity analyses, which may introduce significant errors. In addition, the California risk assessment method does not utilize already established relevant/applicable ambient standards which could speed up the risk analysis.

Conclusions

The California Site Mitigation Tree is useful in its evaluation of multiple exposure pathways (air and water), additive toxic effects, and environmental fate modeling. Toxic effects are evaluated for both carcinogens and noncarcinogens, while the environmental fate modeling considers soil adsorption, bioconcentration, and emission rates.

Weaknesses of this methodology include the fact that exposures via soil ingestion or dermal absorption are not evaluated. In addition, there are no guidelines for assessment of complex mixtures of chemicals, such as petroleum products. Since every chemical in a complex mixture cannot be evaluated, a means of determining the most hazardous substances is necessary.

If NOAELs are unavailable from scientific studies they must be derived from TLVs or structure-activity analyses, which, again, can introduce serious errors. The use of relevant, ambient standards, when available, is not suggested in the California methodology as a means of streamlining the risk assessment process. Therefore, the California risk assessment plant tends to be time-consuming and not applicable to exposures resulting from soil ingestion or dermal absorption (Table 2).

Table 2.   Summary of the California Site Mitigation Decision Tree Methodology with Respect to the Evaluation Criteria.

HAZARD ANALYSIS

Toxicity value	NOAEL[a]
	—epidemiologic
	—chronic
	—subchronic
	TLV
	LD_{50}
	Structure activity analysis
Noncarcinogenic acute toxicity	Yes
Noncarcinogenic chronic toxicity	Yes
Carcinogenic chronic toxicity	Yes
Assessment of mixtures	No
Body weight factor	Yes
Pharmacokinetic factor	Yes

EXPOSURE ANALYSIS

Air inhalation	Yes
Dust inhalation	Yes
Water ingestion	Yes
Soil ingestion	No
Dermal absorption	No
Crop uptake	Yes
Livestock uptake	Yes
Fish uptake	Yes
Environmental fate	Yes
	—soil adsorption
	—bioconcentration
	—air transport
	—water transport
Half-life factor	No
Site-specific factors	Yes

RISK ANALYSIS

Additivity of toxic effect	Yes
Multi-media exposure	Yes
Synergistic effects	No

[a]See text for description of toxicity values.

AN ENVIRONMENTAL FATE MODEL LEADING TO PRELIMINARY POLLUTANT LIMIT VALUES FOR HUMAN HEALTH EFFECTS[2]

Summary

Preliminary Pollutant Limit Values (PPLVs) are acceptable concentrations of a pollutant at a site. PPLVs are derived from Single Pathway PPLVs (SPPPLVs) for a specific contaminant. SPPPLVs represent linear exposure pathways from the site to the human receptor (e.g., soil → water → human or soil → plant → human, etc.). SPPPLVs are based upon the acceptable daily dose of the pollutant (D_T), partition coefficients, standard intake factors, and a body weight factor.

The hazard posed by a specific contaminant is represented by D_T which is based on already established standards, Allowable Daily Intake (ADI), MCL, or TLV, or derived from the no effect level (NEL) or LD_{50} from animal studies. The exposure level is represented by partition coefficients which may include octanol-water partition coefficient (K_{OW}), soil organic carbon adsorption coefficient (K_{OC}), water solubility, saturation vapor density, and others. All SPPPLVs which represent critical exposure pathways are factored into the calculation of PPLVs to determine the cleanup level for a specific contaminant.

Strengths and Weaknesses

The PPLV is a temporary, nonregulatory value that is based on information available in the literature. The PPLV model is flexible and can be modified to account for site-specific conditions. It represents an acceptable cleanup level for a contaminant, and the derivation of this value involves the following major steps.

- Pollutants and pathways are identified.
- An acceptable daily dose of pollutant, D_T, and partition coefficients are determined.
- SPPPLVs are calculated for all potential exposure pathways.
- Critical pathways for each pollutant are selected.
- The PPLV is derived by normalization of SPPPLVs.

In the initial step, pollutants and pathways are identified; however, further guidelines on this step are not given. In addition, guidelines are not offered on how to evaluate complex mixtures of chemicals.

The acceptable daily dose of toxicant, D_T, may be obtained from six sources. Listed in order of preference, these values are: (1) ADI; (2) MCL for drinking water; (3) TLV; (4) Lifetime No Effect Level (NEL_L); (5) Ninety day No Effect Level (NEL_{90}); and (6) Acute Toxicity (LD_{50}).

The TLV must be converted from 5 days/week exposure to 7 days/week exposure by dividing by (7/5 = 1.4). A safety factor of 100 is used to protect sensitive members of the population who would normally not work. The breathing rate, RB, for a 70-kg person doing light work, 12.1 m 3/8 hours, is multiplied times the TLV. Therefore, the TLV is converted as follows: $D_T = (TLL \times RB)/1.4 \times Body\ Weight \times 100$.

The No Effect Level from a lifetime animal study, NEL_L, is adjusted by a safety factor of 100 to account for interspecies differences. The No Effect Level from a 90-day subchronic study, NEL_{90}, requires a safety factor of 1000 due to the shorter period of exposure and interspecies differences. The LD_{50} from an acute toxicity animal study is used when none of the other values are available. A conversion factor is multiplied times the LD_{50} to yield a safe limit for continuous body concentration.

The D_T value based on one of the latter six factors (ADI, MCL, TLV, NEL_L, NEL_{90} or LD_{50}) represents the acceptable daily intake of a noncarcinogenic threshold pollutant. It is unclear how D_T should be determined for a carcinogenic nonthreshold pollutant. In addition, D_T values for ingestion and inhalation exposure are assumed to be the same. This assumption may introduce significant error for some pollutants.

Each exposure pathway involves a linear compartment model through which the pollutant passes (e.g., soil → air → human). The PPLV model assumes that between two adjacent media, the pollutant is partitioned in a constant manner and all compartments are assumed to be at equilibrium.

One assumption of the PPLV model is that the pollutant is conserved in all compartments. Half-life or biodegradation factors are not considered and, therefore, overestimates of risk may be predicted for chemicals with short half-life factors.

Partition coefficients may not be available for some pollutants or may have to be estimated on scanty data. The value of a partition coefficient may also vary with local site conditions such as soil type, climate, etc. If no data is available to estimate the partition coefficient, the method suggests that K = 1.

The PPLV methodology does not specifically address exposure via soil ingestion or dermal absorption. However, new information or different exposure variables can be incorporated into the standard equation.

The derivation of SPPPLV is based on the acceptable daily dose of pollutant (D_T), partition coefficients for a specific exposure pathway, standard intake factors, and a body weight factor. If one or more intermedia transfers occurs in an exposure pathway (e.g., soil → plant → human), the SPPPLV has the following form:

$$C_i = \frac{BW \times D_T}{(Km_1m_2 \times Km_2m_3 \ldots Km_xm_y) \times (f \times DFI)}$$

C_i = acceptable pollutant concentration in one exposure pathway (mg/kg)
BW   = body weight factor (kg)
D_T = acceptable daily dose of pollutant (mg/kg/day)
Km_xm_y = partition coefficients for exposure pathway
$f \times DFI$ = standard intake factor (kg/day) (fraction of total diet represented by food of a given type × daily food intake)

If the pollutant reaches the human receptor without an intermediate compartment (e.g., water→human), the SPPPLV has the following form.

$$C_F = \frac{BW \times D_T}{W_i}$$

W_i = Standard intake factor (e.g., water intake)

Partition coefficients are not needed in direct transfer of pollutant from site to human, as in the case of pollutants originating in drinking water.

Once the critical exposure pathways are identified, the appropriate SPPPLVs are used to determine the PPLV.

$$C_F = \frac{1}{\sum\limits_{i=1}^{n} \frac{1}{C_i}}$$

C_F = final acceptable pollutant concentration for all exposures
C_i = acceptable pollutant concentration in one exposure pathway

The example is given that if C_i was 10 ppm by water ingestion, 5 ppm by fish ingestion, and 20 ppm by crop ingestion, the PPLV would be: $C_F = 1 - (1/5 + 1/10 + 1/20) = 2.86$ ppm. The C_F represents the cleanup level for the specific pollutant, based on all critical exposure pathways.

Conclusions

The PPLV method is easy to use, and evaluates multiroute exposures and intermedia transfers of pollutants. The use of partition coefficients simplifies the mathematical modeling necessary in other methodologies for exposure pathway analysis. The PPLV model is made to be flexible in order to incorporate new information as it becomes available.

An acceptable daily doses of pollutant, D_T, is based on ADI, MCL, TLV, NEL, or LD_{50} values for threshold noncarcinogenic pollutants. The derivation of D_T for carcinogenic pollutants is not detailed. In addition, D_T values are the same for ingestion and inhalation exposures, despite the fact that the hazard level of a pollutant may vary with route of exposure.

The PPLV methodology does not address the problem of how to classify and assess the risks presented by a large, diverse mixture of chemicals at one site. The possibility of additive toxic or carcinogenic effects is also not evaluated. If numerous chemicals at one site exert a cumulative systemic effect, the hazard could be much greater than that represented by the PPLV.

The advantages of use of partition coefficients to determine exposure levels is that this simplifies the need for modeling of release rates of chemicals from a site. Major assumptions which should be noted are (1) all compartments (media) are assumed to be at equilibrium, and (2) no decomposition of the chemical occurs. The first assumption may be violated as a result of local site conditions. At a particular site, the standard partition coefficients may not represent the actual transfer ratio between two media, due to such factors as soil type, weather patterns, etc.

The second assumption could present problems if the chemical of interest has a short half-life. A chemical with a short half-life would present less of a hazard than a chemical with a longer half-life. Therefore, the PPLV could overestimate or underestimate risk.

Partition coefficients may be difficult to locate for some chemicals, and these values may have to be based on a single literature value. Since the environmental fate of the pollutants is based solely on this value, the PPLV cleanup level is very sensitive to errors in determination of partition coefficients.

Exposure via soil ingestion and dermal absorption is not specifically addressed; however, as previously noted, these factors could be incorporated into the standard formula.

The PPLV methodology is site-specific, and evaluates the potential for a pollutant to proceed from its point of origin through specific pathways to a target receptor. The calculations are easy to use; however, some error may be introduced when major assumptions of the model are violated. The authors emphasize, however, that the model is flexible and new data should be incorporated to account for site-specific conditions (Table 3).

Table 3. Summary of the U.S. Army Methodology with Respect to the Evaluation Criteria.

HAZARD ANALYSIS	
Toxicity value	ADI^a
	MCL
	TLV
	NEL_L
	NEL_{90}
	LD_{50}
Noncarcinogenic acute toxicity	Yes
Noncarcinogenic chronic toxicity	Yes
Carcinogenic chronic toxicity	No
Assessment of mixtures	No

Table 3.   (Continued)

Body weight factor	Yes
Pharmacokinetic factor	No
EXPOSURE ANALYSIS	
Air inhalation	Yes
Dust inhalation	Yes
Water ingestion	Yes
Soil ingestion	No
Dermal absorption	No
Crop uptake	Yes
Livestock uptake	Yes
Fish uptake	Yes
Environmental fate	Yes
	—partition coefficients
	—bioconcentration
	—air transport
	—water transport
Half-life factor	No
Site-specific factors	Yes
RISK ANALYSIS	
Additivity of toxic effect	No
Multi-media exposure	Yes
Synergistic effects	No

aSee text for description of toxicity values.

SUPERFUND HEALTH ASSESSMENT MANUAL

Summary

EPA's *Superfund Health Assessment Manual*[3] proceeds from a relatively simple qualitative assessment of available information to a more detailed quantitative risk assessment. The public health evaluation involves three major components:

- Baseline site evaluation
- Public health assessment of the no-action alternative
- Development of design goals and estimation of risk for remedial alternatives

These three components are outlined in the following summary.

Baseline Site Evaluation

The initial step involves determination of the extent of contamination at a site and may be qualitative (Level 1) or quantitative (Level 2). A Level 2 assessment is required at sites in which migration of contaminants is possible, exposure has occurred or is imminent, and/or a large population resides near the site.

Public Health Assessment of the No-Action Alternative

After the baseline site evaluation has indicated whether a Level 1 or Level 2 assessment will be conducted, the regulatory agency must next assess the no-action alternative.

Level 1 assessment of no-action alternative. The Level 1 assessment involves reviewing available information on types and amounts of chemicals, toxic effects, proximity of human populations, likelihood of chemical release and migration, and potential for human exposure. A comprehensive exposure pathway analysis is required which evaluates the following four criteria:

- source and mechanism of chemical release
- environmental transport medium
- human exposure point
- feasible human exposure route

If a complete exposure pathway is found to exist (all four elements are present) then a more detailed Level 2 assessment should be conducted. Both short term and long term exposure pathways should be considered. A Level 1 assessment is conducted when no migration of contaminants is probable over the short term or long term. Remedial actions after a Level 1 assessment generally involve source control measures based on relevant/applicable standards.

Level 2 assessment of no-action alternative. The Level 2 assessment of the no-action alternative involves five analytical steps:

- selection of indicator chemicals
- assessment of exposure pathways
- estimation of human intakes
- toxicity assessment of indicator chemicals
- characterization of human health risks

The procedures required for these five steps are outlined in the following sections.

1. Selection of indicator chemicals: When a site is contaminated with more than 10 chemicals, a subset of indicator chemicals is selected based upon toxicity, mobility, persistence, and concentration. The indicator chemicals are ranked using the following formula:

$$\text{Indicator score}_i = \text{Concentration}_i \times \text{Toxicity constant}_i$$

Toxicity constants for noncarcinogens are derived from the minimum effective dose (MED) for chronic effects, severity of effects rating, standard factors for body weight, and oral or inhalation intake. Toxicity constants for potential carcinogens are based on the dose at which a 10% incremental carcinogenic response is observed (ED_{10}) and standard intake and body weight factors. Intake factors for soil toxicity constants are based on the assumption of 100 mg/day of soil consumed from ages 2 to 6, averaged over a 70-year life span.

Potential carcinogens and noncarcinogens are scored and selected independently due to the fact that indicator scores for these two groups are not on comparable scales. Indicator scores (IS) are summed for each chemical in different media (air, water, soil) to obtain one IS value for each chemical. The IS values are then ranked, and the 10 to 15 top-scoring chemicals within the two groups become the preliminary list of indicator chemicals.

Other chemical properties which are considered prior to selecting the final set of indicator chemicals are water solubility, vapor pressure, Henry's Law constant, organic carbon partition coefficient, and half-life in various media. The final selection should consider the latter properties in the context of the particular exposure pathways at the site.

2. Assessment of exposure pathways: A combination of site monitoring data and environmental modeling results are used to estimate exposure pathways. Because site monitoring data alone will not reveal future conditions, environmental fate modeling is recommended by EPA. Simple environmental fate equations are provided, and more sophisticated computer models are referenced in the manual. EPA suggests that simple models using conservative assumptions be used for Superfund sites.

Exposure pathways must be identified based on the four elements listed in Level 1 assessment (release source, environmental transport media, exposure point, and exposure route). If one of the latter characteristics is missing, the exposure pathway is incomplete.

EPA suggests that chemical releases be quantified in terms of release rates to predict environmental fate. Air, soil, surface water, and groundwater release modeling are described in the manual. Environmental fate and transport modeling is then conducted based on release rates.

The projected concentrations of the indicator chemicals at the exposure points are compared to relevant/applicable standards. If all of the indicator chemicals have relevant/applicable standards, then a full quantitative risk characterization is not required. EPA established a hierarchy of standard values to be applied to Superfund sites:

- relevant/applicable standards for appropriate exposure pathway (EPA considers drinking water maximum contaminant levels (MCLs), state water quality standards, and ambient air quality standards to be the only relevant/applicable standards)
- toxicity values based on EPA's Health Effects Assessment (HEA)
- other standards or criteria such as water quality criteria

If standard values exist for all of the indicator chemicals, the Level 2 assessment of the no-action alternative is complete. The ambient standards can then be used in the development of target concentrations for remedial alternatives. If ambient standards do not exist for all of the indicator chemicals, exposure point concentrations are used to calculate chemical intakes for which risk is estimated.

3. Estimation of human intakes: Subchronic and chronic daily chemical intakes (SDI and CDI) are based on the previously determined short-term and long-term concentrations of the indicator chemicals at the exposure points (step 2). Intake is defined as the amount of contaminant taken into the body per unit body weight per unit time (mg/kg/day). Intakes are calculated separately for each chemical in each medium (air, groundwater, and surface water). Groundwater and surface water intake are summed for each chemical to yield the total oral intake. The following formulas are used to calculate intakes:

$$\text{Subchronic daily intake (SDI)} = \frac{\text{Short-term}}{\text{concentration}} \times \frac{\text{Standard}}{\text{intake factor}}$$

$$\text{Chronic daily intake (CDI)} = \frac{\text{Long-term}}{\text{concentration}} \times \frac{\text{Standard}}{\text{intake factor}}$$

Standard Intake Factors Used by EPA

	Adult	*Child*
Avg. body weight	70 kg	10 kg
Amt. of water/day	2 L	1 L
Amt. of air/day	20 m^3	5 m^3
Amt. of fish/day	6.5 g	—

EPA assumes 100% absorption of the contaminant after intake, due to uncertainties in present data. In addition, formulas for less common exposure routes (soil ingestion, dermal absorption) are not included in the manual. If such an exposure route is important at a Superfund site, EPA suggests that the regulatory agency contact EPA headquarters for guidance on a case-by-case basis.

4. Toxicity assessment of indicator chemicals: For noncarcinogenic indicator chemicals, EPA has determined the acceptable intake for subchronic exposure (AIS) and the acceptable intake for chronic exposure (AIC) for specific media. These values are listed in the appendix of the manual. The AIS and AIC values are derived from data obtained by animal studies and human epidemiological studies, when available. EPA states that these values are designed to protect sensitive populations, but this is not further explained.

For carcinogenic indicator chemicals, carcinogenic potency factors (CPF) have been determined by EPA, and represent lifetime cancer risk per mg/kg/day. The CPF is the estimated upper 95% confidence limit of the carcinogenic potency of a chemical.

The latter critical toxicity values (AIS, AIC, and CPF) are to be used in the risk characterization of the indicator chemicals (step 5). If critical toxicity values are not available for all of the indicator chemicals, EPA suggests that its headquarters be contacted for guidance.

5. Characterization of human health risks: The risk characterization involves a comparison between the estimated daily intakes (SDI or CDI) and the critical toxicity values (AIS, AIC, or CPF) for the indicator chemicals.

For a set of noncarcinogenic indicator chemicals, a hazard index is derived from the summation of the individual risks from each noncarcinogen:

$$\text{Subchronic Hazard Index} = \sum_{i=1}^{n} \frac{\text{subchronic daily intake}_i}{\text{acceptable intake, subchronic}_i}$$

$$\text{Chronic Hazard Index} = \sum_{i=1}^{n} \frac{\text{chronic daily intake}_i}{\text{acceptable intake, chronic}}$$

The hazard indices for inhalation and oral exposure for all noncarcinogenic indicator chemicals are summed to assess the effects of multiple exposure pathways. The hazard index takes into account that multiple subthreshold exposures could result in an overall adverse effect. EPA emphasizes that the hazard index is not a predictor of incidence or severity; rather, it is an indicator of acceptable or unacceptable exposure levels.

EPA recognizes that the application of a hazard index to a mixture of chemicals having varying effects could overestimate risk. Therefore, if the hazard index exceeds 1, the compounds should be segregated by critical effect (described in the manual), and separate hazard indices derived for each critical effect (i.e., neurotoxic, hemotoxic, etc.). If any of the individual hazard indices exceed 1, remedial action will be necessary.

The risk characterization for carcinogenic indicator chemicals is determined using the following formula:

$$\text{Total Carcinogenic Risk} = (\text{chronic daily intake}_i \times \text{carcinogenic potency factor}_i)$$

The carcinogenic risks from various exposure routes (oral, inhalation) are also summed to yield a total carcinogenic risk. The total carcinogenic risk for the site, based on the indicator chemicals, is used to design remedial options based on a target carcinogenic risk.

The Level 2 assessment of the no-action alternative is complete at this point. If remedial action is deemed necessary, the public health evaluation proceeds to the next section (see Figure 1).

Development of Design Goals and Estimation of Risk for Remedial Alternatives

The final state of the health assessment process involves the following steps:

1. Indicator chemicals are reviewed in light of the specific remedial actions to be undertaken. Some indicator chemicals may be treated more easily with certain cleanup methods.
2. Exposure pathways are reviewed for the remedial alternative to be used, because new exposure pathways could be created by the remedial option.
3. Target concentrations are determined at human exposure points.
4. Target release rates are estimated.
5. Chronic risk from noncarcinogens is assessed.
6. Potential short-term health effects of remedial alternatives are assessed.
7. Effects of remedial alternative failure are assessed.

Figure 1.   Superfund Risk Assessment Process

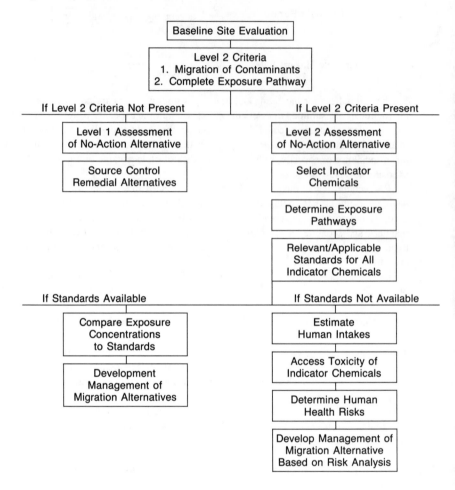

The most important steps for risk assessment are steps 3 and 5. The goal of step 3 is to determine a target concentration range for each indicator chemical. The target concentrations are calculated using relevant/applicable standards or a target risk range. If all of the indicator chemicals have ambient standards, these are used as the basis for the target concentration range. If the indicator chemicals do not have applicable standards, target concentrations are calculated using toxicity and intake values.

Potential carcinogens are evaluated first because their target concentrations are usually lower than those for noncarcinogens. The target total carcinogenic risk at Superfund sites must fall between 10^{-4} to 10^{-7}. Various methods may be used to apportion carcinogenic exposure among multiple carcinogens and multiple exposure routes. One method suggested by EPA is to divide the target carcinogenic risk level (10^{-4} to 10^{-7}) by the number of carcinogenic indicator chemicals. This will yield an individual target carcinogenic risk for each carcinogen. The individual carcinogenic risk is then divided by the carcinogenic potency factor for that chemical to determine the target chronic daily intake level. The target daily intake is then divided by standard intake factors to yield the target long-term concentration. These formulas are outlined below:

$$\frac{\text{Total Target Risk } (10^{-4}-10^{-7})}{\text{Number of Potential Carcinogens}} = \text{Target risk for each chemical}$$

$$\frac{\text{Individual Target Risk}}{\text{Carcinogenic Potency Factor}} = \text{Target chronic daily intake}$$

$$\frac{\text{Target Chronic Daily Intake}}{\text{Human Intake Factor}} = \text{Target concentration}$$

The target concentrations for each carcinogenic indicator chemical are used to design the remedial alternatives for the site.

A second approach suggested by EPA is to let one or two extremely potent carcinogens drive the design process. One or two indicator chemicals may be so potent that exposure to them must be extremely low (e.g., dioxin). In this case, the total carcinogenic risk will fall within the target range for the other carcinogens as a result of remedial actions designed to reduce the most potent carcinogen.

Once remedial alternatives have been assessed for reduction of carcinogenic risk, these remedial alternatives should be checked to ensure that noncarcinogenic risk is reduced as well. Only chronic risk is considered, because long-term remediation is the goal of this process. The hazard index is determined for each critical effect as described in the Level 2 assessment of risk characterization:

$$\text{Hazard Index} = \frac{\text{chronic daily intake}_i}{\text{acceptable intake chronic}_i}$$

The hazard index should be less than 1 but if it exceeds 1 for any endpoint, the remedial design should be altered to reduce this risk.

Strengths and Weaknesses

EPA's Superfund Health Assessment Manual provides a comprehensive step-by-step procedure for assessing the health risks posed by contaminated sites. Since many hazardous waste sites contain a multitude of chemicals, EPA has offered guidelines for selection of the most toxic indicator chemicals which are evaluated in a risk analysis. The number of indicator chemicals can be as high as desired (EPA suggests 5–10), and should include both carcinogens and noncarcinogens, if present at the site.

Chemicals at the site are ranked according to concentration and toxicity values provided by EPA. If no toxicity value has been determined by EPA for a specific chemical, it can be derived using data from animal studies when available (MED or ED_{10}). The top 10–15 scoring chemicals in each class (carcinogens and non-carcinogens) are then further evaluated based on half-life, water solubility, vapor pressure, Henry's Law constant, and organic partition coefficients. These factors are considered in the context of the particular exposure pathway. Therefore, the final indicator chemicals reflect the characteristics of the chemicals in their environment and various toxic effects.

EPA's methodology for selection of indicator chemicals is useful for sites contaminated with a mixture of chemicals, such as petroleum products. Toxicity constants used in selection of indicator chemicals are evaluated for soil ingestion exposure. EPA assumed that 100 mg of soil is ingested per day for children ages 2 to 6, and this value was averaged over a 70-year lifespan. This value could be altered if new information suggests more or less soil ingestion for children and/or adults.

EPA offers guidelines for a comprehensive, yet possibly time-consuming, exposure pathway analysis, including modeling of environmental fate, and transport and release rates for subchronic and chronic exposures. The amount of time needed to complete this portion of the risk assessment may be significant, and could slow down implementation of remedial response measures.

Formulas are provided for estimation of environmental fate and transport and release rates. Computer models which can be used in this step are referenced in the manual. The level of sophistication achieved by an individual regulatory agency may depend on equipment available and familiarity with this type of analysis. It is important that the assumptions of the environmental fate model be evaluated for relevance to the specific site under investigation

For soils contaminated with a complex mixture of substances, analysis of environmental fate and release rates would appear to be very important. Certain chemicals may be released slowly over a lifetime and travel through various exposure pathways (air, water, or soil). For this reason, the site should be evaluated for risks posed over a human lifetime or longer. EPA suggests that simple models with conservative assumptions are sufficient for Superfund sites.

The exposure point concentrations of the indicator chemicals are compared to relevant/applicable standards when available. If standards are not available, guidelines for risk assessment are provided using toxicity data, exposure data, and standard assumptions for intake and body weight. Values for subchronic and chronic daily intake (SDI and CDI) are estimated and multiroute exposures (water ingestion and inhalation) are taken into account.

One problem with the latter procedure is that formulas for less common exposure pathways such as soil ingestion or dermal absorption are not included. This is unfortunate, considering that soil ingestion was taken into account by EPA in the selection of indicator chemicals. EPA stated that there was a desire to include soil toxicity constants in step 1 so that chemicals in the soil could be considered when selecting indicator chemicals. Unfortunately, this was not followed up in the determination of daily chemical intakes to be used in the risk characterization. EPA does emphasize that soil ingestion can be an important exposure pathway for children playing near contaminated sites, yet does not further assess this risk in the manual.

EPA assumes 100% absorption of all chemicals after intake, due to uncertainties in present bioavailability data. If further information on the percent of a chemical absorbed becomes available, EPA suggests that this factor be taken into account. Overestimates of risk can be derived if absorption is less than 100% and, therefore, should be carefully evaluated for specific chemicals and exposure pathways (i.e., some indicator chemicals may bind tightly to soil particles and be less likely to be absorbed in the intestine).

The acceptable daily intakes for subchronic (AIS) and chronic (AIC) exposures have been determined by EPA for many chemicals. It is unclear how EPA derived these values, except that they are based on data from animal studies concerning intake and toxic effect. Carcinogenic potency factors (CPF) have been determined by EPA and represent the lifetime cancer risk per mg/kg/day.

If AIS, AIC, or CPF values are not available for some indicator chemicals at a particular site, the risk analysis cannot be completed. For example, AIS, AIC, or CPF values are not available for benz-(a)-anthracene, cadmium, dichlorobenzene, heptane, hexane, isobutane, isopentene, 1-pentene, and xylene. All of these compounds are major constituents in petroleum products.

The risk characterization involves comparison of intake estimates (SDI or CDI) with acceptable exposure levels (AIS or AIC) or risk based on CPFs for indicator chemicals. Soil ingestion and dermal absorption exposure pathways are not considered in this step.

Additivity of carcinogenic risk from multiple carcinogens is taken into account, as well as additivity of chronic noncarcinogenic effects for a specific end point. The total carcinogenic risk from all routes of exposure must fall within a target range of 10^{-4} to 10^{-7}. It should be noted, however, that the target range for carcinogenic risk involves only the selected indicator chemicals which have CPF values. Other carcinogens cannot be adequately evaluated and may pose a significant health risk.

Conclusion

EPA's *Superfund Health Assessment Manual* provides a useful quantitative risk assessment procedure for sites contaminated with a mixture of chemicals. Major problems with this methodology include lack of consideration of exposures due to soil ingestion or dermal absorption, and values used in the risk analysis (AIS, AIC, and CPF) may not be available for many constituents in petroleum products.

The EPA methodology is useful in its careful selection process for indicator chemicals including relevant factors such as half-life, water solubility and vapor pressure. Carcinogenic and noncarcinogenic chemicals are ranked separately which allows for risk assessment of both carcinogenic effects and noncarcinogenic subchronic or chronic effects. Toxicity constants are provided for soil ingestion exposure in the selection of indicator chemicals.

A comprehensive exposure pathway analysis is detailed by EPA including environmental fate, transport and release rates. Such an analysis, although time-consuming, may be valid for sites in which long-term effects are probable.

The total carcinogenic risk posed by all of the carcinogenic indicator chemicals must fall within a target range of 10^{-4} to 10^{-7}. However, chemicals not having carcinogenic potency factors cannot be adequately assessed. In addition, noncarcinogens cannot be properly evaluated if acceptable intake factors are not available.

The strength of the EPA Superfund manual lies in its long-term exposure pathway analysis, indicator chemical selection process, and risk assessment formulas for carcinogens and noncarcinogens. These portions of the methodology could be utilized in a risk assessment process which does take into account less common exposure pathways such as soil ingestion and dermal absorption. In this manner, the health risk assessment of soils contaminated with petroleum products could be adequately evaluated (Table 4).

Table 4.   Summary of the EPA Superfund Methodology with Respect to the Evaluation Criteria.

HAZARD ANALYSIS	
Toxicity value	AID[a]
	AIC
	CPF
Noncarcinogenic acute toxicity	Yes
Noncarcinogenic chronic toxicity	Yes
Carcinogenic chronic toxicity	Yes
Assessment of mixtures	Yes
Body weight factor	Yes
Pharmacokinetic factor	No
EXPOSURE ANALYSIS	
Air inhalation	Yes
Dust inhalation	Yes
Water ingestion	Yes
Soil ingestion	No
Dermal absorption	No
Crop uptake	No
Livestock uptake	No
Fish uptake	Yes
Environmental fate	Yes
	—air transport
	—water transport
Half-life factor	Yes
Site-specific factors	Yes
RISK ANALYSIS	
Additivity of toxic effect	Yes
Multi-media exposure	Yes
Synergistic effects	No

[a]See text for description of toxicity values.

HEALTH RISK ASSESSMENTS FOR CONTAMINATED SOILS

Summary

Two types of risk analyses were presented in "Health Risk Assessments for Contaminated Soils."[4] The risk assessment formula for acute toxicity utilized the allowable daily intake (ADI) to represent the toxicity of noncarcinogenic contaminants. The ADI was modified by body weight and daily soil intake to determine the soil criteria for a specific contaminant.

The risk analysis formula for chronic toxicity involved dividing an acceptable cancer risk by a unit carcinogenic risk (UCR) established by the U.S. EPA Carcinogen Assessment Group. The factor was then modified by lifetime average soil intake and half-life of the contaminant to yield a soil criteria for a specific carcinogen.

Strengths and Weaknesses

The risk assessment methodology for contaminated soil developed by Ford and Gurba[4] evaluates both acute and chronic toxicity. Since children are believed to ingest the highest levels of soil over a brief time period (ages 2–5) compared to older age groups, they may be a high-risk group for exposure to contaminated soils. For this reason, Ford and Gurba developed a risk analysis formula to evaluate acute toxicity of soil contaminants for children, along with a chronic toxicity formula for carcinogens.

The acute toxicity formula is based on the allowable daily intake (ADI), originally developed to represent the adult intake of a food additive at which no lifetime health effects would occur. Ford and Gurba modified the ADI by body weight and daily soil ingestion for children to yield the following formula:

$$SC = ADI \times \frac{1000}{SI} \times BW$$

SC = soil criteria (mg/kg)
ADI = allowable daily intake (mg/day)
1000 = conversion factor (g/kg)
SI = soil ingestion (g/day)
BW = body weight adjustment (10 kg/70 kg)

The ADI value represents the toxicity of the contaminant, the SI value represents the exposure, and the BW value adjusts the ADI for children.

Since many of the pollutants in contaminated soils may be carcinogens, a different risk analysis was developed for lifetime, chronic toxicity. The term "lifetime allowable daily intake" (LADI) was derived by dividing an acceptable cancer risk (1×10^{-6}) by a unit carcinogenic risk (UCR). UCRs are established by the

U.S. EPA's Carcinogen Assessment Group and are expressed as an excess cancer risk from a lifetime of ingestion of 1 mg/kg/day of a carcinogen. The LADI was modified by lifetime average soil intake and half-life of the contaminant to yield the following formula:

$$SC = LADI \times \frac{1000}{LASI} \times \frac{\frac{t}{2}}{70}$$

LADI = lifetime allowable daily intake (mg/kg/day)

$$= \frac{Risk \ (1 \times 10^{-6})}{Unit \ Carcinogenic \ Risk \ (UCR)}$$

1000 = conversion factor (g/kg)
LASI = lifetime average soil intake (g/kg/day)
t/2/70 = half-life correction factor

The toxicity of a specific carcinogen is represented by the LADI, the exposure is represented by LASI, and a half-life correction factor adjusts for contaminants which may biodegrade significantly over a lifetime.

One limitation of the latter two methodologies is that the values ADI and UCR may not be available for many of the contaminants in petroleum products. In addition, both analyses do not evaluate site conditions, such as geologic or geographic factors, type of soil, and seasonal conditions. Contaminants in soil may react with soil particles by binding more or less strongly and may migrate to drinking water as a result of geologic, geographic, and seasonal influences. The environmental fate of soil contaminants is determined by site conditions which are not taken into account in Ford and Gurba's methodologies.

Soil ingestion is considered to be the most important route of exposure by Ford and Gurba, while other routes of exposure are ignored. One weakness of these risk analyses, therefore, is that the health risk due to exposure via contaminated drinking water, dust inhalation, or dermal absorption is not evaluated.

The effectiveness of the two risk analyses will be reflected by the careful choice of individual constituents which are evaluated in the formulas. For a mixture of pollutants such as petroleum products, the constituents analyzed must represent the overall toxicity of the mixture. If this is not the case, the soil criteria may not establish a safe level of cleanup. A thorough review of the toxicity of constituents in petroleum products should be conducted prior to use of the latter methodologies.

Conclusions

The health risk assessments for contaminated soils developed by Ford and Gurba have included the important factors of acute and chronic toxicity, daily and lifetime soil ingestion, and half-life adjustments for contaminants which biodegrade significantly during a lifetime. Toxicity of the contaminants was evaluated by the terms ADI or UCR, which may not be available for many of the constituents of petroleum products. Both risk analysis formulas failed to evaluate the health risks resulting from exposure via contaminated water, dust inhalation, or dermal absorption. Site conditions, including geologic or seasonal factors, were ignored despite the fact that these variables can strongly influence the environmental fate of soil contaminants. Finally, if the soil contaminant consists of a mixture of chemicals, as is the case for petroleum products, the constituents evaluated in the risk assessment formula must represent the overall toxicity of the mixture.

Table 5. Summary of the Ford and Gurba Methodology with Respect to the Evaluation Criteria.

HAZARD ANALYSIS	
Toxicity value	ADI[a]
	CPF
Noncarcinogenic acute toxicity	Yes
Noncarcinogenic chronic toxicity	Yes
Carcinogenic chronic toxicity	Yes
Assessment of mixtures	Yes
Body weight factor	Yes
Pharmacokinetic factor	No
EXPOSURE ANALYSIS	
Air inhalation	No
Dust inhalation	No
Water ingestion	No
Soil ingestion	Yes
Dermal absorption	No
Crop uptake	No
Livestock uptake	No
Fish uptake	No
Environmental fate	No
Half-life factor	Yes
Site-specific factors	No
RISK ANALYSIS	
Additivity of toxic effect	No
Multi-media exposure	No
Synergistic effects	No

[a]See text for description of toxicity values.

SOIL CLEANUP CRITERIA FOR SELECTED PETROLEUM PRODUCTS

Summary

The New Jersey Department of Environmental Protection's "Soil Cleanup Criteria for Selected Petroleum Products"[5] describes a soil cleanup methodology based on a few individual constituents of petroleum products which pose the greatest threat to public health. The most hazardous constituents were identified to be the carcinogenic polycyclic aromatic hydrocarbons (CaPAHs) and benzene. Acceptable soil contaminant levels (ASCL) were determined based on lifetime soil ingestion, a 1×10^{-6} cancer risk, and carcinogenic potency factors for individual constituents. The ASCL was compared to residual soil levels of CaPAHs and benzene resulting after soil cleanup to 100 ppm total petroleum hydrocarbons, as reported in the literature. For residual soil levels yielding a greater than 1×10^{-6} cancer risk, a lower soil cleanup level was proposed. With the exception of used motor oils over 10,000 km, CaPAHs and benzene levels were below the concentration which would exceed a 1×10^{-6} cancer risk after cleanup to 100 ppm total petroleum hydrocarbons.

Strengths and Weaknesses

The New Jersey Department of Environmental Protection's "Soil Cleanup Criteria for Selected Petroleum Products" examined increased cancer risk as a result of lifetime ingestion exposure to soils contaminated with petroleum products. The risk assessment and soil cleanup objectives were presented only for individual chemical constituents of petroleum products which have the highest toxicity, have the ability to migrate, and/or are present in significant amounts. Petroleum product constituents were reviewed for chemical/physical properties, health effects, environmental effects, and carcinogenic properties. The authors determined that the carcinogenic polycyclic aromatic hydrocarbons (CaPAHs) and benzene were the major constituents of concern.

As noted by the authors, however, the concentrations of CaPAHs and benzene vary depending on the type of crude oil and the fractionation process used. Benzo-(a)-pyrene (BaP) exists in small quantity relative to other CaPAHs, although used petroleum products are often enriched in BaP and other CaPAHs. Information on benzene levels in petroleum products is limited, with the highest levels found in gasoline.

Once the primary chemical constituents were identified, these were utilized in the risk assessment methodology. The authors based their methodology on reports in the literature which examined residual soil levels of CaPAHs and benzene after soil cleanup to 100 ppm total petroleum hydrocarbonds. The residual soil levels were compared to acceptable soil contaminant levels (ASCL) based on a 1×10^{-6} cancer risk, lifetime soil ingestion, and carcinogenic potency factors established for various compounds by the U.S. EPA Carcinogen Assessment Group (CAG):

$$ASCL = \frac{A}{C} \times \frac{1000}{L}$$

A   = acceptable cancer risk$= 1 \times 10^{-6}$ (one in a million)
C   = carcinogenic potency factor (U.S. EPA CAG)
    for BaP $= 11.53$ $(mg/kg/day)^{-1}$
    for benzene$= 0.0052$ $(mg/kg/day)^{-1}$
    1000   = conversion factor (g/kg)
L   = lifetime average daily soil intake

While it is impossible to assess the risk posed by every constituent in petroleum products, the validity of limiting the risk assessment to CaPAHs and benzene should be examined. Care should be taken to adequately and thoroughly assess the toxicity characteristics of the major constituents in petroleum products as a means of identifying the most hazardous elements. The authors of the present methodology did not go into great detail on their rationale for choosing CaPAHs and benzene as the constituents of concern. It is important to explain why the latter two constituents are more hazardous than the many other chemicals in petroleum products.

The carcinogenic potency factor for BaP was used to represent all of the CaPAHs in petroleum products, due to the fact that carcinogenic potency factors do not exist for other CaPAHs. Therefore, the ASCL is based only on BaP, and assumes that all of the other CaPAHs are as toxic as BaP. No justification of this assumption was presented by the authors, except that it was consistent with the U.S. EPA's approach to estimating cancer risk from exposure to mixtures of CaPAHs. It is possible that other CaPAHs may be more toxic or have synergistic effects in the mixture.

The ASCL is based on lifetime soil ingestion; yet the authors contended that it is also based on inhalation of dust. It is unclear how inhalation is taken into account by the model, and this is not explained in the methodology. The method also does not consider the serious threat to public health resulting from contamination of groundwater or surface water via migration of toxic constituents.

The soil cleanup criteria does not take into account biodegradation, volatilization, or half-life of the individual constituents. Considering the fact that benzene is highly volatile and may have a relatively short half-life in soil, this factor should be accounted for in the methodology. Other toxic constituents may also biodegrade in significant amounts over time.

Conclusions

The risk assessment method developed by Stokman and Dime (New Jersey Department of Environmental Protection) focused on increased cancer risk as a result of lifetime ingestion of soil contaminated with petroleum products. Exposure via contaminated drinking water due to migration of toxic constituents was not assessed and is one limitation of this methodology. In addition, the soil cleanup criteria did not take into account biodegradation, volatilization, or half-life of the individual constituents. Half-life can be very important for a constituent such as benzene, which is very volatile.

CaPAHs and benzene were chosen as the major constituents of concern in petroleum products. Although this type of risk assessment simplifies the problem of dealing with numerous compounds in petroleum products, the individual constituents should be chosen carefully and clearly be representative of the overall mixture. In this case, it was not clearly proved that CaPAHs and benzene would be representative of the total mixture, with regard to lifetime toxicity. However, the individual regulatory agency may choose its own representative constituents based on a thorough and complete hazard analysis of petroleum product constituents.

If residual soil levels after cleanup to 100 ppm total petroleum hydrocarbons are below the ASCL for the individual constituents, the risk assessment is complete. However, the effect of residual soil levels of the remaining petroleum compounds will depend to some extent on the type of soil and on geologic, geographic, and seasonal conditions. If environmental factors at the site are not taken into account, and/or the individual constituents were not chosen carefully enough, residual soil levels of other constituents not initially focused on could have long-term toxic effects. In summation, the effectiveness of this methodology depends on how representative the individual constituents used in the ASCL are for the overall mixture and environmental conditions at the site (Table 6).

Table 6.   Summary of the New Jersey Methodology with Respect to the Evaluation Criteria.

HAZARD ANALYSIS	
Toxicity value	CPF[a]
Noncarcinogenic acute toxicity	No
Noncarcinogenic chronic toxicity	No
Carcinogenic chronic toxicity	Yes
Assessment of mixtures	Yes
Body weight factor	No
Pharmacokinetic factor	No
EXPOSURE ANALYSIS	
Air inhalation	No
Dust inhalation	No
Water ingestion	No
Soil ingestion	Yes
Dermal absorption	No
Crop uptake	No
Livestock uptake	No
Fish uptake	No
Environmental fate	No
Half-life factor	No
Site-specific factors	No
RISK ANALYSIS	
Additivity of toxic effect	No
Multi-media exposure	No
Synergistic effects	No

[a]See text for description of toxicity values.

FINAL CLEANUP POLICY

Summary

The Washington Department of Ecology's Final Cleanup Policy[6] involves three levels of cleanup for a contaminated site: (1) Initial Cleanup Levels consist of total cleanup or partial cleanup to eliminate an "imminent" public health threat. Total Cleanup is achievable when site characteristics include well-defined contamination boundaries, concentrated substances, and/or limited extent of contamination. If only partial cleanup is implemented, it is followed by (2) Standard/Background Cleanup Levels, which offer guidelines for soil, water, and air cleanup. The levels for soil and water are based on multiples of the appropriate drinking water quality standard, water quality background, or soil background. The cleanup levels for air are based on OSHA/WISHA limits for air quality, ambient air quality, or air background levels. The feasibility of the latter types of cleanup is evaluated in a Preliminary Technical Assessment based on site-specific characteristics. If Standard/Background Cleanup Levels are unachievable, (3) Protection Cleanup Levels are implemented, based on multiples of the appropriate water quality standard, water quality background, soil background, or site-specific information, followed by predictive modeling. For contaminated soils with a threat to air, "Dangerous Waste Limits" are used to establish cleanup levels. Follow-up includes long-term monitoring to verify that no threat to public health remains (Figure 2).

Figure 2. Washington Department of Ecology's Risk Assessment Process.

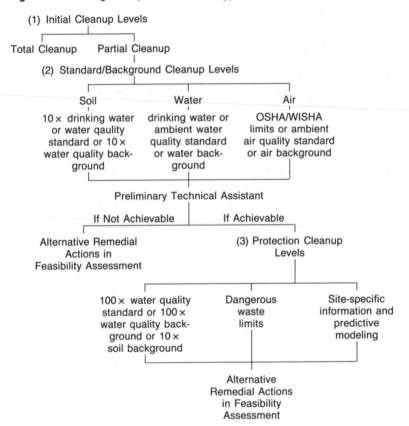

Strengths and Weaknesses

The Washington Department of Ecology's Final Cleanup Policy presents a framework to determine the cleanup levels for a contaminated site which is a threat to public health. The cleanup policy is divided into three levels: (1) Initial Cleanup, (2) Standard/Background Cleanup, and (3) Protection Cleanup.

Initial Cleanup Levels of surface water, groundwater, soil, and air are to be implemented when the contaminated site is an ''imminent threat to public health or difficulty of cleanup increases significantly with time.'' The remedial options within the Initial Cleanup are total cleanup or partial cleanup. Total cleanup is only for situations having well-defined contamination boundaries, concentrated substances, or limited extent of contamination. Except for determination of contamination boundaries, no guidelines are offered for determination of concentrated substances or limited extent of contamination. Once it is determined whether the latter three characteristics exist, no further guidelines are given for total cleanup. Therefore, determination of total cleanup is vague, and no concrete guidelines are offered for risk assessment of the contaminants.

Partial cleanup is the next option if total cleanup is not undertaken to ''eliminate imminent public health and environmental hazards by only removing those portions of the known contamination that represent an immediate hazard or that significantly increase the difficulty of eventual cleanup.'' Neither ''imminent threat'' nor ''immediate hazard'' are further defined, which results in very subjective determinations for this cleanup level. No further guidelines are discussed, so it is unclear what exactly a partial cleanup involves.

The second level of the cleanup policy is Standard/Background Cleanup Levels, which are implemented when total cleanup is unachievable or contaminants pose a threat to public health over the long term. The Standard/Background Cleanup Levels for soil, groundwater, and surface water are based on $10\times$ the appropriate drinking water standard, and $10\times$ the water quality background or soil background. No justification is offered for choosing $10\times$ the drinking water or water quality standard and, therefore, this level appears to be arbitrary.

Soil background levels are used if no water standards exist. This assumes that background levels are already known, which may not be the case for numerous contaminants. A standard procedure for measuring soil background is not given and, therefore, any measurement undertaken after contamination will be variable, depending on the area used for background level measurement and type of equipment available.

The Standard/Background Cleanup Level for air is based on OSHA/WISHA limits for air quality/ambient air quality or background levels. The latter criteria also assume that these standards exist for specific contaminants, or that background levels are known.

The technical feasibility of the Standard/Background Cleanup Levels is evaluated in a Preliminary Technical Assessment, which considers presence of sole source aquifers, barriers to contaminant migration, sorptive properties of soil, contaminant mobility, and depth to groundwater. Unfortunately, these determinations are not elaborated upon for degree of importance and type of measurements, and are not integrated into a risk assessment methodology. The Preliminary Technical Assessment as described in the cleanup policy is neither complete nor quantitative.

The cleanup policy states, "If the Standard/Background Level is achievable . . . [based on the Technical Assessment] . . . it is used to evaluate the alternative remedial actions in the Feasibility Assessment." A Feasibility Assessment is not described in the cleanup policy and, therefore, alternative remedial actions are not offered.

If Standard/Background Levels are not achievable, the third level in the cleanup policy, Protection Cleanup, is implemented. Protection Cleanup Levels are defined using one of the following: (1) specified multiples of the appropriate water quality standard or background (100× water quality standard, 100× water quality background, or 10× soil background); (2) Dangerous Waste Limits; or (3) site-specific information on contaminant migration characteristics, leaching tests, or biologic tests and predictive models.

As previously mentioned, the use of multiples of water quality standards or background is not justified and, therefore, is arbitrary. The term "Dangerous Waste Limits" is not clearly defined, and site-specific tests are vague. Which specific tests should be used and how the results should be weighed is unclear. Predictive modeling is suggested, yet no particular models are mentioned or discussed in terms of applicability to risk assessment. The Protection Cleanup Levels are then used to evaluate the "alternative remedial actions in the Feasibility Assessment," which are not further defined in the cleanup policy.

Conclusions

Overall, the Washington Department of Ecology's Final Cleanup Policy offers a useful outline for a step-by-step general assessment of a contaminated site. The cleanup policy considers the three exposure pathways of soil contamination, water contamination (surface and groundwater), and air contamination resulting from hazardous waste at one site. The above three pathways are important when considering spills of petroleum products in soil. Unfortunately, specific risk assessment guidelines based on justifiable standards are lacking. Many of the guidelines are vague, which forces the regulatory agency to form its own interpretation based on a subjective analysis. The essentials of a risk assessment methodology— exposure analysis, hazard analysis, and predictive modeling—should be based on a quantitative assessment of the contaminated site and justifiable standards (Table 7).

Table 7.   Summary of the Washington Cleanup Methodology with Respect to the Evaluation Criteria.

HAZARD ANALYSIS	
Toxicity value	Multiples (10× or 100×) of water standards, air standards or soil background
Noncarcinogenic acute toxicity	Yes
Noncarcinogenic chronic toxicity	Yes
Carcinogenic chronic toxicity	Yes
Assessment of mixtures	No
Body weight factor	No
Pharmacokinetic factor	No
EXPOSURE ANALYSIS	
Air inhalation	Yes
Dust inhalation	No
Water ingestion	Yes
Soil ingestion	No
Dermal absorption	No
Crop uptake	No
Livestock uptake	No
Fish uptake	No
Environmental fate	No
Half-life factor	No
Site-specific factors	Yes
RISK ANALYSIS	
Additivity of toxic effect	No
Multi-media exposure	No
Synergistic effects	No

LEAKING UNDERGROUND FUEL TANK MANUAL

Summary

The State of California Department of Health Services and State Water Resources Control Board has established procedures for determining whether an underground storage fuel tank site poses a risk to human health. The field manual is a practical extension of the "California Site Mitigation Decision Tree" document produced by the California Department of Health Services, Toxic Substances Control Division.

The guidelines presented in the Leaking Underground Fuel Tank (LUFT) Field Manual apply only to soil and groundwater contamination, rather than surface water contamination or air pollution. In addition, the guidelines only deal with gasoline and diesel fuel products and do not consider waste oil or solvents. The main purpose of the manual is to provide practical guidance to field personnel responsible for dealing with leaking fuel tank problems.

For situations in which gasoline contamination of soil may have occurred, the manual suggests analyzing for benzene, toluene, xylene, and ethylbenzene (BTX and E) and total petroleum hydrocarbons (TPH). For situations in which diesel fuel may have contaminated a site, only TPH is measured.

The manual cautions against measurement of ethylene dibromide (EDB) and organolead for the following reasons. EDB has been in such widespread use that its detection may not be due to gasoline. Concerning organolead, most laboratories have the ability to only analyze for total lead, and cannot distinguish between inorganic and organic lead. In addition, inorganic lead is native to California soil, which may lead to false positive readings unless background levels are known.

The reasons offered by the manual for choosing BTX and E as indicators of gasoline contamination include the following:

1. They are readily adaptable to gas chromatograph detection.
2. They pose a serious threat to human health (i.e., benzene as a carcinogen).
3. They have the potential to move through the soil and contaminate groundwater.
4. Their vapors can be highly flammable and explosive.

Because BTX and E are highly mobile and can migrate from the site, it is also important to analyze for TPH. TPH detection is reported as the total of all hydrocarbons in the samples.

Tank sites are initially classified as one of the following:

- Category I   No Suspected Soil Contamination, i.e., sites in which tanks are being closed for reasons other than a leak.
- Category II   Suspected or Known Soil Contamination, i.e., sites where tanks or lines have failed precision tests, show discrepancies in monitoring records, or show visual evidence of leakage.
- Category III   Known Groundwater Contamination, i.e., sites where tanks or piping have shown a significant loss of product, especially in areas of high groundwater.

When a leaking tank is discovered, immediate safety issues are assessed and information is collected for site categorization, such as inventory records, precision testing records, and repair histories.

To categorize a site, a field TPH test is conducted in which TPH levels are measured in the ambient air, or air drawn from the soil. The result of the TPH test is compared to background levels. Background levels are determined by taking three soil samples from nearby or adjacent properties. If background TPH levels are exceeded, the site is placed in category 2 for further testing.

Under category 2, quantitative lab analyses of soil are conducted for BTX and E, and TPH. Samples are collected from the bottom of the excavation at worst-case locations. The initial trigger levels for BTX and E are the detection limits of the laboratory procedure. According to the manual, the trigger levels should be in the order of 0.3 mg/kg. Cleanup levels for BTX and E are derived from precipitation rates and depth to groundwater, using tables detailed in the manual.

For TPH measurements in category 2, a leaching potential analysis was developed to determine the levels of TPH that can be safely left in the soil. A leaching potential analysis is based on the tendency of TPH to migrate down to groundwater, depending on the features of the site. Four site characteristics which the manual considers important influences on migration of TPH include depth to groundwater, subsurface fractures, precipitation, and man-made conduits. Each characteristic is rated on a scale of high, medium, and low potential for leaching. These three degrees of sensitivity are expressed in terms of TPH that can be safely left in the soil, i.e., high (10 ppm), medium (100 ppm), and low (1000 ppm). The lowest sensitivity level determined for the four characteristics is used as a cleanup level. If a characteristic cannot be rated due to insufficient data, the lowest value (10 ppm) is used as a cleanup level. In addition, other site features may be considered, such as unique site characteristics, actual use of groundwater, and future land use.

If either the TPH levels or the BTX and E levels exceed the allowable limits, additional site analysis is needed. At this point in the evaluation process, the services of a registered geologist, engineer, or environmental chemist are recommended by the manual. In addition, the manual suggests that a general risk appraisal be conducted using environmental fate and chemistry data, and site-specific information. For this stage, computer modeling is used to estimate the concentrations of BTX and E that can be left in place without risking groundwater pollution. The two models recommended for this step are the SESOIL model and the AT123D model. The SESOIL model involves long-term environmental fate simulation of pollutants in the vadose zone, and predicts the amount of pollutants which will enter groundwater. The AT123D model estimates the rate of pollutant transport and transformation in a groundwater system, and predicts groundwater contamination.

For category III sites having groundwater contamination, decisions of site investigations and cleanup measures must be made on a case-by-case basis. This step involves assessing groundwater use, and collecting and analyzing groundwater samples. The appendix of the field manual contains numerous procedures for field measurements of this type. Risk analyses for Category III sites also require consultation with a regulatory agency or professionals in the field. The field manual does not attempt to offer guidelines for an in-depth risk assessment, and instead focuses on site categorization and laboratory analyses of indicator chemicals.

Strengths and Weaknesses

The State of California Leaking Underground Fuel Tank (LUFT) Field Manual is a practical extension of the risk assessment process detailed in the California Site Mitigation Decision Tree. The Field Manual describes the steps to be taken for categorizing sites contaminated from leaking underground storage tanks. The guidelines apply only to sites contaminated with gasoline and diesel fuel, and the risk appraisal focuses only on soil and groundwater contamination.

A site is categorized according to extent of contamination determined by field analysis of indicator chemicals. Levels of total petroleum hydrocarbons (TPH) and benzene, toluene, xylene, and ethylbenzene (BTX and E) are measured to determine extent of contamination. The site is classified based on these findings; however, it is unclear how the trigger levels and the cleanup levels for TPH and BTX and E were derived.

Conclusions

The State of California LUFT Field Manual outlines the step-by-step procedures necessary for field workers to classify sites which may be contaminated by leaking underground fuel tanks. The manual details a classification system which reflects the extent of contamination. Specific testing procedures are explained for the indicator chemicals: benzene, toluene, xylene, ethylbenzene, and total petroleum hydrocarbons. The procedures apply only to leaking underground fuel tanks containing gasoline or diesel fuel products. The risk to human health is evaluated for potential contamination of soil and groundwater.

Environmental fate modeling is encouraged, and several computer programs are detailed for this step. For risk assessment of groundwater contamination, the manual recommends that a regulatory agency and/or professionals in the field be consulted for proper determination of human health risk. Overall, the main goal of the field manual is to provide field workers with the information necessary for accurate analyses of soil and groundwater contamination, and subsequent-characterization (Tables 8 and 9).

Table 8. Summary of the State of California Leaking Underground Fuel Tank Manual with Respect to the Evaluation Criteria.

HAZARD ANALYSIS

Toxicity value	None
Noncarcinogenic acute toxicity	No
Noncarcinogenic chronic toxicity	No
Carcinogenic chronic toxicity	No
Assessment of mixtures	Yes (gasoline and diesel fuel)
Body weight factor	No
Pharmacokinetic factor	No

EXPOSURE ANALYSIS

Air inhalation	No
Dust inhalation	No
Water ingestion	Yes (groundwater)
Soil ingestion	No
Dermal absorption	No
Crop uptake	No
Livestock uptake	No
Fish uptake	No
Environmental fate	Yes
Half-life factor	Yes
Site-specific factors	Yes

RISK ANALYSIS

Additivity of toxic effect	No
Multi-media exposure	No
Synergistic effects	No

Table 9.   Comparison of the Seven Soil Assessment Methodologies for Use in Dealing with Petroleum Contaminated Soil.

	California Site Mitigation	U.S. Army Rosenblatt	EPA Superfund	Ford and Gurba	New Jersey Stokman and Dime	Washington Cleanup Policy	California LUFT
HAZARD							
Toxicity Value	NOAEL—chronic subchronic epidemiologic; TLV; LD_{50}; Structure Activity Analysis	ADI; MCL; TLV; NELL; NEL_{90}; LD_{50}	AIS; AIC; CPF	ADI; CPF	CPF; CPF	Multiples (10× or 100×) of water standards, air standards, or soil background	None
Noncarcinogenic acute toxicity	Yes	Yes	Yes	Yes	No	Yes	No
Noncarcinogenic chronic toxicity	Yes	Yes	Yes	Yes	No	Yes	No
Carcinogenic chronic toxicity	Yes	No	Yes	Yes	Yes	Yes	No
Assessment of mixtures	No	No	Yes	Yes	Yes	No	Yes
Body weight factor	Yes	Yes	Yes	Yes	No	No	No
Pharmacokinetic factors	Yes	No	No	No	No	No	No
EXPOSURE							
Air inhalation	Yes	Yes	Yes	No	No	Yes	No
Dust inhalation	Yes	Yes	Yes	No	No	No	No
Water ingestion	Yes	Yes	Yes	No	No	Yes	Yes
Soil ingestion	No	No	No	Yes	Yes	No	No
Dermal absorption	No	No	No	No	No	No	No
Crop uptake	Yes	Yes	No	No	No	No	No
Livestock uptake	Yes	Yes	No	No	No	No	No
Fish uptake	Yes	Yes	Yes	No	No	No	No

Table 9. (Continued)

	California Site Mitigation	U.S. Army Rosenblatt	EPA Superfund	Ford and Gurba	New Jersey Stokman and Dime	Washington Cleanup Policy	California LUFT
Environmental fate	Yes —soil adsorption —bioconcentration —air transport —water transport	Yes —partition coefficients —bioconcentration —air transport —water transport	Yes —air transport —water transport	No	No	No	Yes
Half-life factor	No	No	Yes	Yes	No	No	Yes
Site-specific factors	Yes	Yes	Yes	No	No	Yes	Yes
RISK ANALYSIS							
Additivity of toxic effect	Yes	No	Yes	No	No	No	No
Multi-media exposure	Yes	Yes	Yes	No	No	No	No
Synergistic effects	No	No	No	No	No	No	No

ACKNOWLEDGMENTS

This work was supported by a grant from the Office of Research and Standards, Massachusetts Department of Environmental Quality Engineering, Boston, MA, and The Environmental Institute, University of Massachusetts, Amherst, MA.

REFERENCES

1. California Department of Health Services, Toxic Substances Control Division, Alternative Technology and Policy Development Section, "California Site Mitigation Decision Tree," Draft Working Document, June 1985.
2. Rosenblatt, D. H., J. C. Dacre, and D. R. Cogley. "An Environmental Fate Model Leading to Preliminary Pollutant Limit Values for Human Health Effects," in *Environmental Risk Analysis for Chemicals,* R. A. Conway, Ed. (New York: Van Nostrand Reinhold Co., 1982).
3. *Superfund Health Assessment Manual,* Office of Emergency and Remedial Response, U. S. Environmental Protection Agency, December 1985.
4. Ford, K. L., and P. Gurba. "Health Risk Assessments for Contaminated Soils," in *Proceedings of the 5th National Conference on Management of Uncontrolled Hazardous Waste Sites,* Washington, D.C., November 1984.
5. Stokman, S. K., and R. Dime. "Soil Cleanup Criteria for Selected Petroleum Products," *Risk Assessment,* 342–345.
6. Department of Ecology, State of Washington, "Final Cleanup Policy—Technical," July 10, 1984.
7. "Leaking Underground Fuel Tank Field Manual: Guidelines for Site Assessment, Cleanup and Underground Storage Tank Closure," State of California, Leaking Underground Fuel Tank Task Force, State of California-Department of Health Services, LUFT Task Force, Draft, May 1987, 96 pp.

CHAPTER 22

Determination of Exposure of Oral and Dermal Benzene from Contaminated Soils

Mohamed S. Abdel-Rahman and Rita M. Turkall

Soil contamination with dangerous toxic chemicals remains one of the most difficult problems in this era. The hazardous chemical may persist in the environment and therefore, the potential for long-term health risk exists. The sources of hazardous chemical wastes are numerous. Industry, agriculture, and institutions such as hospitals and universities are all sources of materials that need to be discarded. People living in proximity to hazardous waste disposal sites or workers at the dump sites are at serious health risk if the sites are poorly managed or improperly designed. Contamination of soil and the leaking of these chemicals to both surface and ground water may lead to long-lasting toxicological problems. As industrial facilities are shut down, all too often they leave behind them heavily contaminated soil. Furthermore, transportation of wastes to dump sites also poses hazards since accidents are an ever-present possibility. If housing, schools, or office buildings are built over these areas, even in the distant future, exposure is likely to occur. Children who play in and around the soil in these areas will receive direct exposure. Children have been estimated to ingest 50–180 mg of soil per day.[1,2]

Paralleling the growth of hazardous wastes, there has been an increasing interest in the development of procedures for assessing public health risks associated with exposures to the hazardous materials. Estimates of health risk following exposure to contaminated soils have largely been based on results of studies performed with pure chemicals. However, the clay, mineral, and organic components of soil form complex, heterogeneous surfaces which are capable of adsorbing organic molecules.[3] The strength of the chemical-soil attractive forces can profoundly affect the reversibility of the adsorptive process. Therefore, the availability and the rate of chemical entering the body, its distribution to tissues, and the rate and amount of excretion may greatly differ than data from pure chemical investigation. Lucier et al.[4] and McConnell et al.[5] suggest that dioxin in soil from Times Beach and Minker Stout sites in Missouri was biologically available, as measured by microsomal enzyme studies in guinea pigs. Umbreit et al.[6] reported that despite the high concentration of dioxin from two manufacturing sites in New Jersey, this soil was unable to produce toxic effects in orally exposed guinea pigs compared with similar amounts of pure dioxin. Tight binding of dioxin to the soil matrix of the New Jersey sites correlated directly with its reduced bioavailability.

Widespread exposure to petrochemicals in dumping sites and groundwater has prompted an evaluation of the kinetics of benzene after oral and dermal treatment. Benzene is a common industrial chemical used for the synthesis of aromatic components.[7,8] It has been identified as the fourth most frequent substance recorded in 818 abandoned dump sites on the U.S. Environmental Protection Agency's 1985 National Priority List for cleanup.

Franz[9] investigated the percutaneous absorption of benzene in animals and men. He reported that less than 0.2% of the applied doses were absorbed in all species studied. Other investigators[10] suggest that workers in tire plants may absorb 4–8 mg of benzene daily through the skin from a rubber solvent mixture containing 0.5% (v/v) benzene.

This study was conducted to compare the extent to which adsorption to either of two different soils (sandy and clay) affects the manner in which benzene is subsequently handled in orally and dermally exposed adult male rats.

MATERIALS AND METHODS

Radioisotopes

All studies were conducted using uniformaly labeled ^{14}C-benzene 50 mCi/mmole (ICN Pharmaceuticals, Irvine, CA) and radiochemical purity was > 98%. Prior to use, dilution with HPLC-grade, unlabeled benzene (Aldrich Chemical Co.) was carried out to reduce specific activity to a workable range.

Soils

Two soils were utilized: a Cohansey Aquifer sandy soil taken from an outcrop site near Chatsworth in southern New Jersey, and a Keyport series clay soil collected from the Woodbury formation near Moorestown in southwestern New Jersey.

Table 1 summarizes the physical, mechanical, sieve, and organic matter characteristics in both soils. Gas chromatography/mass spectrometry analysis indicated that no petrochemical was detected. Because of the Cohansey soil's 90% sand content and the Keyport soil's 22% clay content, these soils will be referred to as sandy and clay, respectively.

Table 1. Soil Properties.

	Cohansey	Woodbury
Sand (%)	90	50
Silt (%)	8	28
Clay (%)	2	22
Organic matter (%)	4.4	1.6
Particle size, mm (%)		
0.05–0.1	22.2	17.0
0.1 –0.25	76.3	65.3
0.25–0.5	1.2	13.6
0.5 –1.0	0.2	3.4
1.0 –2.0	0.1	0.7
>2.0	0	0.1

Animals

Male Sprague-Dawley rats weighing 250–300 g were purchased from Taconic Farms, Germantown, NY, and were immediately quarantined for one week. Animals were housed three per cage at a temperature of 25°C and humidity 50% controlled environment with a 12-hr light/dark cycle. Food and water were provided ad libitum.

Benzene Administration

The oral administration of benzene was performed as follows: 150 μL of ^{14}C-benzene solution (5 μCi) alone, or the same volume of radioactivity added to 0.5 g of soil, was combined with 2.85 mL of aqueous 5% gum acacia and a suspension formed by vortexing. This volume of benzene or benzene soil suspension was immediately administered by gavage to groups of rats which had been fasted overnight. Heparinized blood samples were collected at 5, 10, 14, 18, 20, 22, 30, 45, 60, 90, and 120 minutes by cardiac puncture of lightly ether-anesthetized rats.

In the dermal application, 30 min prior to the administration, five rats/group were shaved on their right costo-abdominal areas. A shallow glass cap (Q Glass Co., Towaca, NJ) circumscribing approximately a 13 cm² area was tightly fixed with Lang's jet liquid acrylic and powder (Lang Dental Manufacturing Corp., Inc., Chicago, IL) on the shaved skin of each animal. Rats were anesthetized with ether during the cap attachment procedure. Either 300 mL of ^{14}C-benzene (40 μCi) alone or with 1 g of soil was introduced by syringe through an opening in the cap, which was immediately sealed. This volume of benzene coated the soil with no excess fluid remaining. Rats were rotated from side to side so that the soil-chemical mixture covered the entire circumscribed area. Volatilization losses during administration were determined. Heparinized blood samples were collected by cardiac puncture under light ether anesthesia at 2, 4, 8, 12, 24, 30, 48, and 72 hr. Samples from both routes of administration were processed and radioactivity was measured by liquid scintillation spectrometry, as previously described.[11,12] Immediately after the collection of the 120-min blood sample in the oral study, rats were sacrificed by an overdose of ether; whole organs or samples of bone marrow, brain, duodenum, adrenal, fat, esophagus, heart, ileum, skin, testes, thymus, thyroid, carcass, stomach, and gastric contents were collected and stored at $-75°C$. Samples of thawed adrenal, bone marrow, thyroid, and thymus 300 mg or smaller were used to determine the distribution of radioactivity as previously reported.[11]

Excretion and Metabolism Studies

In the excretion studies, groups of six rats each were administered benzene or benzene adsorbed to the soil, as described above. Pairs of animals were housed in all-glass metabolism chambers (Bioserve Inc., Frenchtown, NJ) for the collection of expired air, fecal, and urine samples. Expired air was passed through activated charcoal tubes (SKC Inc., Eighty-Four, PA) for the collection of ^{14}C-benzene, then bubbled through traps filled with ethanolamine: ethylene glycol monomethyl ether (1:2 v/v) for the collection of $^{14}CO_2$. Charcoal tubes and trap mixtures were collected at 1, 2, 6, 12, 24, and 48 hr after the administration. Urine samples were collected at 12, 24, and 48 hr, and fecal samples were collected at 24 and 48 hr. Samples were processed and radioactivity was measured as previously described.[11]

At the conclusion of the dermal excretion studies, rats were sacrificed by an overdose of ether. The glass caps were opened, and 1.0 to 1.2 ml of ethyl alcohol was introduced through the cap opening. The animals were rotated from side to side and 100 μL aliquots of ethanol wash were removed to determine the percent of benzene dose remaining on the skin application sites.[12] Then the glass caps were removed from the rats and tissue specimens were collected for the distribution determination.

To determine benzene metabolites produced, urine samples were extracted and analyzed by high performance liquid chromatography, as established in our laboratory.[11]

Data Analysis

Exploratory data analysis was used to summarize replicate data in the plasma time course study. This approach allows a curve to be fitted to all data points, while providing resistance to those points which depart from the primary pattern.[13,14] The curve-fitting procedure which was utilized is called smoothing. For these studies a "4235EH" smoother was used, as described by Velleman and Hoaglin.[14] Each replicate was smoothed over all time points, a median value was calculated from all smoothed replicates at each time point, and a second smooth was applied to these median values. The final smoothed data was used to calculate the rate constants and $t_{1/2}$ of absorption and elimination from plasma by regression analysis and the method of residuals,[15] as well as to determine a maximum concentration, and a time at which the maximum concentration was achieved. Plasma concentrations from 0 min to the time at which maximum concentration was achieved were used for absorption calculations.

For elimination calculations, 45 through 120 min was used in oral route studies, while 12 through 72 hr and 24 through 72 hr were used in dermal route studies in the soil and pure groups, respectively. Since the rate constants and the half-lives are calculated from smoothed data, the standard errors (SE) of the rate constants were determined by the bootstrap method.[16,17] The area under the plasma-time curve (AUC), calculated by the trapezoidal rule using individual replicate data, reflects volatilization losses and is reported as the mean ± standard error of the mean (SEM). All other data are reported as mean ± SEM. Statistical differences between the treatment groups were determined by analysis of variance (ANOVA), F test, and Scheffe's multiple range test. Comparison of slopes were determined using analysis of covariance.

RESULTS

Data showing the absorption and elimination half-lives following administration of ^{14}C-benzene orally and dermally to male rats are given in Table 2. The half-life of absorption into plasma in the presence of either soil for oral or dermal treatment was not statistically altered compared to their respective controls. However, dermal exposure increased absorption half-lives to 25-, 60-, and 44-fold compared to oral exposure in pure, sandy, and clay groups, respectively. In oral treatment, the elimination half-life of the clay group was significantly decreased ($p < 0.05$) compared to either sandy or pure groups. No significant change in elimination half-lives occurred after dermal exposure. In pure and sandy groups, the elimination half-lives of dermal treatment were increased about 2-fold of oral

Table 2.   Absorption and Elimination Half-Lives of Radioactivity in Male Rat Plasma.[a]

Treatment	$t_{1/2}$ (hr)			
	Absorption		Elimination	
	Oral	Dermal	Oral	Dermal
Pure[b]	0.12	3.1	13.4	23.0
Sandy[c]	0.06	3.6	10.8	24.5
Clay[d]	0.10	4.4	1.4[e]	19.4

[a]Values calculated from five or six rats per group. Animals gavaged with an aqueous solution of 5% gum acacia and ^{14}C-benzene alone or adsorbed to soil.
[b][14]C-benzene alone.
[c][14]C-benzene adsorbed to sandy soil.
[d][14]C-benzene adsorbed to clay soil.
[e]Significantly different than treatment with ^{14}C-benzene alone ($p < 0.05$).

treatments, while in the clay group the increase was 13-fold. The AUC for the 2-hr period in oral treatment was increased in both sandy and clay groups; however, only clay soil was significant ($p < 0.05$) compared to the pure group. In dermal treatment both soils significantly ($p < 0.001$) decreased AUC values compared to benzene-alone group during the 72 hr studied (Table 3).

Table 3.   Area Under Concentration-Time Cure of Radioactivity in Male Rat Plasma.[a]

Treatment	Percent Initial Dose—mL/min	
	Oral	Dermal
Pure[b]	1.53 ± 0.06	0.41 ± 0.21
Sandy[c]	2.60 ± 0.19	0.22 ± 0.08[f]
Clay[d]	3.64 ± 0.43[e]	0.17 ± 0.07[f]

[a]Values calculated from five or six rats per group.
[b]^{14}C-benzene alone.
[c]^{14}C-benzene adsorbed to sandy soil.
[d]^{14}C-benzene adsorbed to clay soil.
[e]Significantly different than treatment with benzene alone ($p < 0.05$).
[f]Significantly different than treatment with benzene alone ($p < 0.01$).

Tables 4 and 5 display the pattern of urinary and expired air excretion of radioactivity following oral and topical application of ^{14}C-benzene. In oral treatment the expired air represented the primary excretion route of ^{14}C-activity with lesser amounts eliminated in the urine during the 48 hr following the administration of benzene alone. Expired air and urine represented about equal excretion routes of ^{14}C-activity in the sandy soil treated group, while urine represented the primary route, with lesser amounts eliminated through the expired air in the clay soil group. The precentages of radioactivity in expired air of the clay soil group were significantly lower than those of the pure group at 0–12, 0–24, and 0–48 hr after oral treatment. Unmetabolized ^{14}C-benzene represented 98, 97, and 81% of the total radioactivity collected in the expired air of ^{14}C-benzene, sandy soil, and clay soil groups after oral treatment, respectively, with CO_2 comprising the remainder. With dermal application, the major route of excretion was the urine, and to a lesser extent, the expired air, in all treatment groups. During the 48-hr collection period, 86.2% of the initial dose was recovered in the urine of the pure benzene group. At the same time period, sandy and clay soil significantly decreased urinary excretion to 64.0% and 45.4%, respectively. The expired air recovery in dermal treatment was significantly decreased in the sandy soil group compared with the pure group, while the clay group expired air excretion was without significant change. Less than 1% of the administered dose was expired as $^{14}CO_2$ in all groups.

By comparing the excretion routes for the two different routes of administration, it can be seen that the urinary recovery in the dermal pure group after 48 hr from the administration exceeded the value of the oral pure group (86.2% vs 26.0%), but the other two treatments were almost without change. In the oral route, more than 82% of total radioactivity excreted in the urine of all treatment groups appeared during the 0–12 hr period following administration. However, in the dermal route, the highest amount of radioactivity in urine was recovered in the 12–24 hr interval (Table 4). Table 5 reveals that more than 98% of total radioactivity excreted in expired air of all oral treatment groups appeared during the first 12-hr period following administration. In dermal application, the major portion of radioactivity in expired air (approximately 75%) was also recovered in the 0–12 hr period following the administration. The total activity recovered in expired air in oral administration far exceeded the values of the dermal route in the period studied. During the 48-hr period, the radioactivity in the feces after oral treatment was 0.6, 1.3, and 1.4% of initial dose in pure, sandy, and clay groups, respectively. In the dermal route <0.5% was recovered in all groups during the same time period.

Table 4. Urinary Recovery of Radioactivity Following Oral or Dermal Administration of ^{14}C-Benzene.[a]

Time (hr)	Oral			Dermal		
	Pure	Sandy	Clay	Pure	Sandy	Clay
0–12	23.2±6.9	44.7±21.5	37.9±12.6	9.7±3.8	16.2±0.1	7.2±1.7
12–24	2.3±0.8	5.6±1.9	7.2±1.9	58.8±2.8	31.3±2.8[b]	25.1±3.4[b]
0–24	25.5±7.8	51.6±20.7	45.1±13.4	68.4±2.9	47.4±2.2[b]	32.3±4.0[b]
24–48	0.5±0.5	1.3±0.7	0.8±0.1	17.8±1.8	16.6±1.1	13.1±1.9
0–48	26.0±7.9	52.8±21.4	45.9±13.6	86.2±2.1	64.0±2.8[b]	45.4±4.8[b]

[a]Values represent percentage of initial dose (mean±SEM) of six animals per group.
[b]Significantly different than treatment with benzene alone (p<0.05).

Table 5. Expired Air Recovery of Radioactivity Following Oral or Derman Administration of ^{14}C-Benzene.[a]

Time (hr)	Oral			Dermal		
	Pure	Sandy	Clay	Pure	Sandy	Clay
0–12	58.2±7.2	50.0±7.6	15.6±10.0[b]	9.4±1.0	3.9±0.8[b]	8.5±1.2
12–24	0.1±0.0	0.2±0.1	0.1±0.0	2.5±0.4	0.4±0.1[b]	1.1±0.2[b]
0–24	58.2±7.2	50.2±7.6	15.7±10.1[b]	12.0±1.4	4.3±0.8[b]	9.6±1.3
24–48	0.0±0.0	0.0±0.0	0.1±0.0	0.8±0.2	1.6±0.5	0.5±0.2
0–48	58.2±7.2	50.2±7.6	15.0±10.1[b]	12.8±1.1	5.9±1.3[b]	10.1±1.4

[a]Values represent percentage of initial dose (mean±SEM) of six animals per group.
[b]Significantly different than treatment with pure benzene (p<0.05).

Tissue distribution of radioactivity after oral administration is given in Table 6. Gastric contents contained the highest concentration of radioactivity in all oral groups, with the mean activity (as percentage of initial dose/g) of the clay group (18.7) being about 6-fold higher than that of either the sandy soil (2.8) and pure benzene (2.1) group. Stomach contained the highest tissue concentration of radioactivity, with fat the second highest in all treatments, followed by duodenum and adrenal. No statistically significant differences were detected in the tissue concentrations of radioactivity between the oral treatment groups. In the dermal route, soil-related differences were observed in tissue distribution. The distribution patterns of ^{14}C activity for pure and soil-adsorbed benzene are demonstrated in Table 7. ^{14}C activity 48 hr post-administration of soil-absorbed benzene was greatest in the treated skin followed by kidney, liver, duodenum, spleen, treated fat, and untreated fat, as well as bone marrow in both soil groups. In the pure benzene group, kidney contained the highest amount of radioactivity, followed by liver, treated skin, duodenum, treated fat, and untreated fat as well as bone marrow. Clay soil treatment statistically increased radioactivity (10-fold) in treated skin, while statistically decreasing radioactivity (4-fold) in treated fat compared to benzene alone. It is worth noting that ethanol extraction of treated sites at necropsy contained only 0.1% of the initial dose as loosely retained on the application sites of all groups.

Table 6.   Tissue Distribution of Radioactivity in Male Rat Following Oral Administration of ^{14}C-Benzene.

Gastric Contents[a]	
Pure Benzene	2.1 ± 1.8
Sandy Soil	2.8 ± 0.7
Clay Soil	18.7 ± 10.8[b]

In all treatment groups:
Gastric Contents > Stomach[c] > Fat[c] >
Duodenum[c] > Adrenal[c]

[a]Values represent percent initial dose per gram (mean $\pm$ SEM) from five rats per group, 2 hr following oral administration.
[b]Significantly different than treatment with benzene alone ($p < 0.05$).
[c]No statistical differences between treatment groups.

Table 7.   Tissue Distribution of Radioactivity in Male Rat Following Dermal Administration of ^{14}C-Benzene.

Pure Benzene Group:
Kidney > Liver > Treated Skin > Duodenum >
Treated Fat > Untreated Fat = Bone Marrow

Sandy and Clay Groups:
Treated Skin[a] > Kidney > Liver >
Duodenum > Spleen > Treated Fat[b] >
Untreated Fat = Bone Marrow

[a]Significantly increased in the clay group compared to the pure benzene group ($p < 0.05$).
[b]Significantly decreased in the clay group compared to the pure benzene group ($p < 0.05$).

Data showing the urinary metabolites of ^{14}C-benzene in the male rat after oral and dermal application are given in Table 8. Phenol was the major urinary metabolite detected in the 0–12 hr urines of all treated groups in both routes of administration. Smaller quantities of hydroquinone, catechol, and benzenetriol compared to phenol were also detected. The type and percentage of benzene metabolites produced were not altered in the presence of the soil after oral administration, while in dermal application, hydroquinone was statistically decreased in the 0–12 hr interval of the sandy treatment as compared to the pure group. Similar metabolite percentages were detected in 12–24 hr urines of all treated groups (data not shown). The parent compound was not detected in the urine of any treatment group. Use of acid hydrolysis in the preparation of urinary extract did not permit identification of conjugation products.

Table 8. Urinary Metabolites of ^{14}C-Benzene in the Male Rat.[a]

Metabolite	Oral			Dermal		
	Pure	Sandy	Clay	Pure	Sandy	Clay
Phenol	39 ± 1	41 ± 3	34 ± 2	38 ± 3	44 ± 4	46 ± 3
Catechol	14 ± 1	15 ± 3	18 ± 2	13 ± 1	14 ± 5	12 ± 1
Hydroquinone	17 ± 3	25 ± 3	26 ± 1	19 ± 2	10 ± 1^{b}	25 ± 1
Benzenetriol	12 ± 3	14 ± 3	20 ± 4	5 ± 2	17 ± 7	13 ± 2

[a]Values represent % of total radioactivity in the 0–12 hr collection period from 6 animals per group (mean ± SEM).
[b]Significantly different than treatment with benzene alone ($p < 0.05$).

DISCUSSION

The results of this study revealed that the presence of sandy and clay soil produced qualitative and quantitative differences in the manner of the availability of benzene to the body following oral or dermal treatments. Although the soil group in oral treatment did not significantly alter the rate of benzene absorption ($t_{1/2}$), AUC for 0–2 hr post-administration was increased in both soil groups. Because the density of benzene is less than water, some gavaged material could be vaporized out of the gastrointestinal directly without absorption into the body. Adsorption of benzene to the soil decreased the vaporization of benzene. The detection of the bulk of radioactivity excreted in expired air as unmetabolized benzene during the first 2-hr collection period (data not shown) unassociated with high plasma concentration of radioactivity during the same time period would support this conclusion. The relatively stronger adsorption of benzene to clay soil is supported by significantly increased AUC, as well as significantly decreased excretion in expired air and relatively high concentration of radioactivity in the gastric contents 2 hr after the oral administration.

Dermal exposure of pure and soil-adsorbed benzene produced plasma concentrations of radioactive compound comparable to those generated by oral administration only when rats were exposed to eight times the concentration of ^{14}C-benzene used in the oral route (40 vs 5 μCi). Also, this laboratory reported that peak plasma concentrations after dermal treatments were delayed about 36-fold compared to those which occurred following oral administration.[11,12] The absorption and elimination half-lives in all groups after dermal treatment were much longer compared to their respective oral groups. The result of this study is in agreement with Franz[9] and Susten et al.,[10] which indicated that benzene was not readily absorbed through the skin. After dermal application, both soil groups demonstrated a significantly lower amount of ^{14}C activity in urine compared to the pure group, while radioactivity in expired air was decreased significantly after sandy soil administration.

Routes of excretion and amount excreted by the various routes were changed in the presence of the soils in both routes of administration. Expired air recoveries were decreased in all dermal treatments compared to the oral experiements, while the urine was the primary route of excretion in all dermal groups as well as the oral clay group. The fecal route remains a minor excretion route in the presence or absence of soils.

The quantity and quality of benzene metabolites produced were almost without change, except the amount of hydroquinone in the dermal sandy group was significantly decreased. Phenol was the primary urinary metabolite in all the treatments studied.

Malkinson and Gehlmann[18] reported that the most important factors related to chemical persistence in soil are organic matter and clay content of the soil. In both dermal and oral soil-adsorbed benzene studies, results revealed that a higher percentage of clay rather than organic matter is controlling the retention of benzene in soil and therefore, altering the bioavailability of benzene to the male rat. Particle sizes, and thus surface area, of the two soils were essentially equivalent and do not appear to be a factor in this study.

ACKNOWLEDGMENT

This research was supported as a project of the National Science Foundation/ Industry/University Center for Research in Hazardous and Toxic Substances at New Jersey Institute of Technology, and Advanced Center of the New Jersey Commission on Science and Technology.

REFERENCES

1. Clausing, P., B. Brunekreef, and J. H. van Wignen. "A Method for Estimated Soil Ingestion by Children," *Int. Arch. Occup. Environ. Health.* 59:73–82 (1987).
2. Binder, S., and D. Sokal. "Estimating Soil Ingestion," *Arch. of Environ. Health.* 41(6):341–345 (1986).
3. Hamaker, J. W., and J. M. Thompson, "Adsorption," in *Organic Chemicals in the Soil Environment,* Vol I., C. Goring and J. Hamaker, Eds. (New York: Marcel Dekker, 1972), pp. 49–143.
4. Lucier, G. W., R. C. Rumbaugh, Z. McCoy, R. Hass, D. Harvan, and P. Albro. "Ingestion of Soil Contaminated with 2, 3, 7, 8-Tetrachlorodibenzo-p-dioxin (TCDD) Alters Hepatic Enzyme Activities in Rats," *Fund. Appl. Toxicol.* 6:364–371 (1986).
5. McConnell, E. E., G. W. Lucier, R. C. Rumbaugh, P. W. Albro, D. J. Harvan, J. R. Hass, and M. W. Harris. "Dioxin in Soil: Bioavailability After Ingestion by Rats and Guinea Pigs," *Science* 223:1077–1079 (1984).
6. Umbreit, T. H., E. J. Hesse, and M. A. Gallo. "Bioavailability of Dioxin in Soil from a 2, 4, 5-T Manufacturing Site," *Science* 232:497–499 (1986).
7. Baselt, R. C. *Disposition of Toxic Drugs and Chemicals in Man,* 2nd ed. (Davis, CA: Biomedical Publications, 1982), pp. 71–75.

8. Sandmeyer, E. E. "Aromatic Hydrocarbons," in *Patty's Industrial Hygiene and Toxicology,* Vol. 2B. G. D. Clayton and F. E. Clayton, Eds. (New York: John Wiley & Sons, 1981), pp. 3253–3432.
9. Franz, T. J. "Percutaneous Absorption of Benzene," in *Advances in Modern Environmental Toxicology, Applied Toxicology of Petroleum Hydrocarbons,* Vol. 6. H. N. MacFarland, C. E. Holdworth, J. A. MacGregor, R. W. Call, and M. L. Lane, Eds. (Princeton, NJ: Princeton Scientific Publishers, Inc., 1984), pp. 61–70.
10. Susten, A. S., B. L. Dames, J. R. Burg, and R. W. Niemeir. "Percutaneous Penetration of Benzene in Hairless Mice: An Estimate of Dermal Absorption During Tire-Building Operations," *Am. J. Ind. Med.* 7:323–335 (1985).
11. Turkall, R., G. Skowronski, S. Gerges, S. Von Hagen, and M. Abdel-Rahman. "Soil Adsorption Alters Kinetics and Bioavailability of Benzene in Orally Exposed Male Rats," *Arch. Environ. Contam. Toxicol.* 17:159–164 (1988).
12. Skowronski, G., R. Turkall, and M. Abdel-Rahman. "Soil Adsorption Alters Bioavailability of Benzene in Dermally Exposed Male Rats," *J. Am. Ind. Hyg. Assoc.* Accepted for publication, 1988.
13. Tukey, J. W. *Exploratory Data Analysis* (Reading, MA: Addison Wesley, 1977), pp. 205–235.
14. Velleman, P. F., and D. C. Hoaglin. *Applications, Basics and Computing of Exploratory Data Analysis* (Boston, MA: Duxbury Press, 1981), pp. 159–200.
15. Gibaldi, M., and D. Perrier. *Pharmacokinetics* (New York, Marcel Dekker, 1975), pp. 281–292.
16. Efron, B. *The Jacknife, the Bootstrap and Other Resampling Plans* (Philadelphia, PA: Society of Industrial Applied Math, 1982).
17. Efron, B., and R. Tibshirani. "Bootstrap Method for Assessing Statistical Accuracy," Technical Report 101 (Stanford, CA: Division of Biostatistics, 1985).
18. Malkinson, F. D., and L. Gehlmann. "Factors Affecting Cutaneous Toxicity," in *Cutaneous Toxicity,* V. A. Drill and P. Lazar, Eds. (New York: Academic Press, Inc., 1977), pp. 63–81.

CHAPTER 23

Epidemiological Study to Estimate How Much Soil Children Eat

Edward Calabrese, Charles Gilbert, Paul Kostecki, Ramon Barnes,
Edward Stanek, Petrus Veneman, Harris Pastides, and Carolyn Edwards

INTRODUCTION

It has long been recognized that contaminated soil may present a potential public health concern via the contamination of groundwater, since groundwater is
a significant drinking water source. Recently scientists have become concerned
that consumption of contaminated soil by children may present a significant public health problem. For example, soil levels of lead in certain Boston neighborhoods have been found to be in the range of 2500–7500 μg/g.[1] Consumption of
one-half gram of such soil by a child could result in exposure to 1250 to 3750
μg Pb of ingested lead, or from 62–187-fold greater than permitted by the present
U.S. Environmental Protection Agency (EPA) proposed recommended maximum
contaminant level of 20 μg/L (assuming consumption of 1 L/day).[2] A more widely
discussed health problem from childhood soil ingestion has been the dioxin contamination of soil in Times Beach, Missouri. The Centers for Disease Control
(CDC) derived a theoretical cancer risk associated with levels of dioxin in soil.[3]
In this case, a major exposure component was the assumed consumption by children of soil containing dioxin, a contaminant that is relatively tightly bound in
soil. In light of our concern with lead and dioxin, as well as other soilbound agents
such as PCBs, it is necessary to assess possible human exposure, especially in
children, to contaminants in soil. Some major considerations contributing to an
assessment of exposure to adults and children include ingestion, the bioavailabil-

313

ity of contaminants in soil, dermal absorption from skin contact with soil, and inhalation of soil dust and particulates. This paper describes one methodology to estimate how much soil young children eat.

Techniques to assess the amount of soil consumed by children have been adopted from those used in veterinary science to estimate the soil eaten by animals. The techniques for the estimation of animal soil ingestion are described here to show the evolution of knowledge that can be applied to assess soil consumption in young children and adults.

SOIL INGESTION IN GRAZING ANIMALS

There have been at least eight studies which attempted to determine the extent to which grazing animals ingest soil.[4-11] There were several purposes of such studies, including determining the extent to which mineral intake occurs via soil ingestion and the extent to which pollutants in soil such as DDT may be consumed and retained by such animals. The animals studied included young swine, sheep (ewes and lambs), dairy cows and cattle. The exposure condition of these animals varied widely from studies in confined areas with little grass covering, to vast open ranges with 4–5 in. grass, and over several seasons of the year.

The amount of soil ingested by these animals was estimated through the determination of titanium concentration in feces. Titanium was employed because its concentration in plants consumed by grazing animals was very low (i.e., less than 1 ppm in plant tissue), while the levels in soil ranged from 1000 to 3000 ppm. The results of these different studies indicate that all the grazing animals studied ingest substantial amounts of soil, most notably when the height of the grass had been grazed to shortened stubs. The species displaying the greatest amount of soil ingested were swine, followed by cows and sheep. The reason for enhanced soil ingestion by swine is related to their rooting habit. Their ingestion of soil ranged from 3.3% to 8.0% of dry matter consumed when the swine were maintained on lots of bare soil. The enhanced exposure of swine to soil-laden pollutants is seen in data from Fries et al.[12], in which the average ratios of polybrominated biphenyls (PBB) in body fat to PBB in soil was 0.1:1 for cattle and 2:1 for swine, a 20-fold difference.

It is estimated that the annual soil intake of dairy cow ranges from 200 to 1000 lb, with peak soil consumption occurring in the autumn and winter. This amounts to about 0.5 to 1.5 kg soil/day/animal.[4,5] Soil ingestion by sheep is comparable to the soil ingestion reported for dairy cows.[6]

The issue of animal pollutant exposure via ingestion of contaminated soils has been discussed by Healy[4] and Fries et al.[12] Healy proposes that ingestion of large quantities of soil may contribute significantly to the total amount of DDT consumed by an animal over a year. There is little vertical movement of DDT in the soil, with the bulk of it found in the first inch or two of topsoil, even several years after treatment[7-9,11] Based on this information, Healy speculated that if 2 lb of DDT per acre were applied and concentrated in the top 1/2 in. of soil, and an animal ingested 800 lb of soil per year, this would result in an intake as high as 5 g of DDT per year.[4]

SOIL INTAKE BY HUMANS

Some research has specifically addressed soil consumption in children. Lepow et al. measured a mean amount of 10 mg of dirt on the hands of 22 children.[13] They assumed that a child places his hands into his mouth about 10 times per day. Thus, they calculated a 100 mg/day intake of dirt. The National Research Council estimated that young children average 40 mg/day of ingested street dirt.[14] Day, et al. measured 5–50 mg of dirt transferred from a child's hand to a sticky sweet and estimated a daily intake of 2–20 sweets would result in a dirt intake of 10–1000 mg.[15] CDC scientists developed a speculative model for specific age groups based on observations of child behavior (Table 1).[3] All of these soil estimates, however, have been recently partially refined in a pioneering study by Binder et al.[16] In their study involving 59 children aged 1–3 years, the calculated soil ingestion estimation based on the use of aluminum and silicon as tracers were 121 and 184 mg/day. The soil ingestion estimation increased by 10 times, or 1834 mg/day when titanium concentrations were used to calculate the soil ingested. The authors were unable to resolve the apparent conflict between the estimations based on aluminum or silicon tracers and the estimations based on titanium. It should be emphasized that all three tracers are believed to be poorly absorbed from the GI tract; however, the algorithm will underestimate soil ingestion if young children absorb significant amounts of these tracer elements. It is known that children absorb a number of inorganic elements with greater efficiency than adults in both animal models and their human counterparts. Table 2 summarizes the occurrence of age differences in gastrointestinal absorption for a number of inorganic elements.

The mathematical algorithm used by Binder et al.[16] did not adjust for possible exposure to the tracers from foods orother ingested products such as medicines and toothpaste, since their research did not include collections of duplicate ingestion samples.

One of the most significant sources of titanium to confound estimates of soil ingestion may be the paint chips and dust in the soil surrounding the house. Titanium dioxide is the major pigment in paint and may contribute to substantial soil

Table 1.   Estimated Daily Soil Ingestion Patterns by Age.

Age Group	Soil Ingested
0–9 mo.	0 g
9–18 mo.	1 g
1.5–3.5 yr.	10 g
3.5–5 yr.	1 g
5 yr.	100 mg

Source: Kimbrough et al.[3]

Table 2.   Summary of Difference in G.I. Tract Absorption
Between the Children and Adults.

Element	References	Differences
Lead	Kehoe;[17] Alexander et al.[18]	5-fold humans
Cadmium	Sasser and Jarboe[19]	20-fold guinea pigs
Sr	Taylor et al.[20]	9-fold rats
Ba	Taylor et al.[20]	11-fold rats
Ra	Taylor et al.[20]	24-fold rats
Plutonium	Ballou[21]	85–100-fold rats, guinea pigs
Niobium	Mraz and Eisele[22,23]	485-fold sheep
Cerium	Shiraishi and Ichikawa[24]	10 to 1000-fold rats

Source: Calabrese[25].

concentrations as it weathers from the house and falls to the ground. Estimates have been made of the average adult intakes of aluminum 15 mg/day, silicon 7 mg/day, and titanium 0.3 mg/day.[26,27] Given this adult dietary knowledge, it is probably prudent to consider a child's diet when estimating soil ingestion. Consideration must be given to possible inadvertent cross-contamination of stool specimens from diapers, since titanium dioxide levels in diapers may exceed 1000 ppm.[28] These collective factors may lead to an overestimation of childhood soil consumption.

The process of Binder et al.[16] contains no direct proof that the children are actually consuming soil, nor the amount of soil ingested. Could the presence of the markers in the feces be explained entirely by the presence of these elements in food, toothpaste, and in products that, like diapers, were in contact with the feces? While their approach is reasonable, it is in need of independent attempts at validation.

It is possible to validate the study and algorithm via the comparison of tracer ratios in soil and diet versus feces, assuming that the ratios of the tracers are significantly different in soil compared to food. For example, the estimated ratio of titanium:silicon:aluminum in the overall diet of an adult human is 1:23:50, based on estimated levels of tracer intake of titanium (0.3 mg/day), silicon (7 mg/day), and aluminum (15 mg/day). The estimated ratio of these tracers in soil are 1:10:2, based on the Binder et al. study.[16] Based on these tracer ratios, one would have predicted an approximate fecal tracer ratio of 1:16:26 for 10 mg of soil ingested, 1:10:7 for 100 mg of soil ingested, and 1:9:2.7 for 1000 mg of soil ingested. This mathematical backtracking, while having significant limitations, supports the premise that food consumption could not have explained the presence of tracers in the feces, and that the best estimation based on the tracers was actually 1000 mg/day. Nonetheless, even this validation attempt is still indirect and inconclusive.

Other attempts have been made to estimate the amount of soil consumed by children—one by Clausing et al.,[29] which essentially followed the methodology of Binder et al.,[16] and one by Rabinowitz and Bellinger,[30] which is a pharmacokinetic methodology designed to assess lead ingestion from soil. Clausing et al. reported a possible range of soil ingestion from 127 mg to 1084 mg/day, depending on the marker, with titanium yielding the highest ingestion estimate as in the Binder study. The Clausing et al. study, however, is another indirect attempt to estimate soil ingestion, and suffers from methodologic omissions similar to those of Binder et al., such as no measurement of tracers in food and other ingested products; and presence of tracers in diapers and other materials that may contact the feces.

An interesting aspect of the Clausing et al. study is that they included six children who were hospitalized and bedridden, and assumed to have no soil contact. The hospitalized children had an average soil ingestion of 45 mg per day.

The report of Rabinowitz and Bellinger estimated that children up to the age of two years ingest between 10 and 20 mg of soil/day. The value of this report is highly suspect, due to the use of questionable pharmacokinetic assumptions concerning lead disposition in the individual, the lack of incorporation of environmental measurement into the mathematical algorithm, and the lack of any independent additional markers that could have validated the studies' findings. One of the positive attributes of both the Binder et al. and Clausing et al.studies was the use of tracers with very low absorption rates. Using the heavy metal lead as a marker is problematic because it may be extensively absorbed in the young child, with the rate decreasing as the child ages, and varying due to nutritional status. The body distribution of lead is also affected by age and nutritional status. These variables create significant difficulties in establishing predictable pharmacokinetic constants. The results are also limited since the data did not consider sources of environmental lead other than dirt and breast milk levels.

METHODOLOGY

We used the same conceptual approach as put forth by Binder et al., using tracer elements naturally occurring in the soils and a mathematical algorithm to estimate soil consumption. The present study builds upon the Binder et al. work by addressing identifiable limitations in the design of the study.

Sixty-five children between the ages of 1.0 and 4.0 years were identified through University of Massachusetts-affiliated day care facilities and through word of mouth. The children came from the greater Amherst and greater Springfield areas, and enrolled in the study following parental consent. Children with chronic illness, such as diabetes, heart disease, or gastrointestinal diseases were excluded from participation.

Parents were invited to training sessions to describe the study to them, to invite them to participate in the study, to obtain their consent for their child's participation, and to begin training them in study procedures. The parental training included the creating of duplicate diets, (i.e., estimating the amount of food actually consumed by the child) and the collection of stools and urine. Data collected from the parents included: dates of birth, education, occupation, marital status, and other demographic information; and outdoor and indoor activities reports during the data collection period. All parents were provided with a copy of a description of the study and an informed consent letter. Informed consent was obtained from the parents before the children entered the study.

Total excretory material (feces and urine) was collected from each child starting Monday noon through Friday noon in each of two consecutive weeks by parents and day care staff. All daily excretory output samples were kept separate, so the repeated measures can be used to assess each child's variation and the variation between children. Duplicate food/beverage meal samples were collected for all children by the parents and day care staff. The duplicate meals included all food ingested from breakfast Monday through supper Wednesday over each of the two weekly study periods. All medications ingested by the children were included in the duplicate meals, or were addressed during chemical analysis. Medications not added to the meals by the parents were analyzed for concentration of the eight chemical tracers. None of the medications reported contained excess amounts of the chemical tracers.

Soil samples were collected from the child's day care, and from three play area sites selected by the parents. Dust samples were collected by vacuuming the floors in the living space identified by the parents, where the children usually crawled and played. Dust samples were also collected from the child's day care from areas identified by the staff, where the children crawled or played.

Indirect chemical markers, silicon, aluminum, titanium, barium, vanadium, zirconium, manganese, and yttrium were selected to estimate soil ingestion in these free-living 1- to 4-year-old children via a comprehensive mass-balance exposure assessment.

ADULT VALIDATION

An adult validation study was conducted to evaluate the contribution of known quantities of ingested soil to excretory concentrations of the chemical tracers. During the first week of the study the six volunteers (three male, three female) ingested empty gelatin capsules. Over the subsequent two week periods all six volunteers ingested capsules containing clean soil with breakfast and dinner, Monday through Wednesday. Duplicate meal samples, food, and beverage were collected from the six adults. The duplicate meals included all food ingested from breakfast Monday through the evening meal on Wednesday during each of the three weeks. All medications and vitamins ingested by the adults were included in the duplicate meals or were addressed during chemical analysis. All excretory material, feces and urine, from each volunteer were collected from Monday through Friday midnight during the three week period.

Through the comparison of the ratios of tracers found in the soil, in the food, and in the excretory samples, it was believed possible to assess the extent to which the tracer levels in the ingested soil contribute to the excretory tracer concentrations. These data would constitute proof that soil would contribute to excretory tracer levels and that it was possible to estimate soil ingestion based on the levels found in fecal and urinary specimens. The data would also allow one to infer the comparative bioavailability of tracers from soil versus from foods in these adults.

DISCUSSION

Although the amount of soil consumed by children is quite uncertain, the studies conducted by Binder et al. and Clausing et al. have provided a better understanding of the problem. In our investigation we build upon these efforts, and as a result we have some very ambitious goals. One goal is to determine the range of soil ingestion that occurred in the 1- to 4-year-old children we have studied. Our second goal is that the study will have some generalizability to children of similar socioeconomic backgrounds. A third goal is that, based on the data derived from the adult validation study, we will be able to determine the magnitude of the bioavailability of the tracer elements from soil versus the bioavailability of tracer elements from food. While this investigation uses the indirect approach of naturally occurring chemical tracers, we believe that the controls employed will permit an accurate estimation of the amount of soil ingested by 1- to 4-year-old children. In light of the potential public health significance of soil ingestion by children it is recommended that further research be directed in this area.

REFERENCES

1. Spittler, T. Technical Resource Committee Meeting, Minnesota Pollution Control Agency, Roseville, Minnesota, September 24, 1986.
2. U. S. Environmental Protection Agency, National Primary Drinking Water Standards, 1985.
3. Kimbrough, R. D., H. Falk, P. Stehr, and G. Fries. "Health Implications of 2,3,7,8-TCDD Contamination of Residual Soil," *J. Toxicol. Environ. Health* 14:47–93 (1984).
4. Healy, W. B. "Ingestion of Soil by Dairy Cows," *N. Z. J. Agric. Res.* 11:487–499 (1968).
5. Mayland, H. F., A. R. Florence, R. C. Rosenau, V. A. Lazar, and H. A. Turner. "Soil Ingestion by Cattle on Semiarid Range as Reflected by Titanium Analysis of Feces," *J. Range Management* 28:448–452 (1975).
6. Field, A. C., and D. Purvis. "The Intake of Soil by the Grazing Sheep," *Proc. Nutr. Soc.* 23:24–25 (1964).
7. Gallagher, P.J., and L. Evans. "Preliminary Investigations on the Penetration and Persistence of DDT Under Pasture," *N. Z. J. Agric. Res.* 4:466–475 (1961).
8. Harrison, D.L., J. C. M. Mol, and J. E. Rudman. "DDT and Lindane—New Aspects of Stock Residues Derived from a Farm Environment." *N. Z. J. Agric. Res.* 12:553–574 (1969).
9. Harrison, D. L. "Observations on Background Residues in Animal Fats," *Proc. 21st N. Z. Weed Control Conf.* pp. 225–232 (1968).

10. Harrison, D. L., J. C. M. Mol, and W. B. Healy. "DDT Residues in Sheep from the Ingestion of Soil," *N. Z. J. Agric. Res.* 13:664–672 (1970).
11. Matthews, L. J. "Limitations of DDT on Volcanic Soils," *Proc. 20th N. Z. Weed Control Conf.* pp. 157–159 (1967).
12. Fries, G. F., G. S. Marrow, and P. A. Snow. "Soil Ingestion by Swine as a Route of Contaminant Exposure," *Environ. Toxicol. Chem.* 1:201–204 (1982).
13. Lepow, M. L., L. Bruckman, R. A. Rubin, S. Markowitz, M. Gillette, and J. Kapish. "Role of Airborne Lead in Increased Body Burden of Lead in Hartford Children." *Environ. Health Pers.* 7:99–102 (1974).
14. "Lead in the Human Environment," National Research Council, Washington, DC, 1980.
15. Day, J. P., M. Hart, and M. S. Robinson. "Lead in Urban Street Dust," *Nature* 253:343–345 (1975).
16. Binder, S., D. Sokal, and D. Maughan. "Estimating the Amount of Soil Ingested by Young Children Through Tracer Elements," *Arch. Environ. Health* 41, 341–345 (1987).
17. Kehoe, R. A. "The Metabolism of Lead in Man in Health and Disease. II. The Metabolism of Lead Under Abnormal Conditions," *J. Roy. Inst. Pub. Health Hyg.* 24:101(1961).
18. Alexander, F.W., B. E. Clayton, and H. T. Delves. "Mineral and Trace-Metal Balances in Children Receiving Normal and Synthetic Diets," *Quart. J. Med.* 43:89(1974).
19. Sasser, L. B., and G. E. Jarboe, "Intestine Absorption and Retention of Cadmium in Neonatal Pigs Compared to Rats and Guinea Pigs," *J. Nutr.* 110:1641–1647(1980).
20. Taylor, D. M., P. H. Bligh, and M. H. Duggan. "The Absorption of Cadmium, Strontium, Barium, and Radium from the Gastrointestinal Tract of the Rat," *Biochem. J.* 83:25–29(1962).
21. Ballou, J. E. "Effects of Age and Mode of Ingestion in Absorption of Plutonium," *Proc. Soc. Exp. Biol. Med.* 98:726–727(1958).
22. Mraz, F. R., and G. R. Eisele. "Gastrointestinal Absorption, Tissue Distribution, and Excretion of ^{95}Nb in Newborn and Weanling Swine and Sheep," *Radiat. Res.* 72:533–536(1977).
23. Mraz, F. R., and G. R. Eisele. "Intestinal Uptake and Whole-Body Retention of ^{141}Ce by Suckling Rats," *Health Phys.* 22:169–175(1977a).
24. Shiraishi, Y., and R. Ichikawa. "Absorption and Retention of ^{141}Ce and ^{95}Zr-^{95}Nb in Newborn, Juvenile, and Adult Rats," *Health Phys.* 22:373–378(1972).
25. Calabrese, E. J. *Age and Susceptibility to Toxic Substances* (New York: John Wiley & Sons, 1986).
26. Schroeder, H. A., J. J. Balassa, and I. H. Tipton, "Abnormal Trace Elements in Man: Titanium," *J. Chron. Dis.* 16, 55–69 (1963).
27. Bowen, H. J. M., *Environmental Chemistry of the Elements* (New York: Academic Press, 1979).
28. Barnes, R. Personal communication, Department of Chemistry, University of Massachusetts, Amherst, 1986.
29. Clausing, P., B. Brunekreff, and J. H. van Wijnen. "A Method for Estimating Soil Ingestion in Children," *Int. Arch. Occup. Environ. Med.* 59:73(1987).
30. Rabinowitz, M. B., and D. Bellinger, "How Much Dirt Do Urban Infants Eat?" Personal communication, 1986.

CHAPTER 24

A Site-Specific Approach for the Development of Soil Cleanup Guidelines for Trace Organic Compounds

B. G. Ibbotson, D. M. Gorber, D. W. Reades, D. Smyth, I. Munro,
R. F. Willes, M. G. Jones, G. C. Granville, H. J. Carter, and C. E. Hailes

INTRODUCTION

The decline in demand for petroleum products has resulted in the closure of several oil refineries in Canada. In the fall of 1978, Texaco Canada Inc. (Texaco) stopped production at its Port Credit Plant in Mississauga, Ontario with the exception of petrochemical products which continued to be produced until 1985. In July 1983, Shell Canada Limited (Shell) stopped processing crude oil at its refinery in Oakville, Ontario and announced that the refinery would be dismantled. Both sites are near residential areas and there is a general desire to redevelop parts of both properties for residential housing.

At designated sections of each site, oily wastes were applied to the surface soil and tilled to provide mixing and to encourage the degradation of organic compounds through chemical and biological processes. This method, commonly referred to as landfarming, has been used in Canada and other countries for many years and is generally considered to be an effective, economic method of treatment and disposal for many oily wastes. Some constituents of organic (oily) wastes undergo relatively rapid and complete degradation; others degrade slowly, and the low levels of metals often present do not degrade.

In Ontario, as in the rest of North America, the formal decommissioning of industrial sites in general and landfarm areas in particular is relatively new, and few regulations or guidelines are in place. Two Ontario regulations are relevant:

1. Section 45 of the Ontario Environmental Protection Act restricts alternative land use on any lands which received wastes for a period of 25 years from the time of disposal. The restriction remains in effect unless permission is received from the Environment Minister for a change in land use.
2. Regulation 309 of the Government of Ontario lists generic wastes which are designated as "hazardous." The list includes five wastes from the petroleum refining industry—dissolved air flotation float, slop oil emulsion solids, heat exchanger bundle cleaning sludge, API separator sludge, and leaded gasoline tank scale. Soils which received any of these wastes must be shown to be acceptable prior to leaving the soil in place and developing the site.

The Ontario Ministry of the Environment (MOE) drafted "Guidelines for the Decommissioning (Shut-down) of Major Industrial Sites in Ontario" in 1984. The document outlines a general framework for decommissioning, and currently is being revised and expanded. As part of its decommissioning efforts, the MOE has recommended soil cleanup levels for some general indicator parameters (such as total oil and grease) and some heavy metals, but not for any specific organic compounds.

Shell and Texaco thus were faced with the situation in which various organic compounds were known or suspected to be present in site soils, but without the criteria to determine if conditions were acceptable for redevelopment. When the MOE indicated that criteria for specific compounds could take several years to produce, the companies offered to sponsor the development of an approach for establishing site-specific soil cleanup guidelines for these compounds. The consulting team of Golder Associates, SENES Consultants Limited, and CanTox Inc. was retained to conduct the necessary studies. After discussions between the companies and the consulting team it was decided to develop an approach capable of identifying site-specific cleanup guidelines based primarily on human health effects and "acceptable" levels of exposure.

DESCRIPTIONS AND HISTORIES OF THE TWO SITES

Texaco Port Credit Plant

The Port Credit Plant site is situated in Mississauga, Ontario, approximately 30 km west of Toronto on the north shore of Lake Ontario (see Figure 1). The 89 hectare (ha) property is divided into three major areas: (a) the 30 ha Process Area and South Tank Farm; (b) a 5 ha Bulk Station and Administration Area (also referred to as the Marketing Area); and (c) a 54 ha area which includes the North Tank Farm, the Land Farm, and an Agricultural Area.

The region surrounding the site is a relatively flat plain that is drained by small creeks and ditches that flow to Lake Ontario. Overburden consists of a generally

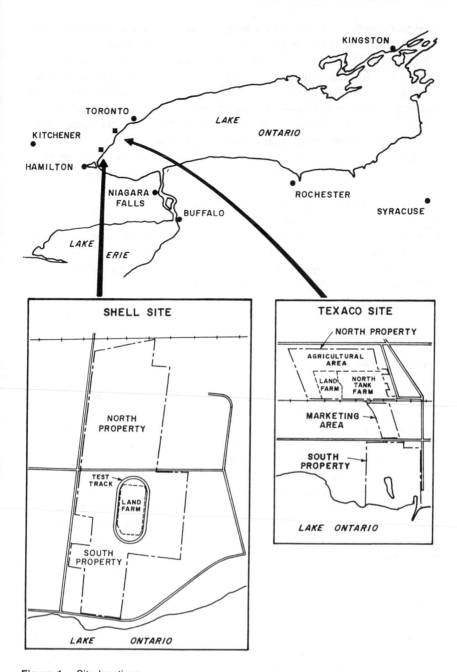

Figure 1. Site locations.

thin layer of glaciolacustrine sands and silts. The regional bedrock consists of grey Dundas shale. Anticipated groundwater yields from overburden and bedrock wells in the area are low (less than 0.1 L/s), thereby precluding wells as an important water supply. Most of the water for domestic, commercial, and industrial uses in the Port Credit area is obtained from Lake Ontario.

The original refinery complex at the Port Credit site started operating in 1933 and had an initial throughput of 160 m³ of crude oil per day. The capacity was increased several times over the next 30 years as the refinery went through several changes of ownership and expansions. Texaco purchased the refinery in 1957 and by 1961, the throughput was 8,000 m³/day. In 1978, Texaco brought another refinery complex into production in Ontario, and many of the refining units at Port Credit were shut down. Between 1978 and 1985, the Port Credit Plant served primarily as an extraction and fractionation unit for benzene, xylene, toluene, and hexane.

Sections of the site were used for waste disposal purposes at various times during the operating life of the refinery. The Land Farm was licensed for the treatment of refinery-derived wastes between 1972 and 1978. Wastes were generally applied to the surface soil and tilled into the upper layers of soil. Tilling was discontinued in 1982. Wastes spread in the Land Farm included leaded and unleaded tank bottoms, API separator sludge, oily waste, catalysts, and light cycle gas oil filter wastes.

Some wastes (including leaded and unleaded tank bottoms) also were disposed in several bermed areas of the South Tank Farm.

Shell Oakville Refinery

The Shell Oakville refinery site is located in Oakville, Ontario, 50 km west of Toronto. The site is bounded by another refinery to the east, Lake Ontario to the south, and residential and agricultural land to the west and north, respectively. The site comprises some 182 ha of land (see Figure 1).

The region surrounding the site is quite similar to that around the Texaco Port Credit Plant. The relatively flat plain is imperfectly drained by a network of small ditches and creeks which generally flow south towards Lake Ontario. The geology of the site consists of a thin layer of overburden underlain by weathered shale bedrock. The water table range is relatively close to the surface, but groundwater quality in the region is only marginally suitable as a domestic water supply. Houses in the area are supplied via the municipal water system.

The Oakville site is divided into northern and southern sections. The Northern Property contained all of the manufacturing facilities. This included the processing units, tank farm, wastewater treatment facilities, marketing terminal and administration buildings, as well as some of the licensed landfarm areas. The Southern Property was largely undeveloped, with the exception of some access roads, an automobile research test track, and the main licensed landfarming area.

When it came on-stream in 1963, the Oakville refinery was considered to be state-of-the-art, and had many features to minimize environmental impacts. Examples included an activated sludge treatment plant for wastewater, steam coils in tanks to minimize sludge accumulation, and no underground piping except sewers. The refinery was capable of processing 7000 m³/day of crude.

Approximately 11 ha at the Oakville site were used for the disposal of refinery wastes. Four sections of the land farm were used for the treatment of oil and biological sludges, and one was used for leaded gasoline tank scale.

THE ORGANIC COMPOUNDS STUDIED

A list of 43 organic compounds was provided to the companies by the MOE. The list was originally established by the U.S. Environmental Protection Agency in 1985 as a compilation of substances most likely to be present, and considered to be of potential environmental health concern in petroleum refining wastes. The list was established to determine those substances which should be examined when delisting such sites.[1] Very few regulatory agencies have established criteria for organic compounds in soil and no criteria have ever been set for most of the 43 compounds.

Preliminary site data suggested that not all of the 43 compounds were likely to be present at the two sites, nor was it necessary (or efficient) to investigate all of the compounds to develop an approach for establishing cleanup criteria. As a result, an early objective of the study was to generate a "short list" of compounds that would provide sufficiently broad ranges of environmental and toxicological parameters to illustrate the approach being developed, and to demonstrate that the approach could be applied to other compounds if required.

Several types of information were sought in the literature for all 43 compounds to determine which were best suited for the short list. To provide an orderly approach to compiling the information, the compounds were arranged into groups based on similarities of chemical structure. Using generally accepted definitions and nomenclature, the compounds were organized into the six chemical groups, plus a seventh miscellaneous category, shown in Table 1.

Information was collected relating to the physical and chemical nature of each compound. These data included aqueous solubility, octanol-water partition coefficient (K_{OW}), and Henry's Law constant. Relatively soluble compounds also tend to be more mobile in the environment. The K_{OW} of a compound (often expressed as a logarithm) is an indicator of its affinity to become bound to organic material. Compounds with relatively high K_{OW} values are easily adsorbed onto the organic components of soil, thus reducing their environmental mobility. Henry's Law constant indicates the rate at which an organic liquid will diffuse or volatilize from the soil into the air.

To consider the persistence of compounds in soil, half-life values were gathered. Organic compounds in soils are affected by various processes including adsorption, volatilization, chemical reactions, photolysis, and microbial degradation. Site-specific factors that affect behavior include physical, chemical, and biological

Table 1. Original List of 43 Compounds.

Monocyclic Aromatics	Phenolics
benzene	phenol
toluene	cresols
xylenes	2,4-dimenthylphenol
ethylbenzene	4-nitrophenol
styrene	2,4-dinitrophenol
Polycyclic Aromatics	**Phthalate Esters**
naphthalene	dimethtl phthalate
1-methyl naphthalene	diethyl phthalate
anthracene	di(n)butyl phthalate
phenanthrene	di(n)octyl phthalate
benzo(a)anthracene	bis(2-ethylhexyl) phthalate
7,12-dimethylbenzo(a)anthracene	butyl benzyl phthalate
chrysene	
methyl chrysenes	**Halogenated Aliphatics**
pyrene	chloroform
fluoranthene	1,2-dichloroethane
dibenz(a,h)anthracene	ethylene
benzo(a)pyrene	
benzo(b)fluoranthene	**Halogenated Cyclics**
benzo(k)fluoranthene	chlorobenzene
dibenz(a,h)acridine	dichlorobenzenes
indene	
	Miscellaneous
	benzenethiol
	carbon disulfide
	1,4-dioxane
	methyl ethyl ketone
	pyridine
	quinolines

characteristics of the soil, the concentration of the compound, the length of time the compound has been in the soil, and the local groundwater regime. The half-lives of most organic compounds in soil are not well documented. For those that have been studied, half-lives range from approximately one day to several hundreds of days.

Information was gathered on toxicological effects. Each substance was matched with those health effects that can result from either short- or long-term exposures. These effects range from carcinogenicity to skin irritation. A qualitative, subjective ranking of the overall level of concern associated with the potential health effects of each compound relative to the hazards presented by the 43 was offered.

Although not tabulated, other types of information were sought during the review of the literature. Information for some compounds was found concerning their presence and abundance in refinery wastes. For others, data about plant toxicity and epidemiological information were reviewed.

All the collected information was used in the development of the short list. Meetings initially among the consultants only, then between the consultants and the companies, and finally with government representatives, resulted in the selection of the 10 compounds shown in Table 2, which also includes key reasons underlying the selection of each compound. Figure 2 uses the data that was gathered to demonstrate that the compounds on the short list reflect the ranges

Table 2.   Short List of Selected Compounds.

benzene
—physico-chemical properties similar to other monocyclic aromatics
—environmental behaviour well established
—volatilization is an important environmental process
—only proven human carcinogen on the original list of 43 compounds
—high level of toxicity concern

naphthalene, phenanthrene, chrysene, benzo(a)pyrene
—these four substances represent the range of physico-chemical properties and health
  considerations present in the polycyclic aromatic hydrocarbons and include a 2-ring,
  a 3-ring, a 4-ring and a 5-ring PAH
—medium-to-low solubility and volatility; medium-to-very slow rates of degradation[a]
—sorption is a key process
—relatively high level of toxicity concern for B(a)P, medium for chrysene, low for naphthalene and phenanthrene

cresols
—physico-chemical properties similar to other phenolics
—high solubility, medium volatility, low K_{OW}, high degradability
—degradation is an important process
—strong skin irritant

bis(2-ethylhexyl) phthalate
—the most persistent, least soluble of the phthalate esters
—mid-range in terms of solubility and volatility[a]
—sorption is an important process
—medium level of toxicity concern

ethylene dibromide
—physico-chemical properties similar to other halogenated aliphatics
—relatively slow to degrade; volatilization is a key process[a]
—high level of toxicity concern
—strict regulatory requirements anticipated in near-future

1,4-dioxane
—highly soluble, very low K_{OW}[a]; degrades in water
—medium level of toxicity concern

methyl ethyl ketone
—highly soluble, low K_{OW}[a]
—medium level of toxicity concern

[a]Relative to the range of characteristics present in the 43 compounds.

of environmental behavior and health concerns presented by the original 43 compounds.

DETERMINING ACCEPTABLE EXPOSURE LEVELS

General Approaches to Assessing Health Risks

Before an "acceptable" concentration in soil of a compound can be determined,
it is first necessary to ascertain the acceptable level of exposure to that compound.
The assessment of "acceptable" exposures and risks to human health is an

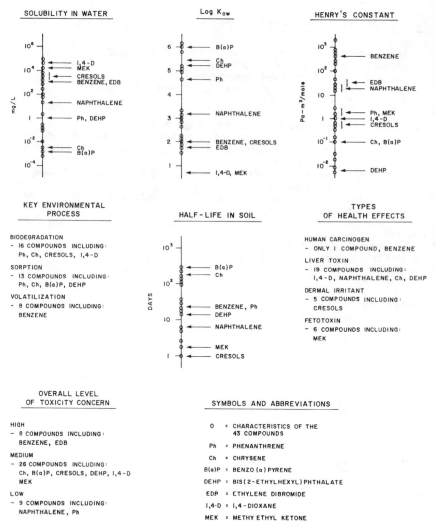

Figure 2.   Ranges of characteristics of the 43 compounds with emphasis on the 10 short list compounds.

inexact exercise, largely based upon theoretical assumptions concerning the extrapolation of animal data to humans. The evaluation of health concerns associated with chemicals in the environment requires an assessment of the potential types of health effects, exposures at which the effects occur, and the comparison of the risks with those of other human activities and conditions. Each of these aspects of health risk assessment includes an element of uncertainty.

To evaluate potential types of health effects, it is usually necessary to extrapolate down several orders of magnitude to evaluate the implications of relatively low exposures, such as those of interest in this study. One way to achieve the necessary extrapolation is to identify a no-observed-adverse-effects-level (NOAEL) and divide that level by an appropriate safety factor (often a factor of 100, 1000, or more) to calculate an acceptable daily intake (ADI). The safety factor approach assumes that there exists an exposure threshold, below which no adverse effects occur. Many North American regulatory agencies prefer to apply this approach to noncarcinogens only. On the other hand, the World Health Organization (WHO) has endorsed this type of approach for assessing carcinogens, and the WHO procedure is used by authorities in many countries.[2]

Extrapolation of low-dose effects can also be accomplished using mathematical models that have been developed for this purpose. All of the models attempt to predict the number of test animals that will respond at low exposure levels, based on the number of responses observed at high dose levels. The models reveal little about predicted human response at any exposure level and require judgments based on broad biological assumptions. Many of the models do not consider the possibility of a threshold response to a chemical because of the statistical problems encountered in the determination of NOAEL values. The assumed absence of a threshold means that absolute safety (or zero risk) cannot be achieved unless exposure is zero. Inherent in the use of such mathematical models, therefore, is the concept of a virtually safe dose (VSD) associated with some suitably low or acceptable level of risk.

The determination of whether to use the ADI approach or one of the mathematical models to calculate a VSD is influenced by many factors, including the types(s) of toxicological information available, and the type(s) of health effects that have been reported. Accordingly, the available literature was reviewed and a toxicological profile was prepared for each of the 10 compounds on the short list. Particular attention was paid to long-term studies designed to demonstrate the relationships between the dose applied and the responses observed in the subjects, and having adequate numbers of experimental subjects to definitively characterize the toxicity of the compound in question. All of the published studies were evaluated based on the assumption that the administered dose (amount of substance given to the subject) is totally absorbed by the subject, irrespective of the route of exposure (method of application). This may considerably overestimate the risks for volatile compounds applied dermally (to the skin) or given by inhalation, or for those compounds that are poorly absorbed.

Calculation of Acceptable Daily Intake (ADI) Values

An ADI is defined as the amount of substance that can be taken into the body daily over a lifetime without producing measurable adverse health effects. The procedure used to calculate ADI values involves dividing the highest NOAEL by a safety factor to account for possible differences insusceptibility between experimental subjects and the general population. A NOAEL is the highest dose of a series of doses tested in experimental animals or human subjects that did not produce any measurable adverse effect. The magnitude of the safety factor depends on the severity of the adverse health effects induced and the quality of the published data available for analysis.

ADI values for nine of the compounds on the short list are presented in Table 3. There are no suitable data for deriving an ADI for ethylene dibromide, while there are sufficient data to calculate separate ADI values for adults and children for benzene and B(a)P.

Calculation of Virtually Safe Dose (VSD) Values

Models that are used to extrapolate to health effects at low doses can assign a level of risk to any exposure, since they do not consider the possibility of a threshold response. A VSD is defined as the exposure at which the level of risk is so small as to be considered "virtually safe," of no consequence, or acceptable. To determine VSD values requires that a model be selected and an acceptable level of risk be established.

Many models have been endorsed by various health authorities to calculate VSD values for certain applications. For this study, the U.S. EPA Global−82 program was selected. The program uses a linearized, multistage model for low dose assessment and is generally perceived by toxicologists to produce extremely conservative (low) VSD estimates.

Just as many models are available for extrapolation; so, too, have several levels of risk been used to establish VSD values. While there is no consensus as to what constitutes an acceptable level of risk, most estimates place the value in the range of 1×10^{-4} to 1×10^{-7} for the probability of death in a lifetime. For this evaluation, the VSD was defined as the lower 95% confidence level of the dose corresponding to the lifetime risk of 1×10^{-6} (one-in-a-million) as calculated by the Global-82 program.

VSD values for the five carcinogens on the short list are presented in Table 3. Sufficient data are available to calculate separate VSD values for benzene based on animal and human data.

Table 3. Acceptable Daily Exposures for Compounds on the Short List.

Compound	VSD[a] (in μg/kg)	ADI (in μg/kg)
benzene	0.034 and 0.32[b]	0.28 (adult)
naphthalene	not applicable	140
phenanthrene	not applicable	40
chrysene	not applicable	10
benzo(a)pyrene	8.7×10^{-5}	2.8×10^{-4} (adult)
cresols	not applicable	51
bis(2-ethylhexyl) phthalate	1.5	106
ethylene dibromide	2.4×10^{-5}	no suitable data
1,4-dioxane	4.6	9.6
methyl ethyl ketone	not applicable	46

[a]Acceptable exposure is defined as the 95% lower confidence limit of the VSD at a lifetime risk of 10^{-6} as calculated using the U.S. EPA Global-82 program.
[b]The VSD of 0.32 μg/kg for benzene is based upon a U.S. National Toxicology Program bioassay. The VSD of 0.034 μg/kg for benzene is based on human data.

Hazard Evaluation for Plant Life

Little has been published regarding the phytotoxicity of the 10 organic compounds. In most cases, no objective data exist to evaluate dose-response relationships and assess risks to plant life. Studies of soil containing elevated levels of polynuclear aromatic hydrocarbons (PAH) have demonstrated that vegetablegrowth is not inhibited by B(a)P concentrations of up to 15,000 g/kg (15 ppm) in soil.[3] Apparently normal vegetables have been grown adjacent to a highway in Ontario even though the PAH concentrations were elevated above background levels.[4] Studies with phthalate esters have demonstrated that concentrations in soil need to exceed 200 ppm before phytotoxic effects are observed. It was estimated that such relatively high concentrations would not be found after the cleanup of either refinery site, in view of the low "acceptable" concentrations that the approach being developed would identify to satisfy human health concerns. On that basis, it was considered that risks to plant life would be remote and need not be considered further.

REFINERY SITE PATHWAYS MODEL

Either refinery site (or portions thereof) eventually may be converted to residential land use. In such a setting, future residents could be exposed to compounds present in site soils through various pathways (routes of exposure) such as the inhalation of dust or vapors, direct ingestion of dirt, and the ingestion of plants grown in local soil.

Each pathway is influenced by many factors related to the environment, the compound being evaluated, and the characteristics of the person being exposed. Often these factors are combined into mathematical equations. The factors and equations used to estimate exposures can be arranged into a "pathways model." The EXPOSE (*EXP*osure to *O*rganic *S*ubstances in the *E*nvironment) model has been developed to assess potential exposures of future residents for these two sites.

The factors and equations developed for the pathways model are based, where possible, on the results of other investigations of exposures under similar circumstances, and on assumptions about the lifestyles of future site residents. When estimating possible exposures, it often is necessary to select specific values for some factors from ranges of values that have been reported. It is general practice in pathways analysis to make such selections so that exposure estimates are likely to be higher than actual exposures. In turn, this should tend to lead to actions that overprotect individuals, rather than provide inadequate protection. Taken to extremes, such a "conservative" approach can lead to gross overestimations of exposures, or portray people with unrealistic lifestyles. For this study, a conservative approach has been followed but tempered by estimating exposures to residents whose lifestyles are likely and appropriate in the context of suburban communities in southern Ontario.

The Receptors

The pathways model is directed toward estimating the exposures to two types of future residents (also referred to as "receptors"): an adult and a young child. The adult and the child are assigned behaviors and living conditions such that they represent the greatest cumulative intakes for individuals living at either site. For example, the receptors are assumed to live in a house onsite, never leave the site, and eat produce from their garden.

The adult receptor has been assigned the characteristics of a man in the 20- to 39-year age bracket. These include a weight of 70 kg, a daily breathing rate of 23 m^3, and the consumption of 0.6 kg of produce daily. The adult is assumed to spend all of his time either in the house or in the yard. During the summer, he spends some time working in the garden. During the winter, he spends all of his time in the house.

The young child receptor has been assigned the characteristics of a child about 2 to 3 years old. These include a weight of 10 kg, a daily breathing rate of 5 m^3, and the consumption of 0.3 kg of produce daily. Like the adult, the child is assumed to spend all of his time in the house or in the yard. During the summer, the child spends five days a week playing in the yard. These are referred to as "active" summer days. The other two days a week are spent totally in the house and are referred to as "passive" summer days. During the winter, all of his time is spent in the house.

The pathways by which the receptors could realize exposures in this setting include direct ingestion of local soil and indoor dust, inhalation of vapors and particulate matter when indoors and outdoors, dermal absorption from soil and dust on exposed skin, and the ingestion of plants grown in local soil. All the pathways considered in the model are indicated in Figure 3.

The Environment

The outdoor environment is based upon average meteorological and air quality conditions experienced in the metropolitan Toronto area, although various simplifying assumptions are made where greater detail would not meaningfully improve exposure estimates. For example, the year has been divided into two seasons only, summer and winter, each lasting six months.

The receptors are assumed to live in a single-story house with a full, concrete block basement. Relevant aspects of indoor air quality include the indoor concentration of total suspended particulate (TSP) matter and vapor, dustfall rates, average dust coverings on indoor surfaces, and the origins of indoor dust. Studies of these conditions in U.S. and Canadian cities were reviewed recently as part of an exposure assessment for residents of a section of Niagara Falls, New York.[5] Several findings of that review were adopted for this assessment.

Concentrations of organic compounds in site soil are assumed to be constant both vertically and horizontally and not to change with time. Actual site conditions regarding concentrations would be quite different, but these assumptions

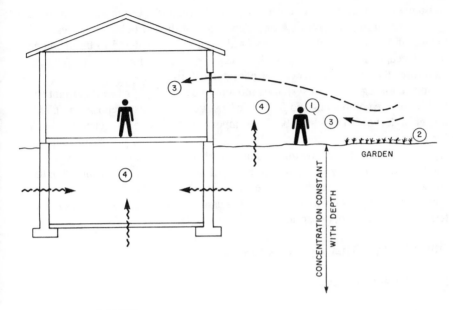

POTENTIAL PATHWAYS:

DIRECT INGESTION OF SOIL
(1) DERMAL EXPOSURE TO SOIL
INHALATION OF PARTICULATE MATTER

(2) INGESTION OF GARDEN PRODUCE

DIRECT INGESTION OF DUST
(3) DERMAL EXPOSURE TO DUST
INHALATION OF PARTICULATE MATTER

(4) INHALATION OF VAPOURS (BOTH OUTDOORS AND INDOORS)

Figure 3. Pathways in the "expose" pathways model.

help minimize model complexity and are conservative; that is, they should lead to overestimations of exposure.

Garden Activities and Products

The adult is assumed to maintain a garden in the yard which provides fruits and vegetables to be consumed by the adult and the child. The amount of garden produce that the receptors consume is influenced by the size of the garden, the yields of the crops grown, and the preferences of the receptors. Based on these types of considerations, various garden scenarios were investigated. It was decided to assume that the receptors' garden provides 1% to 20% of all the produce they consume.

Relatively little has been published about the uptake by plants of the compounds on the short list. Much of the available data concerns uptake in crops sprayed with pesticides or herbicides, or uptake in plants onto which metals or nonvolatile substances such as PCBs have been deposited. The three prime mechanisms of uptake are through the root system, deposition of dirt particles on leaves, and uptake of vapors through leaf pores. Although it is considered to be of minor importance, no information is available on uptake of vapors for the 10 compounds; therefore, that mechanism was not considered further.

Uptake through the root system is thought to be relatively more important for highly soluble compounds and for root crops such as carrots. Deposition of TSP matter on leaves is thought to be more important for less volatile chemicals and for leafy plants such as lettuce. Other factors that influence uptake include the type of plant, length of growing season, soil characteristics, crop yield, volatilization, and washoff by precipitation. Many of these factors are considered in the equation used to estimate plant uptake. As a simplifying and conservative assumption, the type of plant grown in the garden is assumed to have the combined characteristics of both a root crop and a leafy vegetable.

Other Factors That Affect Exposures

Direct Ingestion Exposures

Direct ingestion exposures stem from the inadvertent or deliberate ingestion of dirt and dust. This pathway is particularly relevant to young children who have a predilection for eating foreign material, such as dirt, and frequently put their fingers in their mouths. The adult may transfer dirt and dust from the hands to the mouth through activities such as eating or smoking.

Direct ingestion of dirt and dust by the child is assumed to be 250 mg of dirt while outdoors, plus 50 mg of dust while indoors on active days. On passive summer days and throughout the winter, the child is indoors all day and is assumed to ingest 100 mg of dust. The adult is assumed to ingest dirt from the inside surfaces of his hands on those days when he works in the garden, and dust from his hands on days spent indoors.

For both the adult and the child, it is assumed that 50% to 100% of any organic compound in the ingested soil or dust is absorbed. Actual absorption rates are compound specific and probably include values less than 50%.

Dermal Exposures

Dermal exposures result from the presence of soil or dust on the skin and the subsequent dermal absorption of compounds associated with the soil or dust particles. This pathway is most relevant to outdoor activity during the summer. Dermal absorption rates have been studied by applying pure compounds to the skin of laboratory animals and humans. While the rates vary according to compound, an average of 0.5%/hr has been recommended as a simplifying assumption for adults.[5] For children, a rate of twice that for adults, or 1%/hr, has been used elsewhere and is used in this analysis.

Inhalation Exposures

Inhalation exposures occur continuously as a result of the presence of TSP matter and vapors in the air. Exposures are estimated from the amount of air breathed and the concentrations of TSP matter and vapors either outside or indoors. For both the adult and the child, it is assumed that 75% of inhaled TSP matter is retained in the lungs and that 100% of a compound associated with the retained inhaled soil or dust particles is absorbed. It is also assumed that 50% to 100% of inhaled vapors are absorbed.

Pathways Model Equations

Based on the data and considerations noted above, equations to estimate exposures via each pathway were developed and assembled into the pathways model. For each pathway, an average daily dose is calculated and is expressed in units of mg of soil per kg of receptor body weight per day. For example, exposures from the direct ingestion of dirt or dust are calculated as:

$$\begin{array}{c} \text{amount ingested} \\ \text{daily} \end{array} \times \begin{array}{c} \text{absorbed} \\ \text{fraction} \end{array} \times \begin{array}{c} \text{percentage of days} \\ \text{annually this occurs} \end{array} \div \begin{array}{c} \text{receptor's} \\ \text{weight} \end{array}$$

Similar expressions were prepared to calculate the exposures from dermal contact, the inhalation of TSP, the inhalation of vapors, and the ingestion of food. Exposures via all pathways are converted to average daily dose estimates and summed to produce total dose estimates.

Total Dose Estimates

The total dose estimates for the 10 compounds are presented in Table 4. The ranges reflect the assumptions that absorption of compounds from directly ingested soil and dust ranges from 50% to 100%, that absorption of inhaled vapors ranges from 50% to 100%, and that the backyard garden can provide from 1% to 20% of the fruit and vegetables consumed by the receptors. The estimates are expressed in units of mg of local soil per kg of receptor body weight per day. These values must be multiplied by the concentration of a compound in the soil to produce exposure estimates for that compound. For example, if the concentration of benzene in soil is 3 mg/kg (3×10^{-6} g/g), the Table 4 values for benzene are multiplied by 3×10^{-6} to produce exposure estimates of 0.001 and 0.002 mg/kg/day for the child.

Table 4.   Total Dose Estimates for Compounds on the Short List.

Compound	Adult	Child
benzene	240.4–503.2	349.7–777.7
naphthalene	1.6– 4.9	10.0– 25.9
phenanthrene	0.7– 2.6	8.7– 21.4
chrysene	0.7– 2.5	8.7– 21.2
benzo(a)pyrene	0.7– 2.5	8.7– 21.2
cresols	1.7– 17.6	11.6– 72.8
bis(2-ethylhexyl)phthalate	0.7– 2.5	8.7– 21.2
ethylene dibromide	71.1–200.6	114.8–433.6
1,4-dioxane	71.0–295.7	125.7–789.2
methyl ethyl ketone	104.6–362.9	173.1–884.0

All total dose estimates are in mg of local soil per kg of receptor's body weight per day and must be multiplied by the concentration of a compound in local soil to determine the exposure estimate for that compound.

Ranges reflect the assumptions that absorption from directly ingested soil, ingested dust, and inhaled vapours range from 50 to 100% and that the backyard garden can provide 1 to 20% of the fruits and vegetable consumed by the receptors.

INTERPRETATION OF PATHWAYS MODEL RESULTS

Major Pathways and Key Parameters

The results of the pathways model (summarized in Table 4) illustrate the extents to which exposures vary with compound and the influences of specific pathways on total exposure. One common feature of all the total dose estimates is that the child would experience the highest doses. That the child is the "critical" receptor is frequently encountered in exposure pathways analysis. Contributing factors include the child's relatively low weight, the relatively large amounts of soil and dust assumed to be ingested, and the amount of time spent outdoors in summer, which promotes several exposure mechanisms. Because the child is the critical receptor in this analysis, and the acceptable exposures presented earlier for the child are as low as or lower than those for the adult, the adult was not considered further in deriving acceptable soil concentrations.

A formal sensitivity analysis was not used to characterize the influence of specific parameters, but several key aspects can be derived from inspection of the model results. Although the contribution of each pathway varies from compound to compound, the results of the pathways model indicate that the total exposure estimates are dominated by one or two pathways. For a highly volatile compound such as benzene, the inhalation of vapors is the major pathway. For highly soluble compounds such as 1,4-dioxane and MEK, the ingestion of garden produce dominates. The direct ingestion of soil and indoor dust account for the largest portions of the exposure estimates for relatively long-lived and immobile compounds such as B(a)P. The inhalation of TSP matter and dermal absorption contribute very little to total exposure estimates for all types of compounds.

The domination of the inhaled vapors pathway for volatile compounds largely results from the estimated indoor exposures. Vapor inhalation is estimated to be greatest during the six months of winter, during which time the receptors are assumed not to leave the house. Like other pathways, the estimated exposures are based on an assumption that the concentrations of the compounds in the soil do not decline with time. This is unrepresentative of actual conditions, particularly for volatile compounds which by their nature decline relatively rapidly.

The relative importance of the ingestion of garden produce to the estimates of total dose is strongly influenced by the calculated plant uptake factors and the amount of produce derived from the garden. Two plant uptake mechanisms are estimated in the model. The first mechanism, uptake via the roots, is estimated to increase when the parameter K_{OW} decreases. Thus compounds with relatively low K_{OW} values such as MEK and 1,4-dioxane are estimated to have high uptake factor values with the resulting dominance of the garden produce pathway. For compounds with large K_{OW} values such as B(a)P, chrysene, and DEHP, root uptake is estimated to be small. The second uptake mechanism, foliar deposition, predominates for these latter types of compounds, but does not equal the root uptake estimated for the more mobile, soluble compounds with low K_{OW} values.

Considering the assumed productivity of the garden; the assumption that all garden produce consists of a hypothetical combined leafy root crop; the deliberate overlooking of any losses of substances during food preparation (such as might result from peeling or boiling); the availability of the wide variety of fresh and prepared fruits and vegetables available from commercial outlets year-round; and other conservative assumptions, the treatment of this pathway has been such that estimated exposures should exceed actual exposures. For five of the compounds, the direct ingestion of soil or dust is the major contributor to exposure estimates. For the compounds with high K_{OW} values (and thus low plant uptake), this pathway contributes as much as 90% of the exposure estimates of the young child. Most young children tend to ingest relatively large amounts of soil and indoor dust, and the 250 mg/day of soil ingested by the child assumed for this pathway is toward the middle of the range of values that have been reported for children aged 2 to 6 years. There are records of rare children (and adults) who have demonstrated morbid appetites for unusual substances such as dirt (a condition referred to as pica). Such individuals have been reported to eat 5 g per day and more. Had a higher ingestion value been assumed, the exposure estimates for this pathway could have been increased. On the other hand, the absorption of substances into the body from dirt ingested by individuals most likely occurs at less than the 50% to 100% assumed in this study. The use of 50% to 100% absorption in this model reduces the potential differences between assuming that the child has pica, and the way in which the child is portrayed in the model. To have based the direct ingestion pathway on a child with pica would have been inconsistent with the stated objectiveof determining typical exposures.

Acceptable Onsite Soil Concentrations

Acceptable concentrations of compounds in soils were calculated from exposure levels described in Table 3, divided by the total dose estimates from the sum of the pathways presented in Table 4. For example, if the acceptable exposure of a substance is 5 μg/kg/day and the dose estimate is 2 mg (of soil)/kg/day, the acceptable soil concentration is 2500 μg/g (i.e. $5 \div 2 \times 1000$ to convert to units of μg/g).

Because both the acceptable exposure levels and the total dose estimates have been presented as ranges, acceptable concentrations of compounds in refinery site soils were calculated based on two sets of conditions. Condition "A" is based on the ranges of acceptable exposure levels from Table 3 and the upper values from the ranges of total dose estimates summarized in Table 4. Condition "B" is based on the lowest acceptable exposure level for each compound from Table 3 and the ranges of total dose estimates summarized in Table 4. By definition, the lower value in each scenario is the same. Most values for both scenarios have been rounded off slightly to avoid giving the impression of undue accuracy. Acceptable concentrations for all 10 of the compounds are presented in.

Exposure from Other Environmental Pathways

The pathways included in the model are not the only ways in which future residents would be exposed to organic compounds. Regardless of where a person lives, many of the 10 compounds are present in the air, drinking water, and food as a result of natural processes, man's activities, or both. While there is insufficient information to estimate exposure for many compounds, there is sufficient information for two of the carcinogenic compounds on the short list. These data can be used to put the estimated exposures of the receptors into perspective.

Typical Ontario residents are estimated to be exposed to 4.9 μg/kg/day of benzene and 1.2×10^{-2} of B(a)P. These values indicate that a resident of either site with these two compounds present at the ranges of values indicated in Table 5 would have exposures about 0.7% to 2% greater than a resident living elsewhere. Recognizing the conservative natures of the pathway model and the acceptable

Table 5.   Acceptable Onsite Soil Concentrations Based on Two Sets of Conditions.

Compound	Condition "A"	Condition "B"
benzene	0.04 to 0.13	0.04 to 0.1
naphthalene	5400	5400 to 13960
phenanthrene	1870	1870 to 4600
chrysene	470	470 to 1150
benzo(a)pyrene	0.004 to 0.005	0.004 to 0.01
cresols	700	700 to 4390
bis(2-ethylhexyl)phthalate	70 to 5000	70 to 12200
ethylene dibromide	0.00006	0.00006 to 0.0002
1,4-dioxane	5.8 to 12.2	5.8 to 36.6
methyl ethyl ketone	52	52 to 270

All values are in μg/g.

For Condition "A", concentrations are based on the ranges of acceptable exposures presented in Table 3 and the highest total exposure estimate in Table 4.

For Condition "B", concentrations are based on the lowest acceptable exposure presented in Table 3 and the range of total exposure estimates in Table 4.

exposure values identified in Section 4, the exposures associated with the ranges of soil concentrations in Table 5 represent only minimal incremental increases in overall risk from these two compounds.

Implications of Degradation

As noted previously, the concentrations of organic compounds in site soil are assumed in the model to be constant with time. This simplifies the modeling and the interpretation of health risks associated with site-related exposures, but ignores the reality of degradation and the corresponding decrease in exposure likely to occur. For example, a compound with a half-life in soil of 50 days will be reduced by 99% in 350 days. Six of the compounds (benzene, naphthalene, phenanthrene, cresols, DEHP, and MEK) have half-lives of less than 50 days. B(a)P, the compound on the short list with the longest reported half-life at 480 days, will be reduced by 99% in 3360 days, or approximately 10 years. It also must be remembered that reported half-lives may be based on tests in laboratories under conditions such as constant temperature that do not reflect actual site environments, and therefore the results may not represent events that occur on site. By deliberately ignoring degradation in the modeling of receptor exposures, the risks associated with the ranges in Table 5 overstate the actual risks to site residents by considerable margins for most compounds.

Other Considerations in Establishing Acceptable Concentrations

Many aspects of the pathways model and the interpretation of its results tend to produce low acceptable concentrations for site soils. As a result, there are other factors that should be considered prior to the establishment of site cleanup criteria. Three such considerations are the background levels of compounds in soil, phytotoxicity considerations, and analytical limits of detection.

Background or natural concentrations in soil have not been reported widely for many of the compounds on the short list. Shell and Texaco have recently sponsored a study of background levels for all 43 organic compounds in the Oakville and Port Credit areas. Preliminary results indicate the B(a)P levels typically range from 0.005 to 0.1 μg/g. This range is similar to background concentrations reported by others (for example,[6]). The acceptable concentrations noted in Table 5 (0.004, 0.005 and 0.01 μg/g) are generally lower than the background concentrations. Setting acceptable concentrations or cleanup guidelines at or below background levels could prove unworkable and impractical. The data from Oakville and Port Credit also appear to indicate that an upward adjustment may be appropriate for the acceptable range for benzene. These examples illustrate the need to consider background levels in establishing acceptable concentrations.

Conversely, if human health considerations identify guidelines that far exceed background levels, it may be necessary to adjust guidelines downward, if for no reason other than it may be difficult to convince decisionmakers or members of the public that the high levels are indeed "acceptable." Based on the data from Oakville and Port Credit, the concentrations of several PAHs, including naphthalene, phenanthrene, and chrysene as well as DEHP and methyl ethyl ketone, are well below the acceptable concentration ranges.

As noted earlier, it was felt that the approach used in this study would generate "acceptable" concentrations that would be less than the concentrations needed to observe phytotoxic effects. A subsequent review of the literature on phytotoxic effects of di(n)butyl phthalate (DBP) indicates that effects begin to occur at concentrations of 50 to 200 $\mu g/g$ in soil. Although it was not one of the 10 compounds studied, had the approach been applied to DBP, the range of acceptable concentrations would have been little different from that for DEHP of 70 to 12 200 $\mu g/g$. In this instance, the lower end of the range appears to be appropriate, but the upper end of the range may need to be adjusted downward.

In recent years, many analytical techniques have been developed or improved, with the result that analytical detection limits have declined remarkably and continue to do so. While many substances can be measured in the parts per billion or parts per trillion range, the approach used in this study has indicated acceptable soil concentrations that challenge or exceed current analytical techniques. For example, the acceptable soil concentrations for EDB of 0.00006 and 0.0002 $\mu g/g$ (60 to 200 parts per trillion) can be measured, but require sample preparation techniques that limit the confidence of such measurements. This example illustrates the need to consider analytical techniques and detection limits in establishing acceptable concentrations or cleanup guidelines.

SUMMARY AND CONCLUSIONS

A site-specific approach has been developed for establishing cleanup guidelines for organic compounds in soil at two refinery sites where landfarming has been undertaken. The major results and conclusions of the study follow:

- A review of the physical, chemical, and toxicological properties of 43 compounds was undertaken. For some of the compounds, information was also available that described their occurrence in refinery wastes and phytotoxic effects. Based on that information, 10 compounds were selected to a short list for detailed study. These 10 represent the ranges of environmental behavior and health concerns presented by the original 43 compounds.
- "Acceptable" exposure levels were defined for each of the 10 compounds using two methods. Acceptable daily intake (ADI) values were calculated for substances with suitable data following a method recommended by the World Health Organization. For carcinogens, virtually safe dose (VSD) values were also calculated using the U.S. EPA Global-82 program.
- The EXPOSE pathways model was developed to assess potential exposures of future site residents from direct ingestion of local soil and indoor dust, inhalation of vapors, dermal absorption,and the ingestion of plants grown

in local soil. While designed to estimate the exposures of typical, suburban residents, the model also included several conservative factors that should overestimate exposures.

- The relative importance of individual pathways was found to be influenced largely by the physical and chemical properties of the compounds. For example, the inhalation of vapors was the major pathway for volatile compounds such as benzene, while the direct ingestion of soil and indoor dust accounted for the major portions of the exposure estimates for relatively long-lived and immobile compounds such as B(a)P.

- Based on the total exposure estimates of the pathways model and the acceptable exposure levels, ranges of acceptable onsite concentrations were derived (see Table 5). The calculated ranges of acceptable concentrations for benzene, B(a)P, and ethylene dibromide (all carcinogens) were relatively low, while the ranges for naphthalene, phenanthrene, and cresols (which are all of relatively low toxicity), were among the highest.

- Comparisons of the predicted exposures with those of typical Ontario residents living elsewhere to benzene and B(a)P indicated that residents of either site would have exposures only 0.7% to 2% greater.

- Degradation of organic compounds in the soil was deliberately ignored in the pathways model and the subsequent determination of acceptable onsite concentrations. In reality, all of the compounds degrade; some rapidly.

- Considering the various conservative factors and assumptions incorporated into the determination of acceptable exposures and the pathways model, the acceptable onsite concentrations are considered to overstate the actual risks to site residents.

- To establish acceptable soil concentrations or cleanup guidelines, there are factors other than human health concerns that may need to be considered. These include the recognition that it would not be feasible to establish guidelines at or below background levels; it may not be acceptable to establish guidelines that are far above background levels; that phytotoxic effects from some compounds may occur at concentrations less than those associated with human health effects; and that it would be impractical to establish criteria at concentrations that cannot be measured accurately. The acceptable concentrations identified in this study illustrate the need to consider these factors before final decisions are reached.

- While the general approach developed during this study may be applicable to other sites, the pathways model and the assumptions used are site-specific. The acceptable concentrations should not be used at other sites without proper analysis and a review of the appropriateness of the assumptions made.

REFERENCES

1. The Cadmus Group. "Petitions to Delist Hazardous Wastes: A Guidance Manual," prepared for U.S. EPA, Office of Solid Waste and Emergency Response, Waste Identification Branch (1985).

2. World Health Organization (WHO). "Assessment of the Carcinogenicity and Mutagenicity of Chemicals," report of a WHO Scientific Group Tech. Rep. Series No. 546 (1974).
3. Fritz, W. "Model Tests of the Passage of Benzo(a)Pyrene From the Soil into Crops," *Z. Gesamte Hyg. Ihre Greuzgeb.* 29(7) 370–373 (as reported in *Chem. Abstr.*) (1983).
4. Wang, D. T., and O. Meresz, 1983. "Occurrence and Potential Uptake of Polynuclear Aromatic Hydrocarbons of Highway Traffic Origin by Proximally Grown Food Crops," in *Proceedings of the Sixth International Symposium on Polynuclear Aromatic Hydrocarbons: Physical and Biological Chemistry* Cooke et al., Eds. (New York NY: Springer-Verlag Publishers, 1983).
5. Hawley, J. K. "Assessment of Health Risk from Exposure to Contaminated Soil," *Risk Analysis* 5(4) 289–302 (1985).
6. Edwards, N. T. "Polycyclic Aromatic Hydrocarbons (PAH's) in the Terrestrial Environment—A Review," *J. Environ. Qual.* 12(4) 427–441 (1983).

List of Contributors

Abdel-Rahman, Mohamed S. Pharmacology Department, New Jersey Medical School, University of Medicine and Dentistry of New Jersey, Newark, NJ 07103.

Barnes, Ramon. Chemistry Department, University of Massachusetts, Amherst, MA 01003.

Bauman, Bruce. American Petroleum Institute, Washington, DC 20005.

Beck, Barbara D. Gradient Corporation, Cambridge, MA 02138.

Bricka, R. Mark. U.S. Army Engineer Waterways Experiment Station, P.O. Box 631, Vicksburg, MS 39180-0631.

Brown, Stuart M. CH2M Hill, P.O. Box 91500, Bellevue, WA 98009-2050.

Buelt, James L. Pacific Northwest Laboratory, Richland, WA 99352.

Calabrese, Edward J. Division of Public Health, University of Massachusetts, Amherst, MA 01003.

Callahan, Michael A. Office of Research and Development, U.S. Environmental Protection Agency, Washington, DC 20460.

Carter, Howard J. Texaco Canada Inc., 1210 Sheppard Ave. E., Willowdale, Ontario.

Cullinane, M. John, Jr. U.S. Army Engineer Waterways Experiment Station, P.O. Box 631, Vicksburg, MS 39180-0631.

Dragun, James. The Dragun Corporation, 3240 Coolidge Highway, Berkley, MI 48072-1634.

Eastcott, Linda. Department of Chemical Engineering and Applied Chemistry, University of Toronto M5S 1A4.

Edwards, Carolyn. Division of Public Health, University of Massachusetts, Amherst, MA 01003.

Eklund, Karl. American Reclamation Corporation, P.O. Box 263, Ashland, MA 01721.

FitzPatrick, Vincent F. Pacific Northwest Laboratory, Richland, WA 99352.

Fleischer, Edwin J. CH2M Hill, P.O. Box 4400, Reston, VA 22000.

Genes, Benjamin R. Remediation Technologies, Inc., 9 Pond Lane, Concord, MA 01742.

Gilbert, Charles E. Division of Public Health, University of Massachusetts, Amherst, MA 01003.

Gorber, Donald M. SENES Consultants Limited, 52 West Beaver Creek Road, Richmond Hill, Ontario.

Granville, Geoff C. Shell Canada Limited, 400 4th Avenue S.W., Calgary, Alberta.

Hailes, Charles E. Texaco Canada Inc., 1210 Sheppard Ave. E., Willowdale, Ontario.

Hillel, Daniel I. Plant and Soil Science Department, University of Massachusetts, Amherst, MA 01003.

Hills, John J. Hazardous Waste Management Section, Division of Environmental Health, Orange County, CA 92702.

Horton, Holly M. Division of Public Health, University of Massachusetts, Amherst, MA 01003.

Ibbotson, Brett G. SENES Consultants Limited, 52 West Beaver Creek Road, Richmond Hill, Ontario.

Jones, Murray G. Shell Canada Limited, 400 4th Avenue, S.W., Calgary, Alberta.

Kostecki, Paul T. Division of Public Health, University of Massachusetts, Amherst, MA 01003.

Kucharski, William A. Dames & Moore, 9665 Chesapeake Rd., Suite 360, San Diego, CA 92123.

Lynch, John. Remediation Technologies, 9 Pond Lane, Concord, MA 01742.

Mackay, Donald. Department of Chemical Engineering and Applied Chemistry, University of Toronto M5S 1A4.

McLearn, Mary E. Electric Power Research Institute, 3412 Hillview Avenue, P.O. Box 10412, Palo Alto, CA 94303.

Miller, Michael J. Electric Power Research Institute, 3412 Hillview Avenue, P.O. Box 10412, Palo Alto, CA 94303.

Munro, Ian. CanTox Inc., 460 Wyecroft Road, Oakville, Ontario.

Nash, James H. Roy F. Weston, Inc., 1 Weston Way, West Chester, PA 19380.

Noland, John W. Roy F. Weston, Inc. Weston Way, West Chester, PA 19380.

Pastides, Harris. Division of Public Health, University of Massachusetts, Amherst, MA 01003.

Paustenbach, Dennis J. McLaren Environmental Engineering, 980 Atlantic Avenue, Suite 100, Alameda, CA 94501.

Preslo, Lynne M. Roy F. Weston, Inc., 1001 Galaxy Way, Suite 107, Concord, CA 95210.

Reades, Denys W. Golder Associates, 3151 Wharton Way, Mississauga, Ontario.

Shields, Walter J. CH2M Hill, P.O. Box 91500, Bellevue, WA 98009-2050.

Shiu, Wan Ying. Department of Chemical Engineering and Applied Chemistry, University of Toronto M5S 1A4.

Smyth, David J. Golder Associates, 3151 Wharton Way, Mississauga, Ontario.

Stanek, Edward. Division of Public Health, University of Massachusetts, Amherst, MA 01003.

Suyama, Wendell. Southern California Edison, P.O. Box 800, Rosemead, CA 91770.

Timmerman, Craig L. Pacific Northwest Laboratory, Richland, WA 99352.

Traver, Richard P. Releases Control Branch, Hazardous Waste Engineering Research Laboratory, U.S. Environmental Protection Agency, Edison, NJ 08837-3679.

Tucker, Robert K. Office of Science and Research, New Jersey Department of Environmental Protection, Trenton, NJ 08625.

Turkall, Rita M. Pharmacology Department, New Jersey Medical School, University of Medicine and Dentistry of New Jersey, Newark, NJ 07103.

Valentinetti, Richard A. U.S. Environmental Protection Agency, Office of Underground Storage Tanks, Washington, DC 20460.

Velazquez, Luis A. Roy F. Weston, Inc., Weston Way, West Chester, PA 19380.

Veneman, Petrus. Plant and Soil Science Department, University of Massachusetts, Amherst, MA 01003.

Willes, Robert F. CanTox Inc., 460 Wyecroft Road, Oakville, Ontario.

Index

347

histogram, 72, 75
hot mix process for asphalt paving, 193, 196, 199
HPLC, *See* high pressure liquid chromatographic
Hudson River (New Jersey), 51, 52
humans, risk of petroleum in soil, 211
humus, decomposition of, 152
hydration (water absorption), 142
hydrindenes, 95
hydrocarbon contaminated soils, 163–174
hydrocarbons, 15, 41, 73, 78, 79, 118
  dense, 212
  petroleum, 6, 9, 12, 52, 175, 296
  water-soluble, 95
hydrologic conditions, 89
hydrolysis, 214
hydroquinone, 308, 310
hydroxide sludge, 128
hyonic PE, 90, 158

ICN Pharmaceuticals (Irvine, California), 302
immiscible phase, 87, 90
Imperial Oil Company, Inc. (New Jersey), 52
impervious surfaces, 90
  in situ
    biodegradation, 215
    flushing, 213
    recovery techniques, 215
    soil washing, 157, 158, 161
    techniques, 6, 12, 13, 213
    vitrification, 137–155
  application considerations of large-scale, 147
  benefits of, 155
  combustible limits for, 152, 153
  cost estimate technique of, 143–144
  process equipment for large-scale, 139, 140
  process trailers for large-scale, 139–140
  processing operations, 145
  processing rate for large-scale, 140
  results of large-scale testing, 147
incineration, 180
individual component strategy, 93, 99

individual component-mixture strategy, 94, 99–102
infiltration, 81, 90
infrared spectroscopy, 159, 160
ingestion of contaminated soil, 222, 226, 235–236
  adults, 231–233, 334
  children, 227–231, 313–314, 334
inhalation of contaminated soil and dust, 222, 226, 234–237, 335
injection wells, 215
input parameters, 91
interceptor trenches, 215
interference chemicals, 128
iso-alkanes, 93, 95
isoprenoid, 67, 69, 72, 95
isoprenoid hydrocarbons, 71
isopropanol, 45
issolation/containment, 116, 117

Jumping Brook (New Jersey), 49

kepone, 247
kerosene, 175, 193
Keyport (soil), 302, 303
kinetic constant, 2
King Pygmalion, 86
Kirkwood aquifer (New Jersey), 45
Kuehnle reservoir (New Jersey), 47

land surface conditions, 89
land treatment, 7, 116, 117, 163, 214
Lang Dental Manufacturing Corporation, Inc. (Chicago, Illinois), 303
lateral hetrogeneity, 83
leach tests
  EPA's Extraction Procedure Toxicity Test, 142
  Toxic Characteristics Leach Test, 142
leachate, 45, 66, 142, 165, 169, 197–198
leaching, 5, 11, 82, 88, 90, 97, 99, 100, 116, 128, 197, 223, 247
lead, 51, 52, 148, 230, 232, 250
Leaking Underground Fuel Tank Field Manual, 265, 294, 296, 297
Leaking Underground Storage Tank Committee, 28